Mechanisms and Machine Science

Volume 126

This book series establishes a well-defined forum for monographs, edited Books, and proceedings on mechanical engineering with particular emphasis on MMS (Mechanism and Machine Science). The final goal is the publication of research that shows the development of mechanical engineering and particularly MMS in all technical aspects, even in very recent assessments. Published works share an approach by which technical details and formulation are discussed, and discuss modern formalisms with the aim to circulate research and technical achievements for use in professional, research, academic, and teaching activities.

This technical approach is an essential characteristic of the series. By discussing technical details and formulations in terms of modern formalisms, the possibility is created not only to show technical developments but also to explain achievements for technical teaching and research activity today and for the future.

The book series is intended to collect technical views on developments of the broad field of MMS in a unique frame that can be seen in its totality as an Encyclopaedia of MMS but with the additional purpose of archiving and teaching MMS achievements. Therefore, the book series will be of use not only for researchers and teachers in Mechanical Engineering but also for professionals and students for their formation and future work.

The series is promoted under the auspices of International Federation for the Promotion of Mechanism and Machine Science (IFToMM).

Prospective authors and editors can contact Mr. Pierpaolo Riva (publishing editor, Springer) at: pierpaolo.riva@springer.com

Indexed by SCOPUS and Google Scholar.

Ashok Kumar Pandey · Prem Pal ·
Nagahanumaiah · Lena Zentner
Editors

Microactuators, Microsensors and Micromechanisms

MAMM 2022

Springer

Editors
Ashok Kumar Pandey
Department of Mechanical and Aerospace Engineering
Indian Institute of Technology Hyderabad
Sangareddy, Telangana, India

Nagahanumaiah
Central Manufacturing Technology Institute
Bengaluru, Karnataka, India

Prem Pal
Department of Physics
Indian Institute of Technology Hyderabad
Sangareddy, Telangana, India

Lena Zentner
Department of Mechanical Engineering
Technische Universität Ilmenau
Ilmenau, Germany

ISSN 2211-0984 ISSN 2211-0992 (electronic)
Mechanisms and Machine Science
ISBN 978-3-031-20352-7 ISBN 978-3-031-20353-4 (eBook)
https://doi.org/10.1007/978-3-031-20353-4

This Springer imprint is published by the registered company Springer Nature Switzerland AG
The registered company address is: Gewerbestrasse 11, 6330 Cham, Switzerland

Scientific Advisory Committee

A-R, Eihab, Uni. of Waterloo, Canada
Agrawal, Amit, IIT Bombay, India
Amarnath, C., IIT Bombay, India
Ananthasuresh, G. K., IISc Bangalore, India
Bhattacharya, Enakshi, IIT Madras, India
Bhat, Navakant, IISc, India
Bhatt, Vivekanand, RDE, Pune
Corves, Burkhard, RWTH, Aachen University
Chakraborty, Suman, IIT Kharagpur, India
Dutta, Shankar, SSPL Delhi
Ghosal, Ashitava, IISc Bangalore, India
Goel, Sanket, BITS Hyderabad
Gottlieb, Oded, Technion, Israel
Herder, Just, TU Delft, Netherlands
Kar, Jyoti Prakash, NIT Rourkela, India
Kim, C. J. UCLA, USA
Kumar, Navin, IIT Ropar, India
Lukin, Aleksey, Peter the Great St. Petersburg Polytechnic University, Russia
Misra, Sarthak, Uni. of Twente, Netherlands
Nagahanumaiah, CMTI, Bangaluru
Pratap, Rudra, IISc Bangalore, India
Ranganathan, Santhanam, DRDL, Hyderabad
Raman, Karthik TIFR Hyderabad, India
Ray, MC, IIT Kharagpur, India
Sato, Kazuo AIT Toyota, Japan
Sen, Prosenjit, IISc Bangalore, India
Sesia, Aswin, Cambridge University, UK
Sharma, Jaibir, IME Singapore
Singh, Pushparaj, IIT Delhi
Singh, Ravindra, DRDO Lab Jodhpur
Tanaka, Shuji, Tohoku University

Organizing Committee

Asthana, Saket, IIT Hyderabad
Badhulika, Sushmee, IIT Hyderabad
Duraiswamy, Suhanya, IIT Hyderabad
Emani, Naresh Kumar, IIT Hyderabad
Garlapati, Suresh Kumar, IIT Hyderabad
J., Surya Narayana, IIT Hyderabad
Mohanty, Jyoti Ranjan, IIT Hyderabad
Pal, Prem, IIT Hyderabad
Palathingal, Safvan, IIT Hyderabad
Pandey, Ashok Kumar, IIT Hyderabad
Rajagopal, Amirtham, IIT Hyderabad
Sidhardh, Sai, IIT Hyderabad
Singh, Shiv Govind, IIT Hyderabad
S., Suryakumar, IIT Hyderabad
Vanjari, Siva, IIT Hyderabad
Kumar, Prabhat, IIT Hyderabad

Preface

The 6th Conference on Microactuators, Microsensors and Micromechanisms (MAMM) was held at Indian Institute of Technology Hyderabad from December 3 to 5, 2022. The previous editions of the conferences: MAMM 2010, MAMM 2012, MAMM 2014, MAMM 2016 and MAMM 2020 were held in Aachen, Germany; Durgapur, India; Timișoara, Romania; TU Ilmenau, Germany, and again in TU Ilmenau, Germany. Following the tradition of publishing the peer-reviewed papers from the conference, Springer has consistently supported the publishing of proceedings of the MAMM conference. The Editors of 6th Conference on MAMM were happy to receive the same support for publishing the proceedings of the current MAMM22 conference, held at IIT Hyderabad, India, from December 3 to 5, 2022.

Like previous editions, the thematic area of the conference remained nearly same with the addition of few areas of current interest such as microactuators, microsensors and micromechanisms, compliant mechanisms and actuators, microscale power generation, miniaturized energy harvesting, miniature manufacturing machines, mechatronics and control issues in micromachines, smart circuits for micromachines, micromechanical devices and robotics for life science, biomechatronics at small scale, microscale flight and swimming, mechanic diagnostic of diseases, microopto-mechanical devices, 3D and 4D printings and microfabrication techniques for microactuators and micromechanisms, smart materials, metamaterials, lead-free materials for microsystems, micromechanics and nonlinear coupling in microelectromechanical system (MEMS) oscillators, computation and theoretical methods in microsystem designs.

To our call for papers, the papers were received from India, Germany, Vietnam and Russia. All papers were peer-reviewed by the scientific committee members including Prof. Kazuo Sato, Prof. Lena Zentner, Prof. Ananthasuresh, Prof. Prem Pal, Prof. Shuji Tanaka, Prof. Toshiyuki Tschuiya, Prof. Alexey Lukin, Prof. Oded Gottlieb, Prof. J. P. Kar, Prof. Sanket Goel, Prof. Saket Asthana, Prof. Suhanya Duraiswamy, Prof. Suresh Kumar Garlapati, Prof. J. Surya Narayana, Prof. Jyoti Mohanty, Prof. Safvan Palathingal, Prof. Sai Sidhardh, Prof. Prabhat Kumar, Prof. K. P. Venkatesh, Dr. Rahul Shukla, Dr. Shankar Dutta, Dr. Vivekanand, Prof. Kirti Sahu, Prof. Niranjan Ghasis and others whom I might have forgot to mention. Special thanks to Mr. J. Sai Kishore, Dr. Nikul Jani and Dr. Rakesh Kumar Das for handling the review process.

We would like to appreciate all the authors for taking the reviewers' comments positively and submitting the revised paper in time which helped us to complete the review process in order to get the proceedings of the conference ready. The proceedings of the conference consists of 32 chapters. Chapters are ordered first based on the compliant mechanism, its application in accelerometer and gyroscope, computational methods and, finally, fabrication methods and its effect.

Last but not least, we sincerely thank Prof. B. S. Murty, Director, IIT Hyderabad and local organizing committee members for the support related to the organization of the conference, and Dr. Pierpaolo Riva and the team of Springer for all the support for bringing out the proceedings of MAMM 22.

We sincerely hope the reader will enjoy reading the excellent piece of work by different authors. We look forward to receiving the feedback for the future proceedings.

Sangareddy, India
Sangareddy, India
Bengaluru, India
Ilmenau, Germany
December 2022

Ashok Kumar Pandey
Prem Pal
Nagahanumaiah
Lena Zentner

Contents

Free Vibration of Compliant Mechanisms Based on Euler-Bernoulli-Beams 1
Vivien Platl, Leo Lechner, Thomas Mattheis, and Lena Zentner

Towards Topology Optimization of Pressure-Driven Soft Robots 19
Prabhat Kumar

Compliant Finger Gripper Based on Topology Optimization 31
Estefania Hermoza Llanos, Mathias Hüsing, Burkhard Corves, and Anupam Saxena

Development of a Database to Simulate and Adapt Compliant Mechanisms to a Given Characteristic for Improving Energy Efficiency of a Walking Robot 46
Marten Zirkel, Yinnan Luo, Ulrich J. Römer, Alexander Fidlin, and Lena Zentner

Analytical Characterization of Spatial Compliant Mechanisms Using Beam Theory 61
Stefan Henning and Lena Zentner

Study of Curved Beam Based Displacement Amplifying Compliant Mechanism for Accelerometer Design 77
Mithlesh Kumar, P. Krishna Menon, and Ashok Kumar Pandey

Model of a Micromechanical Modal-Localized Accelerometer with an Initially Curvedmicrobeam as a Sensitive Element 94
Nadezhda Mozhgova, Alexey Lukin, and Ivan Popov

Different Beam Configurations for Compliant Mechanism-Based MEMS Accelerometer 119
Nikul Jani, Mithlesh Kumar, P. Krishna Menon, and Ashok Kumar Pandey

Design and Analysis of the Two-Level Accelerometer 136
Hoang Bao Thien Nguyen, Thanh Lam Tran, and Ngoc Dang Khoa Tran

Microchannel Induced Tailoring of Bandwidth of Push-Pull Capacitive MEMS Accelerometer 147
Nidhi Gupta, Siva Rama Krishna Vanjari, and Shankar Dutta

A Comparison of Ring and Disk Resonator Gyroscopes Based on Their Degenerate Eigenmode Shapes 152
Kandula Eswara Sai Kumar, Karthik Raveendranath, Sudhanshu Sekhar, and G. K. Ananthasuresh

Parametric Tuning of Natural Frequencies of Tuning Fork Gyroscope 162
Rakesha Chandra Dash, Rakesh Tirupathi, P. Krishna Menon, and Ashok Kumar Pandey

Design and Analysis of Single Drive Tri-Axis MEMS Gyroscope 172
Pradnya Chabbi, Anant Khedkar, and Venkatesh K. P. Rao

Analysis of the Equilibrium of a Magnetic Contactless Suspension 183
Pavel Udalov, Alexey Lukin, Ivan Popov, and Dmitriy Skubov

Frequency Analysis of Microbeam with Axial Pretension Using MSGT 191
Sai Kishore Jujjuvarapu, Indrasena Reddy Erravelly, and Ashok Kumar Pandey

Nonlocal Nonlinear Analysis of Functionally Graded Nano Plates Used in MEMs Devices 215
Dhaladhuli Pranavi and Amirtham Rajagopal

Finite Element Analysis of Squeezed Film Damping on Trapezoidal Microcantilever Resonators at Different Pressure Levels 223
Varun P. Sharma and Rahul Shukla

Two-Dimensional Hydrodynamic Forces in an Array of Shape-Morphed Cantilever Beams 232
Lalsingh Devsoth and Ashok Kumar Pandey

Design and Performance Analysis of TiO_x Based MEMS Bolometer Pixel 244
Isha Yadav, Sudha Gupta, A. Vishwakarma, S. Dutta, and Ratnamala Chatterjee

FEM Analysis of Split Electrode IDTs Designed Lithium Tantalate-Polyaniline SAW Gas Sensor 250
Dhananjaya Panda and Koteswara Rao Peta

Thermal Study of Thin-Film Heater for PCR Reaction-Based Applications 266
Andleeb Zahra, Giampiero de Cesare, Domenico Caputo, and Zia Abbas

Etching Characteristics of Si{110} in NaOH Based Solution 275
Soumya Purohit, Veerla Swarnalatha, and Prem Pal

Effect of IPA on Micromachining Characteristics of Silicon in KOH-Based Solution 281
Avvaru Venkata Narasimha Rao and Prem Pal

Deep Grooves in Borofloat Glass by Wet Bulk Micromachining 290
Adarsh Nigam, Swarnalatha Veerla, and Prem Pal

Micro-piezo Actuator for Cell Lysis 296
Sushama Agarwalla and Suhanya Duraiswamy

Wax-Printed Microfluidic Paper Analytical Device for Viscosity-Based Biosensing in a 3D Printed Image Analysis Platform 301
Mrunali D. Wagh, S. B. Puneeth, Subhendu Kumar Sahoo, and Sanket Goel

Detection of Volatile Organic Compounds Using Solution Processed Organic Field-Effect Transistors 310
Suresh Kumar Garlapati, Sheida Faraji, Daniel Tate, Aiman Rahmanudin, Palaniappan Valliappan, Alessandro Patti, Krishna Persaud, and Michael Turner

Thickness Dependent Chlorpyrifos Sensing Behavior of Silver Doped ZnO Nanowires 323
S. P. Ghosh, D. Pradhan, S. Pattnaik, and J. P. Kar

Investigation of $AuCl_3$ Doped MoS_2 Based IR Detector with the Variation of Annealing Temperature 332
Diana Pradhan, Anurag Gartia, Kiran K. Sahoo, Surya P. Ghosh, and Jyoti P. Kar

Investigation of Structural and Electrical Properties of Ta_2O_5 Thin Films with Sputtering Parameters for Microelectronic Applications 342
Kiran K. Sahoo, D. Pradhan, S. P. Ghosh, A. Gartia, and J. P. Kar

Optimization of Controllable Pulsed LASER Deposition Parameters for the Fabrication of Lead Free $Ba(Zr_{0.15}Ti_{0.85})O_3$ Thin Films 351
M. L. V. Mahesh, Prem Pal, V. V. Bhanuprasad, and A. R. James

Resistive Switching Behavior of TiO_2 Nanostructures Grown by Dip Coating Process 357
S. Roy, P. K. Sahu, D. Pradhan, N. Tripathy, and J. P. Kar

Correction to: Detection of Volatile Organic Compounds Using Solution Processed Organic Field-Effect Transistors C1
Suresh Kumar Garlapati, Sheida Faraji, Daniel Tate, Aiman Rahmanudin, Palaniappan Valliappan, Alessandro Patti, Krishna Persaud, and Michael Turner

Free Vibration of Compliant Mechanisms Based on Euler-Bernoulli-Beams

Vivien Platl(✉), Leo Lechner, Thomas Mattheis, and Lena Zentner

Fachgebiet Nachgiebige Systeme, Technische Universität Ilmenau, Max-Planck-Ring 12, Ilmenau 98693, Germany
vivien.platl@tu-ilmenau.de
https://www.tu-ilmenau.de/nsys

Abstract. This paper presents an analytical approach for computing the natural frequencies of planar compliant mechanisms consisting of any number of beam segments. The approach is based on the Euler-Bernoulli Beam theory and the transfer matrix method (TMM), which means there is no need for a global dynamics equation, but instead low-order matrices are used which result in high computational efficiency. Each beam segment is elastic, thin, has a different rectangular cross-section or a different orientation and is treated as an Euler-Bernoulli beam. The approach in principle does not differentiate between the flexure hinges, and the more rigid beam sections, both are treated as beams. The difference in stiffness solely results from the changes in the cross sections and length. A finite element analysis (FEA), as often used in practical applications, has been carried out for various geometries to serve as state-of-the-art reference models to which the results obtained by the presented analytical method could be compared. Various test specimens (TS) consisting of concentrated and distributed compliance in various degrees of complexity were produced and measured in free- and forced vibration testing. The results from experiments and the FEA compared to those of the proposed method are in very good correlation with an average deviation of 1.42%. Furthermore, the analytical method is implemented into a readily accessible computer-based calculation tool which allows to calculate the natural frequencies efficiently and to easily vary different parameters.

Keywords: Compliant mechanism · Vibration frequency · Transfer matrix method · Bernoulli beams · Free vibration · Forced vibration

1 Introduction

Compliant mechanisms gain their mobility partially or exclusively from the compliance of its flexible members rather than from traditional joints. Nowadays this compliance is no longer seen as just a disadvantage but is used in a targeted manner in many areas of application. While lots of progress has been made in the static analysis of such systems in recent years, their dynamic behaviour has been

A. K. Pandey et al. (Eds.): MAMM 2022, 126, pp. 1–18, 2023.
https://doi.org/10.1007/978-3-031-20353-4_1

subject to little research to date. Dynamic analysis is indispensable especially for systems that are exposed to high dynamic processes.

So far, pseudo-rigid-body models (PRBM), FEA and beam models are the most common approaches for calculating the natural frequencies of compliant mechanisms. For example, Lyon et al. [10] and Yu et al. [14] used PRBM to predict the first modal response of compliant mechanisms. Liu and Yan [9] used a modified PRBM approach by considering also the nonlinear effects and Vedant and Allison [12] propose a general pseudo-rigid body dynamic model for n-links. A hybrid compliant mechanism with a flexible central chain and a cantilever is examined through a PRBM by Zheng et al. [16].

As for the FEA Li et al. [8] and Wang et al. [13] both propose an approach based on it for the dynamic analysis of compliant mechanisms.
Concerning the classical Euler-Bernoulli beam theory, an approach is proposed for example by Vaz and de Lima Junior [3], where multi-stepped beams with changing cross sections, material properties or different boundary conditions were considered.

A different approach to obtain the vibration frequencies for uniform or non-uniform beams is the transfer-matrix-method (TMM). It is commonly used, for instance by Boiangiu et al. [2] to calculate the natural frequencies and observe the relation between them and the geometric parameters for multi-step beams. Khiem et al. [5,6] as well as Attar [1] use this method for crack detection and investigation of damaged beams. Obradovic et al. [11] represents an analytical treatise, where the TMM is used on rigid bodies in conjunction with elastic beam sections and Hu et al. [4] propose a new closed-form dynamic model for describing the vibration characteristics of actual compliant mechanisms with serial and parallel configuration by using the TMM.
Despite extensive research, the authors are not aware of any widely applicable analytical method or standalone tool for the calculation of the natural frequencies for compliant mechanisms.
Therefore, the purpose of this paper is the development of an analytical method (Sect. 2) and its implementation into a time efficient, intuitively operable tool (Sect. 3) for determining the transverse vibration of planar compliant mechanisms with minimum input data. This will enable the analysis of the dynamic behaviour of compliant mechanisms from the design phase to its application. Thus, the natural frequencies can easily be taken into account which can safe a lot time and money. The method is validated through several experiments and FEA (Sect. 4), the results are presented and discussed (Sects. 5 and 6) and finally conclusions are drawn (Sect. 7).

2 Analytical Method

2.1 Differential Equations of Motion

Different forms of the following compliant mechanism, as shown in Fig. 1, are evaluated in this paper, consisting of n beam segments and joints.

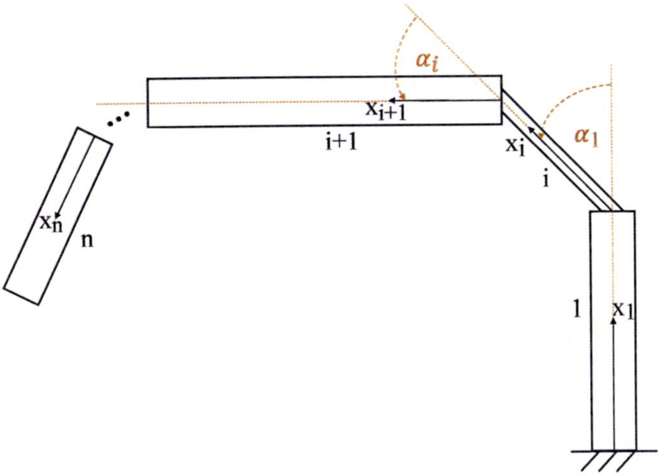

Fig. 1. Compliant mechanism consisting of n beam and joint segments

The partial differential equations of motion of an elastic Euler-Bernoulli beam can be expressed as

$$\text{transversal} \quad k_i^4 \frac{\partial^2 w_i(x,t)}{\partial t^2} + p_i^2 \frac{\partial^4 w_i(x,t)}{\partial x^4} = 0, \qquad i = 1, \ldots, n \tag{1a}$$

$$\text{axial} \quad k_i^4 \frac{\partial^2 u_i(x,t)}{\partial t^2} - p_i^2 \frac{\partial^2 u_i(x,t)}{\partial x^2} = 0, \qquad i = 1, \ldots, n \tag{1b}$$

$$\text{with the relations} \quad k_i^4 = \frac{\rho_i A_i}{E_i I_x(i)} \omega^2 \quad p_i^2 = \frac{\rho_i}{E_i} \omega^2. \tag{1c}$$

The following notations are used: E_i-Young modulus, $I_x(i)$-second moment of area, ρ_i-density, A_i-area surface, ω-frequency, α_i-angle between x_i and x_{i-1} in a positive sense, $w_i(x,t)$-transversal displacement and $u_i(x,t)$-axial displacement of the ith beam segment at axial coordinate x and time t, x_i-coordinate axis of the beam segments. The considered compliant mechanisms are monolithic structures; therefore, all segments are from the same material and have the same Young modulus and density. Later in this paper the Lagrange's and Newton's notations are used for differentiation.

Applying Bernoulli's commonly used method of separation of variables and using the approach $W(x) = e^{\lambda x}$, the general solutions for the transversal $W(x)$ and axial displacement $U(x)$ are

$$W_i(x_i) = C_{1i} \cos(k_i x_i) + C_{2i} \sin(k_i x_i) + C_{3i} \cosh(k_i x_i) + C_{4i} \sinh(k_i x_i), \tag{2}$$

$$U_i(x_i) = C_{5i} \cos(p_i x_i) + C_{6i} \sin(p_i x_i), \tag{3}$$

$$\text{with} \quad i = 1, \ldots, n$$

in which $C_{1i} - C_{6i}$ are unknown constants.

2.2 Boundary Conditions

In order to solve these differential equations (2) and (3), boundary conditions are used. The boundary conditions at the beginning describe the slopes and displacement at $x_i = 0$ and $i = 1$, and at the end at $x_i = L_i$ and $i = n$.

Table 1. Boundary conditions, beginning and end

Boundary condition		beginning ($x_1 = 0$)		end ($x_n = L_n$)	
clamped	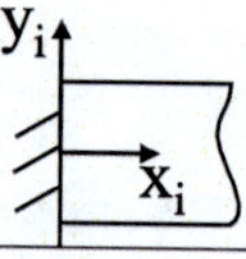	$U_1(0) = 0$ $W_1(0) = 0$ $W_1'(0) = 0$	(4a)	$U_n(L_n) = 0$ $W_n(L_n) = 0$ $W_n'(L_n) = 0$	(4b)
pinned	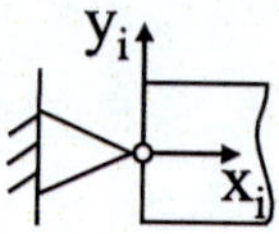	$U_1(0) = 0$ $W_1(0) = 0$ $W_1''(0) = 0$	(4c)	$U_n(L_n) = 0$ $W_n(L_n) = 0$ $W_n''(L_n) = 0$	(4d)
free	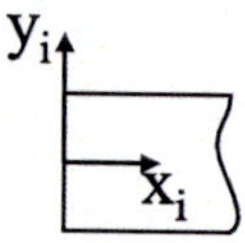	$U_1'(0) = 0$ $W_1''(0) = 0$ $W_1'''(0) = 0$	(4e)	$U_n'(L_n) = 0$ $W_n''(L_n) = 0$ $W_n'''(L_n) = 0$	(4f)

2.3 Continuity Conditions

The continuity conditions describe the relations of force and deformation quantities at the junction of two beam segments, see Fig. 2.

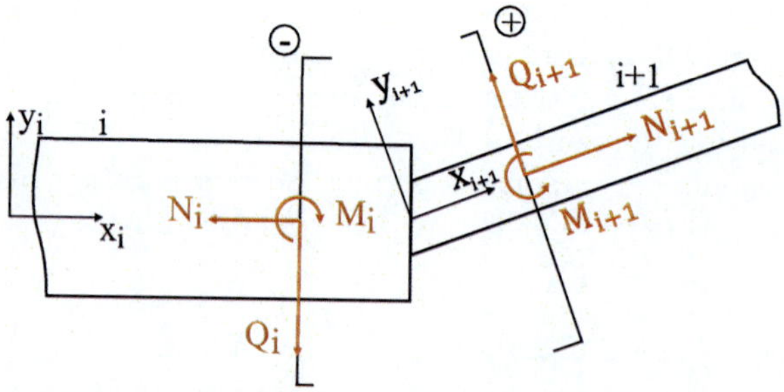

Fig. 2. Connecting point of two adjacent beams with internal forces and moments

All internal forces must occur in collinear pairs, according to Newton's Third Law, and must be equal in magnitude and opposite in direction. Taking further into account the following relations

$$N = U'EA, \quad -Q = W'''EI_z, \quad M = W''EI_z, \tag{5}$$

we obtain the continuity conditions regarding the internal forces and moments at the connecting point:

$$\begin{aligned} N: \quad & U_i'(L_i)A_i = U_{i+1}'(0)A_{i+1}\cos\alpha_i + W_{i+1}'''(0)I_{z,i+1}\sin\alpha_i, \\ Q: \quad & -W_i'''(L_i)I_{z,i} = U_{i+1}'A_{i+1}\sin\alpha_i - W_{i+1}'''(0)I_{z,i+1}\cos\alpha_i, \\ M: \quad & W_i''(L_i)I_{z,i} = W_{i+1}''(0)I_{z,i+1}. \end{aligned} \tag{6}$$

Slopes as well as axial and transversal displacements at the connecting point must be continuous, which leads to the following continuity conditions:

$$\begin{aligned} U_i(L_i) &= U_{i+1}(0)\cos\alpha_i - W_{i+1}(0)\sin\alpha_i, \\ -W_i(L_i) &= U_{i+1}\sin\alpha_i + W_{i+1}(0)\cos\alpha_i, \\ W_i'(L_i) &= W_{i+1}'(0). \end{aligned} \tag{7}$$

2.4 Matrix Form

To describe the motion of a compliant mechanism Eqs. (1a)–(7) can be written in a simple yet ineffective matrix form as follows.

$$\begin{pmatrix} x_{1,1} & x_{1,2} & \cdots & x_{1,n} \\ x_{2,1} & x_{2,2} & \cdots & x_{2,n} \\ \vdots & \vdots & \ddots & \vdots \\ x_{m,1} & x_{m,2} & \cdots & x_{m,n} \end{pmatrix} \begin{pmatrix} C_{11} \\ \cdots \\ C_{61} \\ C_{12} \\ \cdots \\ C_{6n} \end{pmatrix} = \begin{pmatrix} 0 \\ 0 \\ \cdots \\ \cdots \\ 0 \end{pmatrix} \tag{8}$$

$$\mathbf{TC} = \mathbf{0} \tag{9}$$

where $m = n$. The second index of the vector $\mathbf{C}$ indicates the corresponding segment and the first index, ranging from 1–6, symbolises the coefficient. The square matrix $\mathbf{T}$ grows in correlation to the number of segments of the mechanism. The first 3 rows are the initial conditions for the first beam and the last 3 rows for the ending conditions of the last beam, see (5) in Table 1. In between are 6 rows each for the continuity conditions (6) and (7) of the individual connecting points. For each additional segment six new rows and columns are added to the matrix.

In order to find the natural frequency of the mechanism under consideration the non-trivial solutions of its system (8) are calculated. To do this, the determinant of the matrix $\mathbf{T}$ is calculated as a function of the frequency ω. The values for ω, for which the determinant is equal to zero, are the natural frequencies of the mechanism. The determinant calculation of such huge matrices soon exceeds the computing capacity of common computers, which makes this approach unsuitable and impractical for the calculation of compliant mechanisms with already more than three segments.

2.5 Transfer Matrices

To be able to perform the calculation faster and for compliant mechanisms with an arbitrary number of beams and joints, each segment must be considered individually—individual matrices and sets of equations must be derived and then put in relation to each other. Instead of one big matrix for the whole mechanism, the boundary and continuity conditions for each beam and connection point are set up individually and then multiplied together. This always results in one single matrix $\mathbf{T}$ with the size 3×3, regardless of the number of segments within the mechanism. The matrix multiplication and the calculation of the determinant of the 3×3 matrix can be done with high computational efficiency.

First Beam Analysing for instance the motion of the mechanism in Fig. 1, the starting point is the clamped beam 1. Respectively using the boundary conditions (4a) combined with Eqs. (2) and (3), a set of equations is formed and can be written in matrix form as follows:

$$C_{11} + C_{31} = 0, \quad C_{21} + C_{41} = 0, \quad C_{51} = 0, \tag{10}$$

$$\begin{pmatrix} 1 & 0 & 0 \\ 0 & 1 & 0 \\ -1 & 0 & 0 \\ 0 & -1 & 0 \\ 0 & 0 & 0 \\ 0 & 0 & 1 \end{pmatrix} \begin{pmatrix} C_{11} \\ C_{21} \\ C_{61} \end{pmatrix}, = \begin{pmatrix} C_{11} \\ C_{21} \\ C_{31} \\ C_{41} \\ C_{51} \\ C_{61} \end{pmatrix} \tag{11}$$

$$\underset{(6\times3)\cdot(3\times1)\;=(6\times1)}{\mathbf{T}_0\mathbf{C}_0 = \mathbf{C}_1.} \tag{12}$$

$\mathbf{T}_0$ for pinned and free boundary conditions are obtained in the same way. The matrices $\mathbf{T}_i$ are called transfer matrices and the vectors $\mathbf{C}_i$ coefficient vectors.

Last Beam Moving on to the end of the last beam of the mechanism, which is free, Eqs. (2) and (3) are combined with the boundary condition (4f). This results in the following set of equations and can also be written in matrix form:

$$\begin{aligned} -C_{1n}k_n^2 c(k_n l_n) - C_{2n}k_n^2\, s(k_n l_n) + C_{3n}k_n^2 ch(k_n l_n) + C_{4n}k_n^2 sh(k_n l_n) &= 0, \\ C_{1n}k_n^3\, s(k_n l_n) - C_{2n}k_n^3 c(k_n l_n) + C_{3n}k_n^3 sh(k_n l_n) + C_{4n}k_n^3 ch(k_n l_n) &= 0, \\ -C_{5n}p_n s(p_n l_n) + C_{6n}p_n c(p_n l_n) &= 0. \end{aligned} \tag{13}$$

Instead of `sin` and `sinh` it is `s` and `sh`; instead of `cos` and `cosh` it is `c` and `ch`. The set of equations for clamped and pinned boundary conditions are obtained in the same way. These equations, describing the end of the last beam, can also be written in matrix form, such as

$$\underset{(3\times6)\cdot(6\times1)\ =(3\times1)}{\mathbf{T}_n\mathbf{C}_n = \mathbf{0}.} \tag{14}$$

Connecting Point This leaves the evaluation of the two connecting points within the mechanism. Equations (6), (7) and (2), (3) lead to a set of equations, with k is a substitute for $k_i l_i$ and p for $p_i l_i$, which can be written in matrix form as demonstrated in (15)–(17).

$$\mathrm{T}_{iL} = \begin{pmatrix} 0 & 0 & 0 & 0 & -A_i p_i s(p) & A_i p_i c(p) \\ -I_i k_i^3 s(k) & I_i k_i^3 c(k) & -I_i k_i^3 sh(k) & -I_i k_i^3 ch(k) & 0 & 0 \\ -I_i k_i^2 c(k) & -I_i k_i^2 s(k) & I_i k_i^2 ch(k) & I_i k_i^2 sh(k) & 0 & 0 \\ -k_i s(k) & k_i c(k) & k_i sh(k) & k_i ch(k) & 0 & 0 \\ c(k) & s(k) & ch(k) & sh(k) & 0 & 0 \\ 0 & 0 & 0 & 0 & c(p) & s(p) \end{pmatrix} \tag{15}$$

$$\mathrm{T}_{iR} = \begin{pmatrix} 0 & -I_{i+1}k_{i+1}^3 s(\alpha_i) & 0 & I_{i+1}k_{i+1}^3 s(\alpha_i) & 0 & A_{i+1}p_{i+1}c(\alpha_i) \\ 0 & +I_{i+1}k_{i+1}^3 c(\alpha_i) & 0 & -I_{i+1}k_{i+1}^3 c(\alpha_i) & 0 & A_{i+1}p_{i+1}s(\alpha_i) \\ -I_{i+1}k_{i+1}^2 & 0 & I_{i+1}k_{i+1}^2 & 0 & 0 & 0 \\ 0 & k_{i+1} & 0 & k_{i+1} & 0 & 0 \\ c(\alpha_i) & 0 & c(\alpha_i) & 0 & s(\alpha_i) & 0 \\ -s(\alpha_i) & 0 & -s(\alpha_i) & 0 & c(\alpha_i) & 0 \end{pmatrix} \tag{16}$$

Which can be written as

$$\begin{aligned} \mathbf{T}_{iL}\mathbf{C}_i &= \mathbf{T}_{iR}\mathbf{C}_{i+1} \\ \mathbf{T}_{iR}^{-1}\mathbf{T}_{iL}\mathbf{C}_i &= \mathbf{T}_{iR}^{-1}\mathbf{T}_{iR}\mathbf{C}_{i+1} \\ \underset{(6\times6)(6\times1)\ =(6\times1)}{\mathbf{T}_i\mathbf{C}_i} &= \mathbf{C}_{i+1}. \end{aligned} \tag{17}$$

2.6 Equation of the Natural Frequencies

Inserting Eqs. (11), (14) and (17) into one another leads to the following equation

$$\mathbf{T}_n\mathbf{T}_{n-1}\cdots\mathbf{T}_{i+1}\mathbf{T}_i\mathbf{T}_0\mathbf{C}_0 = \mathbf{0} \tag{18}$$

$$\Rightarrow \mathbf{T}\mathbf{C}_0 = \mathbf{0}. \tag{19}$$

$$\det(\mathbf{T}) = 0. \tag{20}$$

The natural frequencies of the mechanism correspond to the values for ω for which $\det(\mathbf{T})=0$.

2.7 First Verification of the Analytical Approach

In order to prove the presented analytical method a numerical verification is carried out through FEA. For this purpose eleven mechanisms with different dimensions and angles are examined, such as the mechanism in Fig. 3. It consists of three beams with the angles $\alpha_{1,2} = \frac{\pi}{6}$ and dimensions $5\times5\times50\,\text{mm}^3$, $1\times1\times10\,\text{mm}^3$ and $5\times5\times40\,\text{mm}^3$. The material of the mechanisms is structural steel with a density of $\rho=7850\,\frac{kg}{m^3}$ and Young's modulus E=200,000 MPa.

Table 2. Natural frequencies of Fig. 3; results and deviation (dev.)

Natural frequency	Mathematica [Hz]	ANSYS [Hz]	Dev. [%]
1	84.4221	84.294	0.15
2	755.874	750.35	0.73
3	2439.97	2396.3	1.79
4	7742.07	7517.4	2.90
5	12737.9	12212	4.13

The Program Wolfram Mathematica 9 is used to calculate the natural frequencies with the given Eqs. (18)–(20) and the numerical analysis is carried out as a modal analysis with FEA-models using ANSYS Workbench 2019 R3.

As one can see in Table 2 the deviations of the first five natural frequencies of the mechanism shown in Fig. 3 between the analytical and numerical calculation are in a range of less than 4%. Similar deviations are obtained for the other calculated mechanisms.

3 Design of the Calculation Tool

The derived analytical method can be easily implemented into a calculation tool, due to the pre defined boundary conditions and the consistent calculation of the matrices.

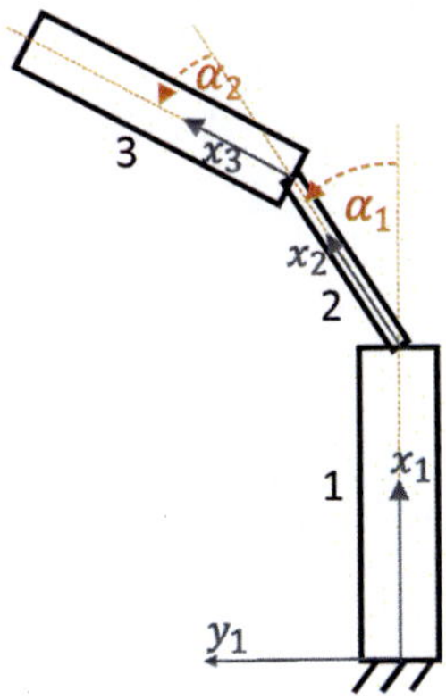

Fig. 3. Exemplary mechanism, beginning clamped, end free

3.1 Programming Language Python

For programming the graphical user interface the programming language Python was chosen with its library Tkinter. It is a universal, usually interpreted, higher programming language, which can be run platform independent. One of the great advantages of Python is the extreme variety of freely available standard libraries, which allow the language's range of functions to be extended at will. The majority of the libraries are also platform independent, so that even larger Python programs run on Unix, Windows, macOS and other platforms without any changes.

The Python program is split into several files, for example the language-dependent terms, standard values, colours of the controls etc., making it easy to make changes to the standard configuration of the program.

3.2 Design of the Graphical User Interface for the Calculation Tool

The concept behind the structure of the user interface is a large drawing area for the two-dimensional graphical representation of the current beam configuration on the right-hand side, with all operating tools of the program on the left-hand side, see Fig. 4.
The two-dimensional drawing of the beams shows the width, length and angle of each beam segment. Furthermore, a graphic representation of the respective boundary conditions at the start and end is implemented. On the left side of the program the parameters are added. The Young's-Modulus and density are defined for the whole mechanisms, whereas the geometric parameters for the beam segments are assigned individually. The program supports entering the angle in both degrees and radians, which can be selected via an associated checkbox. The selection of the beam segment can be changed using a drop-down menu or by clicking the respective segment with the mouse. Changes to the width, length or angle of a segment are shown in real time on the drawing

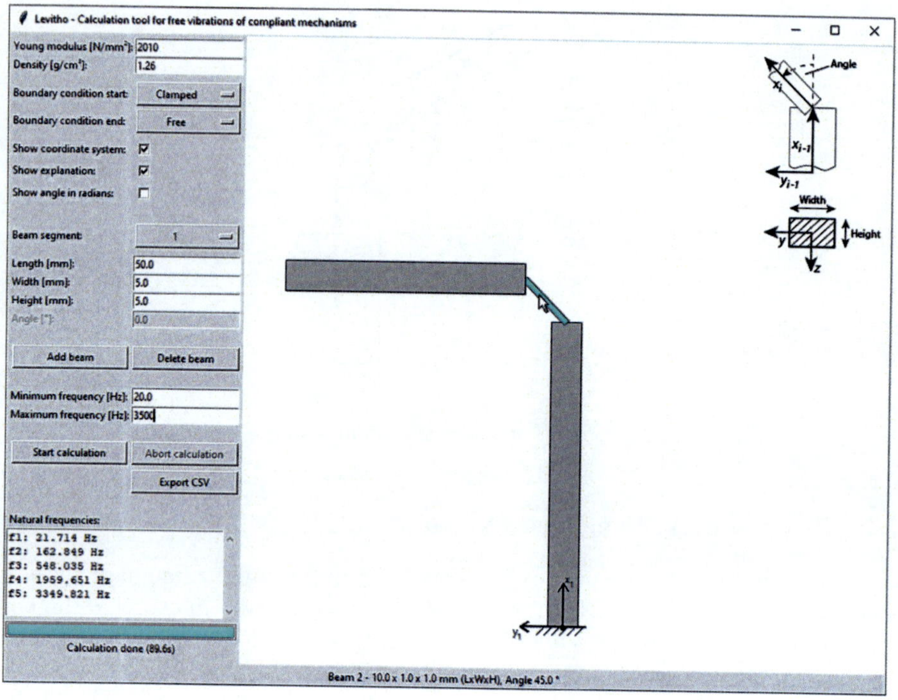

Fig. 4. Screenshot of the calculation tool LeViTho

area. Furthermore, checkboxes for the display of the coordinate system for the first segment as well as a short explanation of the dimensions and numbering of each beam segment are featured. In the lower section of the input area the user can input the frequency range, in which the calculation function will search for the natural frequencies. Below the input fields there is a button to start the calculation and another one to cancel the calculation.

Invalid entries are checked before the actual calculation is started and corresponding error messages are displayed in the result text field. During the calculation, all the solutions found in the given frequency range are displayed in the results text field, while the progress bar below shows the current calculation progress in percent as well as a rough estimate of the remaining calculation time in seconds.

Finally, the program offers the possibility to export the entered parameters (Young's-Modulus, density, boundary conditions, geometry of each beam segment, frequency range) as well as the corresponding natural frequencies resulting from the calculation in form of a .csv file.

4 Validation and Verification

To create an assortment of reference models on which the proposed analytical method can be assessed various Test Specimens (TS) are designed and manufactured. The tests for the validation are carried out as free and forced vibration tests. The verification is then done by FEA and the parameters used can be found in the appendix Table 7.

4.1 Test Specimen Design

The TS are plane symmetric in respect to the XY-plane and the XZ-plane. For compliance with the TMM the TS have a rectangular cross section, though for mechanical and manufacturing reasons the corners of the flexure hinges are filleted, characterized as corner-filleted notch hinges, see Fig. 5.

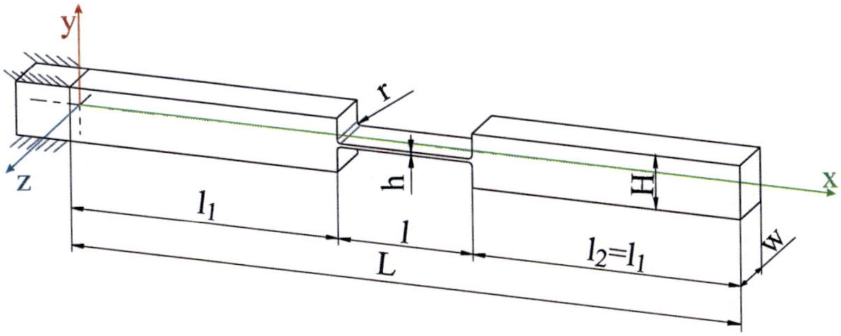

Fig. 5. Visual representation of test specimens parameters, see Table 3

The specimens have a constant width in the Z-direction, so they are biased to oscillate around this axis. The dimensions of the four TS are specified in Table 3. All values are within ±0.020 the nominal value if not stated differently. Due to the immense influence of the notch height of the flexure hinge, the actual height was measured and later used for the calculations with LeViTho and for the FEA.

Table 3. Geometric dimensions of TS: L-length of whole mechanism, l-length of compliant segment, w-width, H-height, h-height compliant segment, r-radius

TS	L [mm]	l [mm]	w [mm]	H [mm]	h [mm]	$h_{measured}$ [mm]	r [mm]
1	140	8	8	$8^{+0.004}_{-0.082}$	$0.8^{+0.004}_{-0.082}$	0.777	0.5
2	140	30	8	$8^{+0.002}_{-0.084}$	$0.8^{+0.002}_{-0.084}$	0.753	0.5
3	100	8	8	$8^{+0.010}_{-0.035}$	$0.5^{+0.010}_{-0.035}$	0.460	0.5
4	100	20	8	$8^{+0.013}_{-0.065}$	$0.5^{+0.013}_{-0.065}$	0.444	0.5

Compliant mechanisms can be categorized to have concentrated or distributed compliance. Specimens with a dimensionless ratio of $\frac{L}{l} \geq 10$ are defined to have Concentrated Compliance (CC), specimens with a ratio of $\frac{L}{l} < 10$ are defined to have Distributed Compliance (DC) [15]. Thus, TS 1 and 3 feature CC, TS 2 and 4 DC.

4.2 Material Choice and Manufacturing

To guarantee the longevity and avoid deterioration of the TS a cold work tool steel, 100MnCrW4 (DIN EN ISO 4957; 1.2510) was picked. To avoid creeping of the TS during the cutting process due to releasing internal stresses the material was annealed twice by the supplier. The Youngs-Modulus was specified by the supplier as E= 193 000 MPa, the density is measured to be 7776 $\frac{kg}{m^3}$. The TS were cut on a Wire Erosion Discharge Machine (WEDM) Charmilles Robofill 240 with a single rough cut (v_c=3 $\frac{mm}{min}$).

4.3 Free Vibration Testing

The test setup for measuring the first natural frequency can be seen in Fig. 6. The TS are treated as cantilever beams. The fixed end is clamped with a set of wedges in a force-fitting manner in a clamping device. The XY-Plane is oriented parallel to the ground to avoid interaction with gravity.

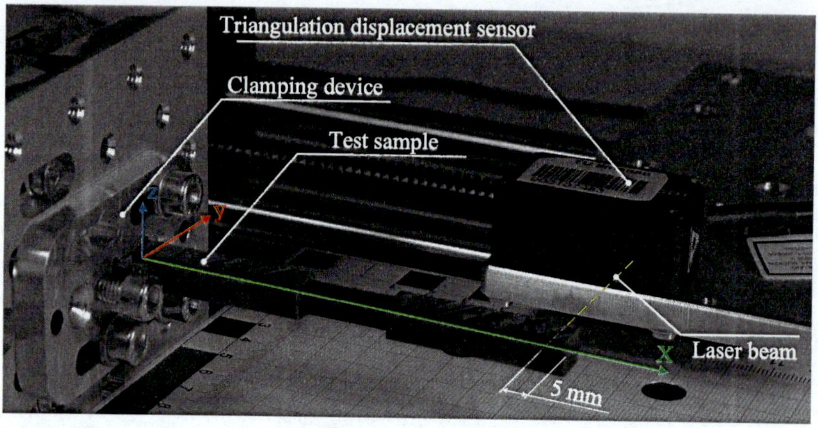

Fig. 6. Free vibration test setup with a triangulation displacement sensor

Every TS is deflected to a specified value and then released to start the oscillation. The procedure is repeated six times for every test set up, the Standard Deviation (SD) is deducted. The amplitudes are recorded over time by the laser triangulation displacement sensor (LTD) Micro Epsilon optoNCDT 1420-100, its parameters can be found in the appendix, see Table 6. The Fast Fourier Transformation (FFT) is applied to the raw data, which shows the amplitudes and corresponding frequencies.

4.4 Dynamic Vibration Testing

To recover the second natural frequency the TS are additionally examined with a Laser-Doppler-Scanning-Vibrometer. The TS are clamped in a cantilever manner to the shaker and are agitated along a band of frequencies while the Vibrometer scans the respond on the surface of the TS. The system is set to a sampling rate of 5.12 kHz at a sweep resolution of 625 mHz. The shaker is set to sweep a frequency band from 1 Hz to 1.6 kHz in a time frame of 4 seconds. Once again, the first natural frequencies are really evident. The frequency of the second planar eigenmode is found for the 2nd and 4th test specimen (DC), see Fig. 7. The 1st and 3rd specimen (CC) showed no clear frequency respond for the second planar eigenmode.

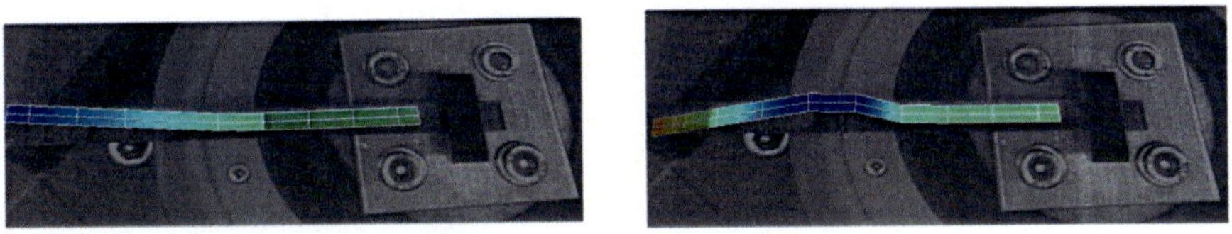

Fig. 7. First (left) and second (right) planar normal mode of test specimen 4, excitation through Vibrometer

4.5 Verification Through Finite Element Analysis

To verify the results obtained by the TMM, the models of the test specimen are analysed with the commercial software SolidWorks with the simulation add-on. The first 5 normal modes are calculated, those corresponding with the transversal vibration on the XY-plane, see Fig. 8, are used as comparative values for the evaluation.

Fig. 8. 1st (left) and 2nd (right) planar natural frequency of TS 4 in FEM

5 Results

The prismatic TS were examined by free vibration testing. The first natural frequency is found by processing the signal from the LTD sensor with FFT. The elastic behaviour of the flexure hinge results in a well-defined peak in the amplitude spectrum, as shown for TS1 in Fig. 9, corresponding to the frequency of the first mode shape of the investigated specimen. The first natural frequency

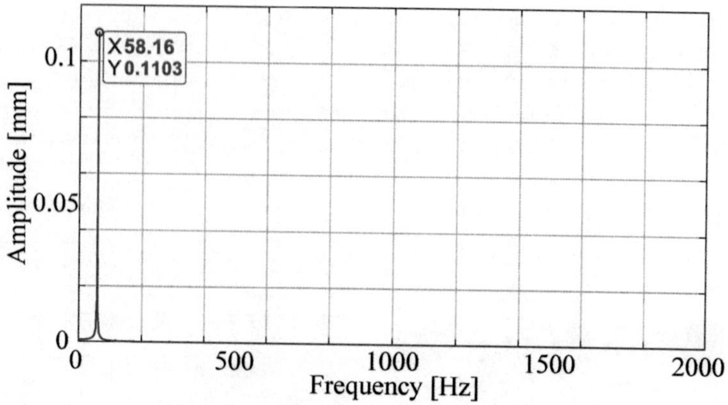

Fig. 9. Frequency spectrum of TS1

is very evident for all specimens, the second natural frequency around the Z-direction (the 3rd or 4th spatial natural frequency) though is very small and not clearly distinguishable from the background noise.

The oscillation of the test specimen is subject to losses which results is the decay of the amplitudes over time. Energy dissipation occurs externally due to air resistance and internally due to internal friction resulting in a dampened system. From the decay of the amplitudes the logarithmic decrement δ is calculated, as shown in (21).

$$\delta = \frac{1}{i} ln \left(\frac{x(t)}{x(t+iT)} \right) \tag{21}$$

i is an integer defining the number of consecutive positive peaks occurring within a decrement of 10 dB, $x(t)$ is the amplitude peak value at time t, T is the period duration. From the logarithmic decrement the damping ratio ξ is found in (22).

$$\xi = \left(\frac{\delta}{\sqrt{4\pi^2 + \delta^2}} \right) \tag{22}$$

The damping ratio as well as the logarithmic decrement for the results of the first natural frequencies of the free vibration tests are listed in the appendix, see Table 5.

To reveal the second natural frequency of the prismatic TS a modal-analysis is carried out with a Vibrometer. The second mode shape for the specimens with DC is identified as well as the related natural frequency. The first two planar natural frequencies are also calculated for all four TS with the LeViTho tool and through FEA.

The results for free and forced vibration tests as well as the FEA and their respective deviations to LeViTho are listed in Table 4.

Table 4. Comparison of results of free and forced vibration test (V.T.) and FEA to LeViTho results; N.F.-natural frequency, Δ-devitation to LeViTho

Test specimen		1	2	3	4
LeViTho	1st N.F. [Hz]	57.09	28.13	43.52	25.93
	2nd N.F. [Hz]	981.37	407.54	1160.42	401.39
Free V.T.	1st N.F. [Hz]	58.09	27.89	43.81	24.27
	Δ [%]	1.74	−0.85	0.66	−6.39
Forced V.T.	1st N.F. [Hz]	54.69	27.50	43.13	24.38
	Δ [%]	−4.21	−2.23	−0.90	−5.97
	2nd N.F. [Hz]	–	391.30	–	390.60
	Δ [%]	–	−3.98	–	−2.69
FEA	1st N.F. [Hz]	56.97	28.20	44.33	26.16
	Δ [%]	−0.21	0.27	1.87	0.88
	2nd N.F. [Hz]	967.63	403.40	1151.2	400.22
	Δ [%]	−1.40	−1.02	−0.79	−0.29

6 Discussion

As can be seen in Table 4 and is visualized in Fig. 10, the results from the experiments and the finite element analysis compared to the results obtained with the proposed analytical method implemented in the programm LeViTho are mostly in very good correlation with an average deviation of 1.42%.

The experimental results from TS 4 show comparably large deviations for the first natural frequency. As mentioned, the TS were cut via WEDM process, though they were only rough cut. This may have led to many imperfections in the surface and dimensional inaccuracy which leads to larger deviations between the experiment and the calculation with the ideal geometry (LeViTho). Another factor which would influence the natural frequency of the TS could be stress within the material, due to manufacturing processes.

Furthermore, the properties of the rim zone that was altered by the cut might have influenced the results to a certain degree, as stated by [7]. The dimensional inaccuracies were compensated in the calculations as good as possible by measuring the deviation from the nominal value, see Table 3. The calculations depend on the geometrical dimensions as well as the physical quantities of Young's-Modulus and the density. The density was calculated from the theoretical volume of the TS and their measured weight. The Young's-Modulus was taken as indicated by the supplier, but it was not independently measured.

Additionally, the different orientation of the TS during the two testing setups may influence the measured natural frequency. In the free vibration test run the vibration axis was oriented in line with the gravitational vector. This was not possible in the forced vibration test run due to the setup of the Vibrometer, the specimen were therefore oriented perpendicular. The somewhat larger deviations

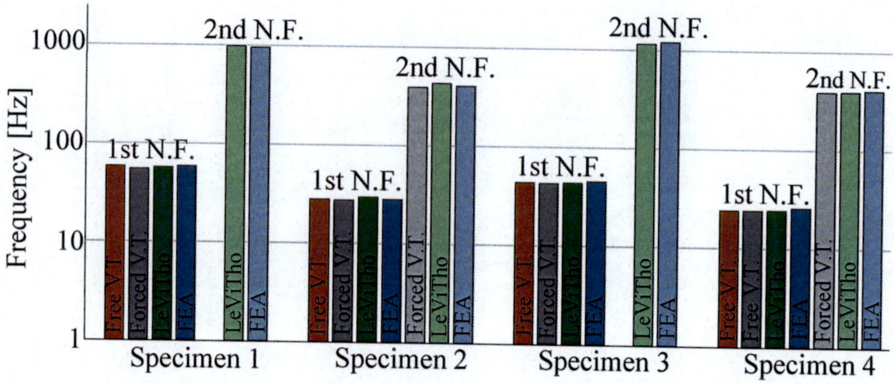

Fig. 10. Comparison of results of free and forced vibration testing, *LeViTho* and FEA: 1st N.F.-first natural frequency, 2nd N.F.-second natural frequency, free V.T.-free vibration test, forced V.T.-forced vibration test

in the forced V.T. might be attributable to the acting gravity. The damping ratio was calculated for the free vibration, it is small enough to be neglected comparing to other influence factors. Furthermore, neither LeViTho nor the FEA take into consideration the filleted corners of the TS.

7 Conclusion and Outlook

This paper shows that the proposed analytical approach can be legitimately applied to compliant mechanisms for calculating their natural frequencies. Several single flexure hinges with CC and DC were designed, manufactured and tested. The frequency values obtained by this method are in very good correlation to the measured frequencies and to the results from the FEA. With the development of LeViTho a handy tool with practical application was created. It might be used in the design stage of compliant machines to prevent harmonic reactions or it could be applied to troubleshoot when existing machines show signs of harmonic interaction. The big advantage beside the reasonably precise calculations, is the fast and intuitive handling of the software, which compared to the FEA has the potential to save a good amount of time. As observed in the testing phase, compliant mechanisms are very sensible to manufacturing tolerances, in this case LeViTho can be used to simulate the consequences deriving from manufacturing variations. The current state of the analytical method implemented in LeViTho is that it is only applicable for planar continuous systems, even though the calculation of merged systems, i.e. the segments can have different material properties, is possible with few modifications. In the future, further investigations will be carried out to calculate spatial systems, systems with branching points and possibly flexure hinges.

8 Appendix

Table 5. Free vibration test results; N.F.-natural frequency, SD-standard deviation, δ-logarithmic decrement, ξ-damping ratio

TS	1st N.F. [Hz]	SD [Hz]	δ	SD_δ	ξ	SD_ξ
1	58.09	0.054	1.84E−02	2.44E−04	2.92E−03	3.89E−05
2	27.89	0.029	6.30E−03	5.87E−05	1.00E−03	9.35E−06
3	43.81	0.049	1.47E−02	1.24E−03	2.34E−03	1.97E−04
4	24.27	0.010	5.86E−03	4.16E−04	9.32E−04	6.62E−05

Table 6. Sensor parameters Micro Epsilon optoNCDT 1420-100

Parameter	Value	Unit
Measuring range	100	mm
Sampling rate	4	kHz
Linearity	$< \pm 0.08$	%
Repeatability	4	µm

Table 7. FEA parameters

Parameter	Value	Unit
Solver	FFEPlus	
Mesh elements	4 Jacobi points	
Max. element size	1	mm
Max. aspect ratio	4.38	
Jacobi mesh quality	1–1.213	
Mesh size at flexure	0.162	mm
Mesh size ratio at flexure	1.9	

References

1. Attar, M.: A transfer matrix method for free vibration analysis and crack identification of stepped beams with multiple edge cracks and different boundary conditions. Int. J. Mech. Sci. **57**(1), 19–33 (2012). https://doi.org/10.1016/j.ijmecsci.2012.01.010

2. Boiangiu, M., Ceausu, V., Untaroiu, C.D.: A transfer matrix method for free vibration analysis of Euler-Bernoulli beams with variable cross section. J. Vibr. Control **22**(11), 2591–2602 (2016). https://doi.org/10.1177/1077546314550699
3. Da Vaz, J.C., de Lima Junior, J.J.: Vibration analysis of Euler-Bernoulli beams in multiple steps and different shapes of cross section. J. Vibr. Control **22**(1), 193–204 (2016). https://doi.org/10.1177/1077546314528366
4. Hu, J., Wen, T., He, J.: Dynamics of compliant mechanisms using transfer matrix method. Int. J. Precis. Eng. Manuf. **21**(11), 2173–2189 (2020). https://doi.org/10.1007/s12541-020-00395-9
5. Khiem, N.T., Lien, T.V.: A simplified method for natural frequency analysis of a multiple cracked beam. J. Sound Vibr. **245**(4), 737–751 (2001). https://doi.org/10.1006/jsvi.2001.3585
6. Khiem, N.T., Lien, T.V., Ninh, V.T.A.: Natural frequencies of multistep functionally graded beam with cracks. Iran. J. Sci. Technol., Trans. Mech. Eng. **43**(1), 881–916 (2018). https://doi.org/10.1007/s40997-018-0201-x
7. Klocke, F., Hensgen, L., Klink, A., Mayer, J., Schwedt, A.: EBSD-analysis of flexure hinges surface integrity evolution via wire-EDM main and trim cut technologies. Procedia CIRP **13**, 237–242 (2014). https://doi.org/10.1016/j.procir.2014.04.041
8. Li, Z., Kota, S.: Dynamic analysis of compliant mechanisms. In: Howell, L.L. (ed.) Proceedings of the 2002 ASME Design Engineering Technical Conferences and Computers and Information in Engineering Conference. pp. 43–50. American Society of Mechanical Engineers, New York, NY (2002). 10.1115/DETC2002/MECH-34205
9. Liu, P., Yan, P.: A modified pseudo-rigid-body modeling approach for compliant mechanisms with fixed-guided beam flexures. Mech. Sci. **8**(2), 359–368 (2017). https://doi.org/10.5194/ms-8-359-2017
10. Lyon, S.M., Erickson, P.A., Evans, M.S., Howell, L.L.: Prediction of the first modal frequency of compliant mechanisms using the pseudo-rigid-body model. J. Mech. Des. **121**(2), 309–313 (1999). https://doi.org/10.1115/1.2829459
11. Obradović, A., Šalinić, S., Trifković, D.R., Zorić, N., Stokić, Z.: Free vibration of structures composed of rigid bodies and elastic beam segments. J. Sound Vibr. **347**(347), 126–138 (2015)
12. Vedant, Allison, J.T.: Pseudo-rigid body dynamic modeling of compliant members for design. In: Proceedings of the ASME International Design Engineering Technical Conferences and Computers and Information in Engineering Conference—2019. The American Society of Mechanical Engineers, New York, N.Y. (2020). 10.1115/DETC2019-97881
13. Wang, W., Yu, Y.: Analysis of frequency characteristics of compliant mechanisms. Front. Mech. Eng. China **2**(3), 267–271 (2007). https://doi.org/10.1007/s11465-007-0046-2
14. Yu, Y.Q., Howell, L.L., Lusk, C., Yue, Y., He, M.G.: Dynamic modeling of compliant mechanisms based on the pseudo-rigid-body model. J. Mech. Des. **127**(4), 760–765 (2005). https://doi.org/10.1115/1.1900750
15. Zentner, L., Linss, S.: Compliant Systems: Mechanics of Flexible Mechanisms, Actuators and Sensors. De Gruyter, Berlin and Boston (2019). 10.1515/9783110479744
16. Zheng, Y., Yang, Y., Wu, R.J., He, C.Y., Guang, C.H.: Dynamic modeling of compliant mechanisms based on the pseudo-rigid-body model. Mech. Mach. Theory **155**, 104095 (2021). https://doi.org/10.1016/j.mechmachtheory.2020.104095

Towards Topology Optimization of Pressure-Driven Soft Robots

Prabhat Kumar(✉)

Department of Mechanical and Aerospace Engineering, Indian Institute of Technology Hyderabad, Telangana 502285, India
pkumar@mae.iith.ac.in

Abstract. Soft robots are made of compliant materials that perform their tasks by deriving motion from elastic deformations. They are used in various applications, e.g., for handling fragile objects, navigating sensitive/complex environments, etc., and are typically actuated by Pneumatic/hydraulic loads. Though demands for soft robots are continuously increasing in various engineering sectors, due to the lack of systematic approaches, they are primarily designed manually. This paper presents a systematic density-based topology optimization approach to designing soft robots while considering the design-dependent behavior of the actuating loads. We use the Darcy law with the conceptualized drainage term to model the design-dependent nature of the applied pressure loads. The standard finite element is employed to evaluate the consistent nodal loads from the obtained pressure field. The robust topology optimization formulation is used with the multi-criteria objective. The success of the presented approach is demonstrated by designing a member/soft robot of the pneumatic networks (PneuNets). The optimized member is combined in several series to get different PneuNets. Their CAD models are generated, and they are studied with high-pressure loads in a commercial software. Depending upon the number of members in the PneuNets, different output motions are noted.

Keywords: Soft robots · Topology optimization · Design-dependent loads · Compliant mechanisms

1 Introduction

Soft robots are constituted by compliant materials and have monolithic lightweight designs [1]. Such robots are actuated primarily by pneumatic/hydraulic (fluidic pressure) loads and use motion obtained from elastic

Science & Engineering Research Board, Department of Science and Technology, Government of India under the project file number RJF/2020/000023.

A. K. Pandey et al. (Eds.): MAMM 2022, 126, pp. 19–30, 2023.
https://doi.org/10.1007/978-3-031-20353-4_2

deformation to perform their tasks. Nowadays, they are being used in a wide range of applications, e.g., to handle fragile objects, fruits, and vegetables, in sensitive and unstructured environments for navigation, etc. In addition, they provide high power-to-weight ratios and help achieve complex motions [1]. Therefore, interest in designing them for different applications is constantly growing. The pneumatically/hydraulically (pressure loads, air/water) actuated soft robots are sought the most and are used relatively more. In general, soft robots are designed manually using heuristic methods because of the lack of systematic approaches. Heuristic methods greatly depend upon the designers' knowledge and experience and may require many resources/iterations. Therefore, the goal of this paper is to present a systematic approach using topology optimization for designing pressure-driven soft robots. Figure 1 displays a schematic diagram of a soft robot with a bellow-shaped pressure loading chamber. When pneumatic/hydraulic loads inflate the chamber, it is desired that the output point P moves in a bending motion, as shown by the red curved arrow.

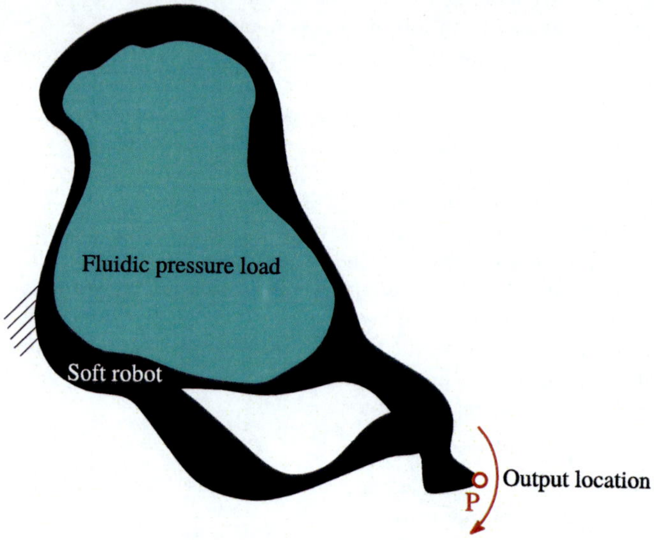

Fig. 1. A schematic diagram for a soft robot.

Topology optimization (TO) provides an optimized material distribution for a design problem by extremizing the desired objective under the given physical and geometrical constraints. In a typical TO setting, the design domain is described by finite elements (FEs, cf. [2]), and each element is assigned a design variable $\rho \in [0, 1]$. $\rho = 0$ and $\rho = 1$ indicate an element's void and solid states, respectively. The applied loads/actuating forces depending upon the applications soft robots are designed for, can be either design-dependent, e.g., fluidic pressure load [3], or constant. A design-dependent load changes its location, magnitude, and direction as TO advances and thus, poses several distinctive challenges [3].

Hiller and Lipson [4] used the evolutionary topology optimization approach to design soft robots. Chen et al. [5] developed a soft cable-driven gripper using the level-set topology optimization method. Zhang et al. [6] developed a soft pneumatic gripper by maximizing the output displacement. Pressure loads always act normal to the boundaries of the design domain, which alter with TO iterations. Therefore, including design-dependent behavior within the optimization formulation is essential for fluid pressure loads, which is not considered in Refs. [4–6]. In addition, Refs. [7,8] mentioned that compliant mechanisms (CMs) optimized using TO suffer from point (single-node) connections, and thus, they either become challenging to realize or require post-processing. Further, the performance of the post-processed designs may not be the same as that of their numerical counterparts. Herein, we use the robust formulation presented in [9] to circumvent this issue, considering the design-dependent characteristics of pressure loads for designing soft robots. Readers can refer to Refs. [10–13] and references therein for designing various CMs for different applications using TO.

Hammer and Olhoff [14] were the first to consider the design-dependent nature of the pressure loads in TO while designing structures. Chen et al. [15] used a fictitious thermal model to design pressure-actuated compliant mechanisms. Sigmund and Clausen [16] presented the mixed-finite element-based approach. Panganiban et al. [17] employed a non-confirming FE method in their approach for pressure-actuated CMs. The solid isotropic material with penalization (SIMP) and moving isosurface threshold schemes are used by Vasista and Tong [18]. de Souza and Silva [19] used the method presented in Ref. [16] with a projection filter. Kumar et al. [3] presented a novel approach using the Darcy law for pressure field modeling. The method uses the standard FE method and works fine for designing 3D CMs [20]. Thus, we adopt the method to model the pressure load. The prime goal herein is to design a member (soft robot) of the pneumatic network (PneuNets, cf. [21]) to achieve the specified motion. The optimized member is further connected in several series to get different output motions with high-pressure loads.

The remainder of the paper is structured as follows. Section 2 summarizes pressure load modeling using the Darcy law in brief. TO formulation is provided in Sect. 3. Section 4 presents an optimized design for a member of the PneuNet soft robot. The optimized design is extracted, and its CAD model is made. The different networks are generated from the CAD model and further analyzed with higher pressure loads in commercial software to achieve complex motions. Lastly, the conclusions are drawn in Sect. 5.

2 Pressure Load Modeling

In this section, we present the pressure load modeling using the Darcy law in brief herein for completeness. One can refer to [3] for a detailed description.

As TO advances, the material states of the associated FEs evolve. We already have known boundaries with input pressure and zero pressure load at the initial stage of TO, i.e., a pressure difference across the domain. Therefore, using the

Darcy low to determine the pressure field while assuming elements as porous is natural. Given Darcy law, the flux $\boldsymbol{q}$ is determined as

$$\boldsymbol{q} = -\frac{\kappa}{\mu}\nabla p = -K(\bar{\rho})\nabla p, \tag{1}$$

where ∇p is the pressure gradient. κ and μ represent the permeability of the medium and the fluid viscosity, respectively. $\bar{\rho}$ indicates the physical design variable. $K(\bar{\rho})$ is called the flow coefficient. For element e, the flow coefficient is defined as [3]

$$K(\bar{\rho_e}) = K_v\left(1 - (1-\epsilon)\mathcal{H}(\bar{\rho_e},\, \beta_\kappa,\, \eta_\kappa)\right), \tag{2}$$

where $\mathcal{H}(\bar{\rho_e},\, \beta_\kappa,\, \eta_\kappa) = \frac{\tanh(\beta_\kappa\eta_\kappa)+\tanh(\beta_\kappa(\bar{\rho_e}-\eta_\kappa))}{\tanh(\beta_\kappa\eta_\kappa)+\tanh(\beta_\kappa(1-\eta_\kappa))}$, and $\epsilon = \frac{K_s}{K_v}$ is called flow contrast [20]. K_s and K_v are the flow coefficient of solid and void phases of element e respectively. $\{\eta_\kappa,\, \beta_\kappa\}$ are termed flow parameters which define respectively the step position and slope of $K(\bar{\rho_e})$. To get the meaningful pressure distribution in a TO setting, drainage, Q_drain, is conceptualized [3, 20]. The balanced equation of Eq. 1 with the drainage can be determined as [3]:

$$\nabla\cdot\boldsymbol{q} - Q_\text{drain} = 0. \tag{3}$$

where $Q_\text{drain} = -D(\bar{\rho_e})(p - p_\text{ext})$ with $D(\bar{\rho_e}) = D_\text{s}\mathcal{H}(\bar{\rho_e},\, \beta_d,\, \eta_d)$. $\{\eta_\text{d},\, \beta_\text{d}\}$ are the drainage parameters. In view of the fundamentals of the finite element formulations, Eq. 3 transpires to [3]

$$\mathbf{Ap} = \mathbf{0}, \tag{4}$$

in case both external pressure load and surface flux are set equal to zero, which is the case herein considered. $\mathbf{A}$ and $\mathbf{p}$ are the global flow matrix and pressure vector, respectively. Equation 4 gives the pressure field distribution within the design domain as TO advances. We find the consistent nodal loads from the pressure field distribution as [3]:

$$\mathbf{F} = -\mathbf{Tp} \tag{5}$$

where $\mathbf{F}$ is the global force vector, and $\mathbf{T}$ is the transformation matrix [3]. To summarize, with TO iterations, Eqs. 4 and 5 are used to determine the pressure field and corresponding nodal force vector.

3 Topology Optimization Formulation

We use the robust formulation [9], i.e., the eroded, intermediate, and dilated descriptions of the design field, to find the optimized designs for the soft robots. The worst objective of the eroded, intermediate, and dilated designs is minimized with the given volume fraction. Mathematically, the optimization formulation can be written as [22]

$$\left.\begin{array}{l}\min_{\boldsymbol{\rho}} : \max : \left(f_0(\bar{\boldsymbol{\rho}}^d),\, f_0(\bar{\boldsymbol{\rho}}^i),\, f_0(\bar{\boldsymbol{\rho}}^e)\right) \\ \text{Subjected to :} \\ \boldsymbol{\lambda}_m^1 : \qquad \mathbf{A}_m\mathbf{p}_m|_{m=d,i,e} = \mathbf{0}, \\ \boldsymbol{\lambda}_m^2 : \qquad \mathbf{K}_m\mathbf{u}_m = \mathbf{F}_m = -\mathbf{T}_m\mathbf{p}_m \\ \boldsymbol{\lambda}_m^3 : \qquad \mathbf{K}_m\mathbf{v}_m = \mathbf{F}_\text{d} \\ \mu : \qquad V(\bar{\boldsymbol{\rho}}^d(\boldsymbol{\rho})) - V_d^* \le 0 \\ \quad \mathbf{0} \le \boldsymbol{\rho} \le \mathbf{1} \\ \text{Data: } V_d^*, \Delta\eta, r_\text{fill}, P_\text{in}, \mathbf{F}_\text{d}\end{array}\right\}, \tag{6}$$

where f_0 is a multi-criteria objective, which is defined by $-s\frac{MSE}{SE}$ [23]. MSE indicates the mutual strain energy, and SE represents the strain energy. $MSE = \mathbf{v}^\top\mathbf{K}\mathbf{u}$ and $SE = \frac{1}{2}\mathbf{u}^\top\mathbf{K}\mathbf{u}$. s is a consistent scale factor. As per [9], the volume fraction is applied using the dilated field and is updated with TO iterations. $\mathbf{u}$ and $\mathbf{v}$ are the global displacement vectors obtained in response to the actual and dummy forces, respectively. $\mathbf{K}$ is the global stiffness matrix. $\bar{\boldsymbol{\rho}}$ is the physical design vector. $\bar{\rho}_j$ of element j is determined as [9]:

$$\bar{\rho}_j(\tilde{\rho}_j,\, \beta,\, \eta) = \frac{\tanh(\beta\eta) + \tanh(\beta(\tilde{\rho}_j - \eta))}{\tanh(\beta\eta) + \tanh(\beta(1-\eta))}, \tag{7}$$

where $\eta \in [0,\, 1]$ defines the threshold, and $\beta \in [0,\, \infty)$ controls the steepness of the projection function. Generally, β is increased from 1 to a large finite value using a continuation scheme. $\tilde{\rho}_j$, the filtered design variable of element j, is defined as [24]:

$$\tilde{\rho}_j = \frac{\sum_{k=1}^{N_e} v_k\rho_k w(\mathbf{x}_k)}{\sum_{k=1}^{N_e} v_k w(\mathbf{x}_k)} \tag{8}$$

where N_e indicates the total number of elements used to parameterize the design domain, and v_k is the volume of neighboring element k. $w(\mathbf{x}_k) = \max\left(0,\, 1 - \frac{d}{r_\text{fill}}\right)$, is the weight function, wherein $d = ||\mathbf{x}_j - \mathbf{x}_k||$ is a Euclidean distance between centroids $\mathbf{x}_j$ and $\mathbf{x}_k$ of elements j and k, respectively. One finds the derivative of $\bar{\rho}_j$ (Eq. 7) with respect to $\tilde{\rho}_j$ as:

$$\frac{\partial\bar{\rho}_j}{\partial\tilde{\rho}_j} = \beta\frac{1 - \tanh(\beta(\tilde{\rho}_j - \eta))^2}{\tanh(\beta\eta) + \tanh(\beta(1-\eta))}. \tag{9}$$

Likewise, the derivative of $\tilde{\rho}_j$ (Eq. 8) with respect to ρ_k can be evaluated as

$$\frac{\partial\tilde{\rho}_j}{\partial\rho_k} = \frac{v_k w(\mathbf{x}_k)}{\sum_{i=1}^{N_e} v_i w(\mathbf{x}_i)}. \tag{10}$$

Finally, using the chain rule one can find the derivative a function f with respect to ρ_k as

$$\frac{\partial f}{\partial\rho_k} = \sum_{j=1}^{Ne} \frac{\partial f}{\partial\bar{\rho}_j}\frac{\partial\bar{\rho}_j}{\partial\tilde{\rho}_j}\frac{\partial\tilde{\rho}_j}{\partial\rho_k}, \tag{11}$$

We use the modified SIMP material scheme [25] for finding Young's modulus of element j as

$$E_j = E_0 + (\bar{\rho_j})^\chi (E_1 - E_0), \qquad \bar{\rho_j} \in [0,\, 1] \tag{12}$$

where χ is the SIMP parameter. E_0 and E_1 are Youngs' moduli of the void and solid states of an FE, respectively. $\frac{E_0}{E_1} = 1\mathrm{e}^{-6}$ is set in this paper.

The Method of Moving Asymptotes (MMA, cf. [26]), a gradient-based optimizer, is used to solve the optimization problem (Eq. 6). Thus, we need the sensitivities of the objective and constraint with respect to the design variables, which are determined using the adjoint-variable method. Complete detail on sensitivity analysis can be found in [3,22].

4 Numerical Results and Discussions

In this section, a member of PneuNets (Fig. 1(B) of [21]) is designed using the presented method. Different PneuNets are constructed using the optimized design to achieve complex motions with high-pressure loads, which we analyze in a commercial software.

The design domain is displayed in Fig. 2. $L_x = 0.1$ m and L_y =0.15 m indicate respectively the dimension in $x-$ and $y-$directions. The out-of-plane thickness is set to 0.001 m. The input pressure load of 1 bar is applied at the center of the domain, as shown in Fig. 2. Edges of the domain experience zero pressure load. Half the length of the left edge is fixed (Fig. 2). Workpiece of stiffness k_{ss} $= 1 \times 10^4 \mathrm{Nm}^{-1}$ is used. The domain is parameterized using quadrilateral FEs herein. $N_\mathrm{ex} \times N_\mathrm{ey} = 100 \times 150$ bi-linear quadrilateral FEs are used to describe the design domain, where FEs in $x-$ and $y-$directions are represented via N_ex and N_ey, respectively. Each element is assigned a design variable that is considered constant within the element. The external move limit for the MMA optimizer is set to 0.1. The maximum number of the MMA iterations is fixed to 400. The filter radius $r_\mathrm{fill} = 6.0 \times \min\left(\frac{L_x}{N_\mathrm{ex}}, \frac{L_y}{N_\mathrm{ey}}\right)$ is set. A void region of dimension $\frac{L_x}{2} \times \frac{L_y}{10}$ exists within the domain, as displayed in Fig. 2. We consider the plane stress and small deformation finite element formulation assumptions. The SIMP parameter $\chi = 3$ is taken.

The permitted volume fraction is taken to 0.20. The robust parameter $\Delta\eta =$ 0.15 is set. β (Eq. 7) is varied from 1 to 128, which is doubled at each 50 MMA iterations, and once it reaches its maximum value of 128, it remains so for the rest of the optimization iterations. The dilated volume is updated every 25 MMA iterations. The higher β helps achieve a solution close to 0–1 that is measured herein using the discreteness scale M_nd defined in [25] as

$$M_\mathrm{nd} = \frac{\sum_{e=1}^{Ne} 4(\bar{\rho_e})(1 - \bar{\rho_e})}{Ne}, \tag{13}$$

For the mechanism, it is expected that as it is inflated with pressure load, the bottom right corner of it should move down, i.e., motion in the negative $y-$direction

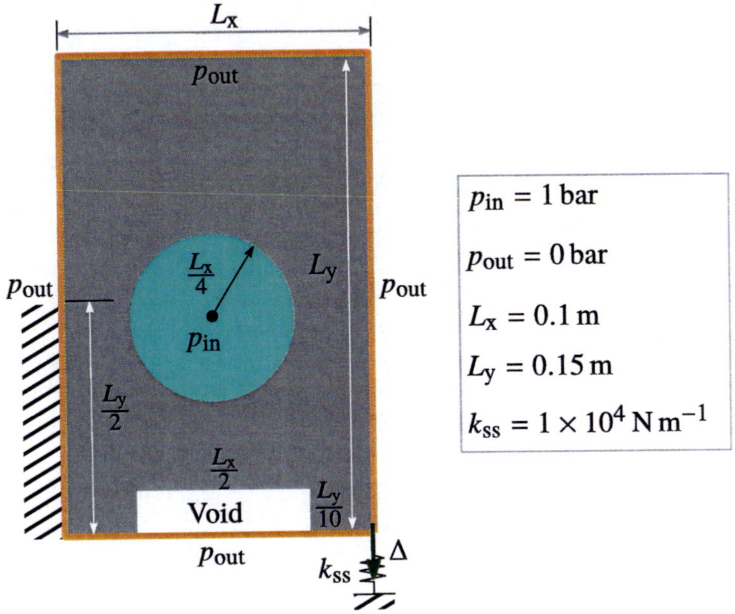

Fig. 2. Design domain of a pressure-driven soft robot. $L_x = 0.1$ m and $L_y = 0.15$ m are set, where L_x and L_y represent dimensions in $x-$ and $y-$directions, respectively. The input pressure load of 1 bar is applied using a circular chamber of radius $\frac{L_x}{4}$, as shown in the figure. The desired output motion is indicated by a green arrow. k_{ss} indicates the work-piece stiffness. A predefined void region of dimension $\frac{L_x}{2} \times \frac{L_y}{10}$ is shown within the domain.

is sought (Fig. 2). The flow contrast $\epsilon = 1\times\ 10^{-7}$ is used. Young's modulus of material $E_1 = 100$ MPa, and Poisson's ration $\nu = 0.40$ are set. The flow parameters $\{\eta_\kappa,\ \beta_\kappa\} = \{0.20,\ 10\}$ and the drainage parameters $\{\eta_d,\ \beta_d\} = \{0.30,\ 10\}$ are taken.

The optimized eroded, intermediate, and dilated designs with respective optimum pressure fields are displayed in Fig. 3. The optimized PneuNet design gets an arbitrary-shaped chamber to contain the fluidic pressure load in the optimized designs (Fig. 3), and that is expected to enhance the performance of pneumatic networks. On the other hand, when PneuNets are designed by the heuristic method, they typically have regular-shaped pressure chambers [21].

The eroded, intermediate, and dilated optimized mechanisms have identical topologies. The intermediate design is taken for the blueprint/fabrication. The eroded designs contain relatively thinner members, whereas the dilated designs are thicker. As per the value of M_{nd}, the optimized designs are very close to 0–1, i.e., black-white designs are obtained.

A member of PneuNet soft robot is designed with $P_{in} = 1$ bar while considering small deformation finite element assumptions in Fig. 3. However, soft robots typically experience finite deformation while performing their tasks. Including

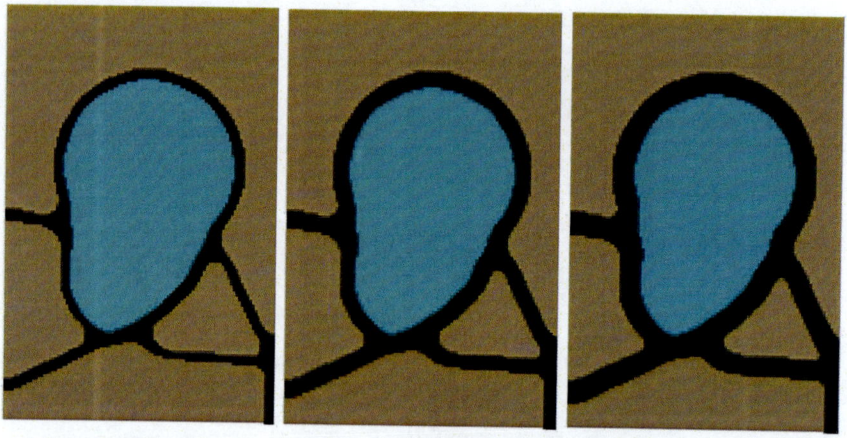

(a) $\Delta = -0.231\,\text{mm}$ $M_{\text{nd}} = 0.42\%, V_f = 0.135$ (b) $\Delta = -0.224\,\text{mm}$ $M_{\text{nd}} = 0.38\%, V_f = 0.2$ (c) $\Delta = -0.209\,\text{mm}$ $M_{\text{nd}} = 0.76\%, V_f = 0.26$

Fig. 3. Optimized design of a member of the PneuNet. Filter radius $r_{\text{fill}} = 6.0h$ is employed. $h = \min\left(\frac{L_x}{N_{\text{ex}}}, \frac{L_y}{N_{\text{ey}}}\right)$. (a) Eroded design, (b) Intermediate design, and (c) Dilated design.

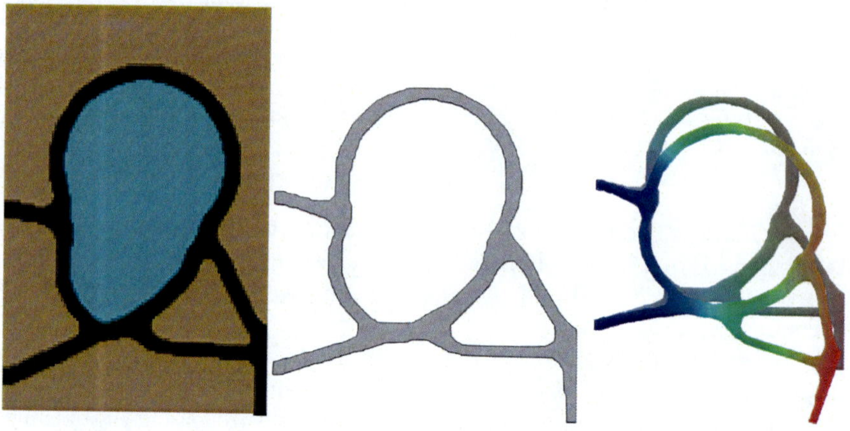

(a) (b) (c)

Fig. 4. (a) The optimized PneuNet intermediate design, (b) The CAD model, (c) The deformed and undeformed profiles. Red and blue colors indicate maximum and minimum deformation locations. The chamber is pressurized by a 100 bar load.

nonlinear mechanics formulation within the proposed design approach poses several other challenges, as mentioned in [22], and is out of the scope of the current study. However, we extract the optimized intermediate design using the technique mentioned in [22] (Fig. 3b) and study its behavior with high-pressure loads using a neo-Hookean material model in a commercial software. In addi-

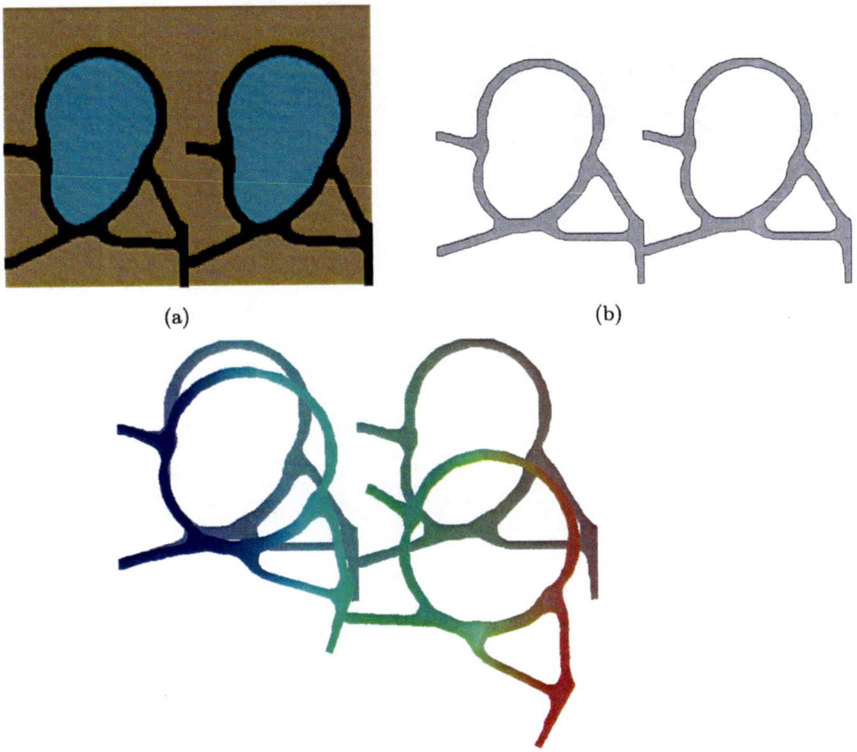

(a) (b) (c)

Fig. 5. (a) The optimized PneuNet design with two inflating members, (b) The CAD model, (c) The deformed and undeformed profiles. Each chamber is pressurized by a 100 bar load.

tion, we also form different PneuNets from the optimized design displayed in Fig. 3b and demonstrate their output deformation profiles. In [27] also a single member is optimized first and combined to get a pneumatic network. They use polygonal/hexagonal [2] FEs to represent the design domain.

The intermediate optimized design of the soft robot is analyzed in ABAQUS with $P_{\text{in}} = 100$ bar. The optimized design (Fig. 4a), its CAD model (Fig. 4b), and deformed-undeformed profiles superimposed on each other (Fig. 4c) are shown in Fig. 4. One can note that we get bending motion for the output node with high-pressure loading, which is expected.

Next, we combine two (Fig. 5a) and four (Fig. 6a) members of the optimized design to get different PneuNets (Fig. 5b and Fig. 6b). The pressure chambers of the designs are inflated using a 100 bar pressure load. The superimposed deformed and undeformed profiles for these robots are displayed in Fig. 5c and Fig. 6c, respectively. As the number of members increases in the network, the final robot deforms relatively more and provides bending motion. With more

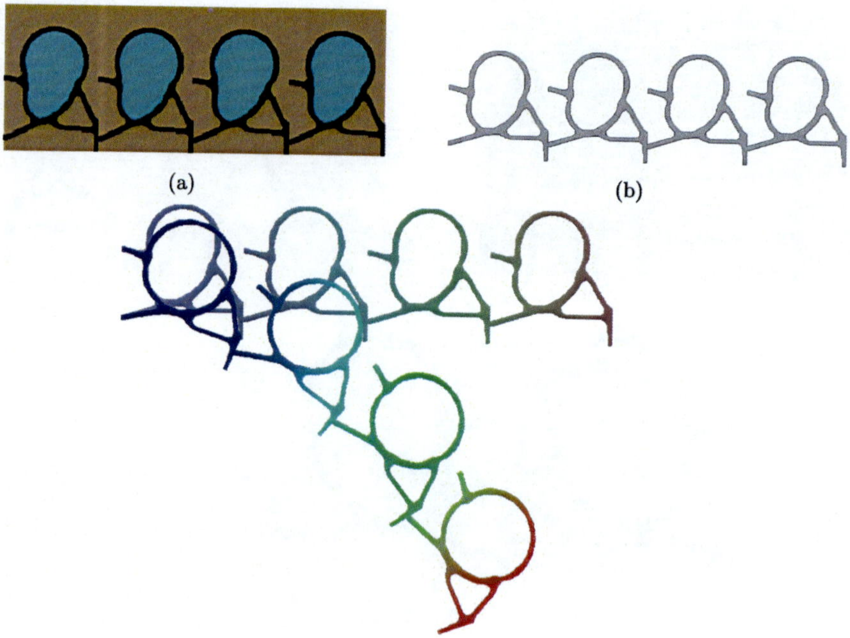

(a) (b) (c)

Fig. 6. (a) The optimized PneuNet design with four inflating members, (b) The CAD model, (c) The deformed and undeformed profiles. Each chamber is pressurized by a 100 bar load.

pneumatic members and high-pressure loads, the branches may interact and come in contact, i.e., situations for self-contact may occur [28,29] and thus, pose another set of challenges for designing such robots using TO. It can be noted that the pressurized chambers become close to circular shapes in their deformed profiles (Figs. 4c, 5c, and 6c); this is due because the pressure load acts normal to the surface. In addition, as the number of PneuNet members increases, one gets the complex output motions from the PneuNets, as depicted in Figs. 4c, 5c, and 6c.

5 Closure

This paper presents a density-based topology optimization approach to designing soft robots while considering the design-dependent nature of the actuating (fluidic pneumatic) loads. The robust formulation is employed to subdue point/single-node connections. To model the design-dependent character of the pressure load, we use the Darcy law with a conceptualized drainage term per [3]. The consistent nodal loads are determined from the obtained pressure field using the standard finite element method.

The proposed approach is demonstrated by designing a PneuNet soft robot. A min-max optimization problem is solved wherein the objective is determined using the multi-criteria formulation. The Method of Moving Asymptotes is used to solve the optimization problem. The optimized PneuNet gives the desired motion and has an arbitrary-shaped pneumatic chamber. The optimized design is combined to form numerous PneuNets, which provide different movements with high-pressure loads. Typically, soft robots experience finite deformation during their performance. Thus, it is necessary to include nonlinear finite element modeling within the optimization formulation, which forms the future direction for this research.

References

1. Xavier, M.S., Tawk, C.D., Zolfagharian, A., Pinskier, J., Howard, D., Young, T., Lai, J., Harrison, S.M., Yong, Y.K., Bodaghi, M., et al.: Soft pneumatic actuators: a review of design, fabrication, modeling, sensing, control and applications. IEEE Access (2022)
2. Kumar, P.: HoneyTop90: A 90-line MATLAB code for topology optimization using honeycomb tessellation. Optimization and Engineering, pp. 1–18 (2022)
3. Kumar, P., Frouws, J.S., Langelaar, M.: Topology optimization of fluidic pressure-loaded structures and compliant mechanisms using the Darcy method. Struct. Multi. Optim. **61**(4), 1637–1655 (2020). https://doi.org/10.1007/s00158-019-02442-0
4. Hiller, J., Lipson, H.: Automatic design and manufacture of soft robots. IEEE Trans. Robot. **28**(2), 457–466 (2011)
5. Chen, F., Xu, W., Zhang, H., Wang, Y., Cao, J., Wang, M.Y., Ren, H., Zhu, J., Zhang, Y.: Topology optimized design, fabrication, and characterization of a soft cable-driven gripper. IEEE Robot. Autom. Lett. **3**(3), 2463–2470 (2018)
6. Zhang, H., Kumar, A.S., Fuh, J.Y.H., Wang, M.Y.: Design and development of a topology-optimized three-dimensional printed soft gripper. Soft Robot. **5**(5), 650–661 (2018)
7. Yin, L., Ananthasuresh, G.K.: Topology optimization of compliant mechanisms with multiple materials using a peak function material interpolation scheme. Struct. Multi. Optim. **23**(1), 49–62 (2001)
8. Sigmund, O.: On the design of compliant mechanisms using topology optimization. J. Struct. Mech. **25**(4), 493–524 (1997)
9. Wang, F., Lazarov, B.S., Sigmund, O.: On projection methods, convergence and robust formulations in topology optimization. Struct. Multi. Optim. **43**(6), 767–784 (2011)
10. Kumar, P., Fanzio, P., Sasso, L., Langelaar, M.: Compliant fluidic control structures: concept and synthesis approach. Comput. Struct. **216**, 26–39 (2019)
11. Zhu, B., Zhang, X., Zhang, H., Liang, J., Zang, H., Li, H., Wang, R.: Design of compliant mechanisms using continuum topology optimization: a review. Mech. Mach. Theory **143**, 103622 (2020)
12. Kumar, P., Schmidleithner, C., Larsen, N., Sigmund, O.: Topology optimization and 3D printing of large deformation compliant mechanisms for straining biological tissues. Struct. Multi. Optim. **63**(3), 1351–1366 (2021)

13. Kumar, P., Sauer, R.A., Saxena, A.: On topology optimization of large deformation contact-aided shape morphing compliant mechanisms. Mech. Mach. Theory **156**, 104135 (2021)
14. Hammer, V.B., Olhoff, N.: Topology optimization of continuum structures subjected to pressure loading. Struct. Multi. Optim. **19**(2), 85–92 (2000). https://doi.org/10.1007/s001580050088
15. Chen, B.C., Silva, E.C., Kikuchi, N.: Advances in computational design and optimization with application to mems. Int. J. Numer. Methods Eng. **52**(1–2), 23–62 (2001)
16. Sigmund, O., Clausen, P.M.: Topology optimization using a mixed formulation: an alternative way to solve pressure load problems. Comput. Methods Appl. Mech. Eng. **196**(13–16), 1874–1889 (2007)
17. Panganiban, H., Jang, G.W., Chung, T.J.: Topology optimization of pressure-actuated compliant mechanisms. Finite Elem. Anal. Des. **46**(3), 238–246 (2010)
18. Vasista, S., Tong, L.: Design and testing of pressurized cellular planar morphing structures. AIAA J. **50**(6), 1328–1338 (2012)
19. de Souza, E.M., Silva, E.C.N.: Topology optimization applied to the design of actuators driven by pressure loads. Struct. Multi. Optim. **61**(5), 1763–1786 (2020). https://doi.org/10.1007/s00158-019-02421-5
20. Kumar, P., Langelaar, M.: On topology optimization of design-dependent pressure-loaded three-dimensional structures and compliant mechanisms. Int. J. Numer. Methods Eng. **122**(9), 2205–2220 (2021)
21. Shepherd, R.F., Ilievski, F., Choi, W., Morin, S.A., Stokes, A.A., Mazzeo, A.D., Chen, X., Wang, M., Whitesides, G.M.: Multigait soft robot. Proc. Natl. Acad. Sci. **108**(51), 20400–20403 (2011)
22. Kumar, P., Langelaar, M.: Topological synthesis of fluidic pressure-actuated robust compliant mechanisms. Mech. Mach. Theory **174**, 104871 (2022)
23. Saxena, A., Ananthasuresh, G.K.: On an optimal property of compliant topologies. Struct. Multi. Optim. **19**(1), 36–49 (2000)
24. Bruns, T.E., Tortorelli, D.A.: Topology optimization of non-linear elastic structures and compliant mechanisms. Comput. Methods Appl. Mech. Eng. **190**(26), 3443–3459 (2001)
25. Sigmund, O.: Morphology-based black and white filters for topology optimization. Struct. Multi. Optim. **33**(4–5), 401–424 (2007)
26. Svanberg, K.: The method of moving asymptotes-a new method for structural optimization. Int. J. Numer. Methods Eng. **24**(2), 359–373 (1987)
27. Caasenbrood, B., Pogromsky, A., Nijmeijer, H.: A computational design framework for pressure-driven soft robots through nonlinear topology optimization. In: 2020 3rd IEEE International Conference on Soft Robotics (RoboSoft), pp. 633–638. IEEE (2020)
28. Kumar, P., Saxena, A., Sauer, R.A.: Implementation of self contact in path generating compliant mechanisms. In: Microactuators and Micromechanisms, pp. 251–261. Springer (2017)
29. Kumar, P., Saxena, A., Sauer, R.A.: Computational synthesis of large deformation compliant mechanisms undergoing self and mutual contact. J. Mech. Des. **141**(1), 012302 (2019)

Compliant Finger Gripper Based on Topology Optimization

Estefania Hermoza Llanos[1(✉)], Mathias Hüsing[1], Burkhard Corves[1], and Anupam Saxena[2]

[1] Mechanical Engineering, Institute of Mechanism Theory, Machine Dynamic and Robotics, RWTH Aachen University, Eilfschornsteinstraße 18, 52062 Aachen, Germany
{Hermoza-Llanos,huesing,corves}@igmr.rwth-aachen.de
[2] Mechanical Engineering, Indian Institute of Technology Kanpur, Kanpur 208016, India
anupams@iitk.ac.in

Abstract. We present the design of a compliant finger based on the topology optimization method through Hill Climber stochastic search. The compliant finger is composed of an arrangement of beams, whose behavior is studied under the formulation of large deformation co-rotational beam elements. The design optimization seeks to maximize deflection of the tip of the compliant finger with the lowest possible input torque and the mechanical advantage via the contact force between the compliant finger and a free-form shaped object. Large tip deformation is sought along a desired path. The actual path traced by the candidate mechanism and the desired path are compared through their Fourier descriptors. Computation of the contact force is performed between the surfaces of Euler-Bernoulli beam and a prescribed free-form shape.

Keywords: Topology optimization · Compliant mechanism · Compliant finger · Hill climber stochastic search · Fourier shape descriptors · Large deformation finite element analysis

1 Introduction

Compliant mechanisms (CMs) are monolithic mechanical entities that can transmit motion and forces through deflection of their flexible parts. These mechanisms are omnipresent in nature as parts of compacts natural machines, e.g., bees' wings, elephant trunks, blooming flowers and many more [1].

With this premise, it is possible to replicate the functionality part of a human finger with such compliant monolithic mechanisms. A human hand has 25 degrees of freedom in total [2] which are coordinated through a complex biological control system, making it difficult to entirely replicate its performance and features. One of the major problems in the artificial hands already created is the mechanical complexity of the design of the fingers, their joints and the actuation method [3]. The goal here is to present the design of a single, fully compliant finger undergoing large deformation that will have one degree of freedom, as observed in the human hand as extension and flexion.

A. K. Pandey et al. (Eds.): MAMM 2022, 126, pp. 31–45, 2023.
https://doi.org/10.1007/978-3-031-20353-4_3

2 Related Work

With the aim to accomplish several types of grasping and manipulation related tasks, the human design process, as starting point, has mostly taken inspiration from the biological hand. When designing a single finger, Topology Optimization (TO) has been a viable method to get a functional, optimized and practical mechanism's topology.

Renghao et al. [4] set the objective to be the maximization of output displacement to design and 3D-fabricate a rigid-compliant parallel exoskeleton; Saurabh et al. [5] use TO based on an approximated geometry of the structure to achieve a monolithic finger design for prosthetic applications; Hongying et al. [6] and Liu et al. [7] propose TO to get prototypes of multi-material compliant fingers, where both aim at maximizing the output displacement. Hongying et al. [6] optimize only the output displacement while Liu et al. [7] minimize the strain energy as well, showing by experimental results that the bi-material finger is a better design to reduce the driving force and to increase the output displacement; Chen et al. [8] propose a soft finger composed of silicone and a hard outside structure to impose constraints. The finger may have two different movements: bending and twisting. They used TO on rational approximation of material properties to obtain an optimal design of the outer layer of the soft fingers. Liu et al. [9] and Petović et al. [10] use TO to get the topology of a compliant finger, which is based first on an initial mesh and, once the final model of a finger is obtained, they seek different configurations, as multi-fingered grippers presented by Dechev et al. [11] and Yang et al. [12], via three contact points to achieve more stable and safer grasping.

It is also pertinent to mention different actuation methods used for compliant fingers. Chen et al. [13] propose a network of inflated dielectric elastomer actuators, interconnected through a chamber; Li et al. [14], Popov et al. [15] and Manti et al. [16] present mechanisms guided by cable wires; Wehner et al. [17] control the CM with microfluidic actuation that autonomously regulates fluid flow and its catalytic decomposition to get a gas that inflates a chamber to actuate the mechanism; Kim et al. [18] use shape-memory alloys; Drotman et al. [19], Çulha et al. [20], Deimel et al. [21] and Polygerinos et al. [22], all, employ hydraulic or pneumatic actuation for their CM designs.

3 Motivation and Organization

We aim at designing a monolithic, single finger unit of a three-finger soft-robotic system that could be employed in several fields, e.g., medical and automation, for a variety of tasks such as assembly, transportation and sample collection, precision manipulation and others. Our topology design methodology for a monolithic finger incorporates large deformation analysis with contact modeling. This is to facilitate the fingertip to trace a prescribed path before the unit makes contact with the work piece. Our approach focuses on maximizing the contact force as well to achieve stable grasp. Previous attempts to design monolithic fingers for flexion-extension using topology optimization did not involve computation/modeling of contact force(s) with the work piece and its maximization for stable grasp. None or few employed large deformation analysis, most performing primarily small deformation based topology optimization.

The rest of the paper is organized as follows. Section 4 describes the problem formulation for a compliant finger. Objective(s) and constraints are detailed herein. Section 5 describes large deformation contact modeling with co-rotational frame elements. Contact with only rigid surfaces (mutual contact) is considered. The Hill Climbing algorithm, and the design variables/parameters used are briefed in Sect. 6. Results are described and discussed in Sect. 7 and finally, conclusions are drawn.

4 Formulation of the Problem, Objective and Constraints

The design of a monolithic, single finger unit involves (i) replication of the extension-flexion movement of the human finger through compliant mechanisms based on co-rotational beams, (ii) maximization of the contact force between the workpiece and the single finger to get a stable grasp and (iii) trace of the fingertip along a predetermined path before it makes contact with the workpiece.

To solve this problem, an objective function based on different characteristics of the design is defined and maximized.

$$F_{obj} = K_1 F_1 + K_2 F_2 \tag{1}$$

where K_1 and K_2 are user chosen constants and, F_1 and F_2 are the sub objectives described in Table 1. The weight factors K_1 and K_2 have a considerable effect on the solution, depending on the weight, the objectives will have relative prominence.

Table 1. Sub-functions to maximize.

Name	Description	Sub-function
F_1 = Path generation	Comparison between the Fourier descriptors and the length of the path created by the candidate mechanism and the desired path	$\frac{1}{F_1} = C_a \sum_{k=0}^{T} \left(a_k^{cand} - a_k^{des}\right)^2$ $+C_b \sum_{k=1}^{T} \left(b_k^{cand} - b_k^{des}\right)^2$ $+C_c \left(L^{cand} - L^{des}\right)^2$
F_2 = Contact force	Modulus of the contact force between the candidate mechanism and the surface	$F_2 = \int_{-1}^{1} \epsilon g_N \left(\mathbf{N}\mathbf{N}^T \mathbf{n}_m\right) l d\xi$

where: C_a, C_b and C_c = Importance coefficients defined by the user. a_k^{cand}, b_k^{cand} = Candidate's Fourier descriptors. a_k^{des}, b_k^{des} = Desired Fourier descriptors. L^{cand} and L^{des} = Length of the candidate and desired paths

The sub-function $\frac{1}{F_1}$ is responsible to minimize the difference between the Fourier descriptors of the path described by the candidate mechanism and the desired curve. Therefore, the fingertip will follow the desired path; this objective was included so that the algorithm would find a topology faster and that a part of the finger would come into contact with the object, so that the contact forces would be maximized. In addition, the sub-function F_2 maximizes the contact force, to provide stability to the grasp. Correspondingly, the weight for the path was chosen less compared to that for the contact forces.

Various design conditions/constraints are also specified, as mentioned in Table 2.

Table 2. Design feasibility constraints/bounds

Constraint/Bound	Description
Force on one node	The point of force application must be a node
Constant out-of-plane thickness	The design is prismatic
Boundary conditions	The mechanism has two nodes fixed along the horizontal and vertical directions, but not in the rotational degree of freedom
Condition of equilibrium	For all candidate designs, the vector of internal forces must be equal to the same of external forces $\mathbf{F}_{int} = \mathbf{F}_{ext}$ Although it is an axiom of mechanics, these nonlinear equality constraints are mentioned explicitly because they are considered in the iterative search process
Presence of important nodes	The following nodes must be present in the candidate model: nodes where boundary conditions are applied, node representing the fingertip and the node where the input force/torque is applied
Elimination of disconnected elements	Elements that are dangling/disconnected must be removed before the analysis. This constraint does not apply to the important nodes mentioned before
Elimination of elements with low strain energy	Elements with relatively low strain energy (almost zero) must be eliminated
Volume	The total volume of the design must be smaller than a volume defined by the user $V \leq V^*$
In-plane widths/nodes	Bounds chosen on the in-plane widths and nodal positions are such that the design is manufacturable

Intermediate designs for which there is no displacement of the node representing the fingertip, or the one in which any essential node is excluded, are penalized.

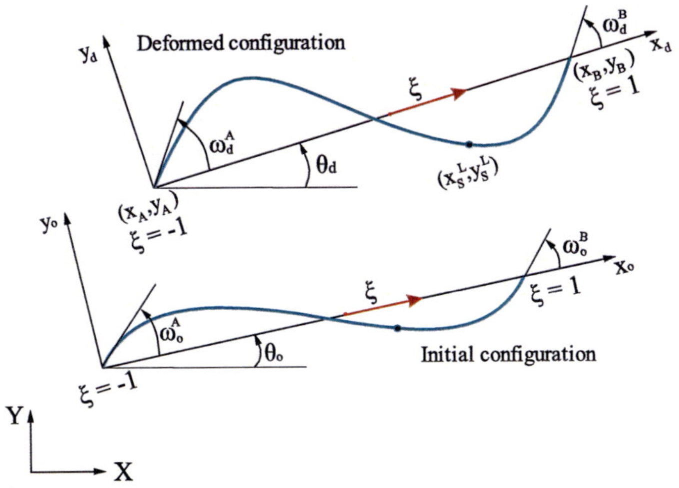

Fig. 1. Initial and deformed configuration of a co-rotational beam.

5 Contact Modeling in Large Deformation Between Beams and External Surfaces

Implementation details of large deformation analysis with Euler-Bernoulli co-rotational beams can be found in [23–26]. This section presents the analysis followed to capture contact behavior between a beam and an external surface. A deformed co-rotational beam in the global XY system is shown in Fig. 1, where (x_d, y_d) is the local coordinate system describing the beam. (x_A, y_A) and (x_B, y_B) are two end nodes, ω^A, ω^B are current slopes, θ_o is the angle of x_o with the global horizontal X and θ_d is the angle of the deformed axis x_d with the same. A point $({x_s}^L, {y_s}^L)$ on the beam can be interpolated cubically as

$$x_s^L = N_1 l = \left(\frac{1+\xi}{2}\right) l, \tag{2}$$

$$y_s^L = N_2 l \omega_d^A + N_3 l \omega_d^B \tag{3}$$

where $\xi \in [-1, 1]$ is the local coordinate.

and

$$N_2 = \left(\frac{1+\xi}{2}\right)^3 - 2\left(\frac{1+\xi}{2}\right)^2 + \left(\frac{1+\xi}{2}\right)^3 \tag{4}$$

$$N_3 = \left(\frac{1+\xi}{2}\right)^3 - \left(\frac{1+\xi}{2}\right)^2 \tag{5}$$

In the global system,

$$x_S = N_1(x_B - x_A) - \left(N_2 \omega_d^A + N_3 \omega_d^B\right)(y_B - y_A) + x_A \tag{6}$$

$$y_S = N_1(y_B - y_A) + \left(N_2\omega_d^A + N_3\omega_d^B\right)(x_B - x_A) + y_A \tag{7}$$

and

$$\begin{aligned} l\cos\theta_d &= x_B - x_A \\ l\sin\theta_d &= y_B - y_A \end{aligned} \tag{8}$$

Change of slope at node A and B

$$\begin{aligned} \omega_d^A + \theta_d - \theta_o &= \beta^A \\ \omega_d^B + \theta_d - \theta_o &= \beta^B \end{aligned} \tag{9}$$

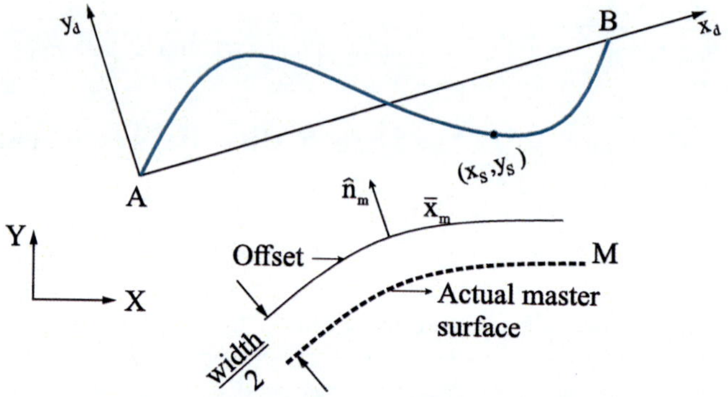

Fig. 2. Deformed beam and surface of contact

The beam (slave surface) interacts with the master, external, surface. The master surface is fixed, rigid and is offset by half of the beam width so that is not the neutral axis, but the beam surface that interacts with the master surface (Fig. 2).

Let $\mathbf{x}_m$ be the closest point on the master surface to $\mathbf{x}_s = \begin{bmatrix} x_s \\ y_s \end{bmatrix}$, a point on the slave surface. Let $\mathbf{n}_m$ be the unit normal defined at $\mathbf{x}_m$. We define the normal gap between the beam (slave) and the master external surface as:

$$g_N = (\mathbf{x}_s - \mathbf{x}_m) \cdot \mathbf{n}_m \tag{10}$$

The contact potential Π_e within the penalty method framework can be written as

$$\Pi_e = \int_{\zeta_s} \frac{1}{2} \epsilon g_{N^2} d\zeta_s = \frac{1}{2} \int_{-1}^{1} \epsilon g_{N^2} l d\xi \tag{11}$$

where: ζ_S = beam surface.

ϵ = penalty parameter (N/m^2).

The contact potential Π_e is calculated only when $g_N < 0$. If $g_N \geq 0$, $\Pi_e = 0$. The contact forces $\mathbf{f}_c$ can be computed via the principle of virtual work such that

$$\delta\Pi_e = \delta\mathbf{u}^T\mathbf{f}_c = \int_{-1}^{1} \epsilon g_N \delta g_N l d\xi \tag{12}$$

where $l = |AB|$. Taking variation of the normal gap, and noting that the master surface is stationary and rigid, we get

$$\delta g_N = (\delta\mathbf{x}_s - \delta\mathbf{x}_m).\mathbf{n}_m + (\mathbf{x}_s - \mathbf{x}_m).\delta\mathbf{n}_m = \delta\mathbf{x}_s.\mathbf{n}_m \tag{13}$$

We take variations in Eq. (6) so that

$$\begin{aligned}\delta x_S &= N_1(\delta u_B - \delta u_A) - \left(N_2\omega^A + N_3\omega^B\right)(\delta v_B - \delta v_A)\\ &\quad - \left(N_2\delta\omega^A + N_3\delta\omega^B\right)(y_B - y_A) + \delta u_A\end{aligned} \tag{14}$$

Note that $\delta u_A = \delta x_A$, $\delta v_A = \delta y_A$, $\delta u_B = \delta x_B$ and $\delta v_B = \delta y_B$, with (u_A, v_A) and (u_B, v_B) being the global displacements at the two nodes. Also, from Eqs. (8) and (9):

$$-l\sin\theta_d\delta\alpha = -l\sin\theta_d\delta\theta_d = \delta x_B - \delta x_A = \delta u_B - \delta u_A \tag{15}$$

$$l\cos\theta_d\delta\alpha = l\cos\theta_d\delta\theta_d = \delta y_B - \delta y_A = \delta v_B - \delta v_A \tag{16}$$

$$\delta\omega^A + \delta\theta_d = \delta\omega^A + \delta\alpha = \delta\beta^A \tag{17}$$

$$\delta\omega^B + \delta\theta_d = \delta\omega^B + \delta\alpha = \delta\beta^B \tag{18}$$

Substitution in Eq. (14) yields

$$\begin{aligned}\partial x_S &= (1 - N_1 + N_2 + N_3)\delta u_A + \left(N_2\omega^A + N_3\omega^B\right)\delta v_A - N_2(y_B - y_A)\delta\beta^A\\ &\quad + (N_1 - N_2 - N_3)\delta u_B + \left(-N_2\omega^A - N_3\omega^B\right)\delta v_B\\ &\quad - N_3(y_B - y_A)\delta\beta^B\end{aligned} \tag{19}$$

Likewise for ∂y_S:

$$\begin{aligned}\partial y_S &= \left(-N_2\omega^A - N_3\omega^B\right)\delta u_A + (1 - N_1 + N_2 + N_3)\delta v_A + N_2(x_B - x_A)\delta\beta^A\\ &\quad + \left(N_2\omega^A + N_3\omega^B\right)\delta u_B + (N_1 - N_2 - N_3)\delta v_B\\ &\quad + N_3(x_B - x_A)\delta\beta^B\end{aligned} \tag{20}$$

Summarizing Eqs. (19) and (20):

$$\begin{pmatrix} \partial x_s \\ \partial y_s \end{pmatrix} = \begin{bmatrix} (1-N_1+N_2+N_3) & (N_2\omega^A+N_3\omega^B) & -N_2(y_B-y_A) & (N_1-N_2-N_3) & (-N_2\omega^A-N_3\omega^B) & -N_3(y_B-y_A) \\ (-N_2\omega^A-N_3\omega^B) & (1-N_1+N_2+N_3) & N_2(x_B-x_A) & (N_2\omega^A+N_3\omega^B) & (N_1-N_2-N_3) & N_3(x_B-x_A) \end{bmatrix} \cdot \begin{bmatrix} \delta u_A \\ \delta v_A \\ \delta\beta_A \\ \delta u_B \\ \delta v_B \\ \delta\beta_B \end{bmatrix}$$

or

$$\delta \mathbf{x}_s = \mathbf{NN}.\delta\mathbf{u} \tag{21}$$

so

$$\delta g_N = \delta\mathbf{u}^T\mathbf{NN}^T\mathbf{n}_m \tag{22}$$

Returning to Eq. (12):

$$\delta\Pi_e = \delta\mathbf{u}^T\mathbf{f}_c = \int_{-1}^{1} \epsilon g_N\left(\delta\mathbf{u}^T\mathbf{NN}^T\mathbf{n}_m\right) l d\xi$$

or

$$\mathbf{f_c} = \int_{-1}^{1} \epsilon g_N\left(\mathbf{NN}^T\mathbf{n}_m\right) l d\xi \tag{23}$$

Equation (23) gives the expression for the contact force vector which is subtracted from the internal force vector and the result is assembled as in conventional finite element analysis. To obtain the contact tangent stiffness matrix, we use the following equations:

$$\mathbf{k}_T^{cont} = \frac{\partial \mathbf{f}_c}{\partial \mathbf{u}}$$

$$\mathbf{k}_T^{cont} = \int_{-1}^{1} \epsilon g_N\left(\mathbf{NN}^T\mathbf{n}_m\right)\left(\frac{\partial g_N}{\partial \mathbf{u}}\right) l d\xi + \int_{-1}^{1} \epsilon g_N\left(\frac{\partial\left(\mathbf{NN}^T\mathbf{n}_m\right)}{\partial \mathbf{u}}\right) l d\xi \tag{24}$$

where:

$$\frac{\partial g_N}{\partial \mathbf{u}} = \mathbf{n}_m^T\mathbf{NN}$$

and

$$\frac{\partial(\mathbf{NN}^T\mathbf{n}_m)}{\partial\mathbf{u}} = \begin{bmatrix} 0 & 0 & -N_2n_{my} & 0 & 0 & -N_3n_{my} \\ 0 & 0 & N_2n_{mx} & 0 & 0 & N_3n_{mx} \\ -N_2n_{my} & N_2n_{mx} & 0 & N_2n_{my} & -N_2n_{mx} & 0 \\ 0 & 0 & N_2n_{my} & 0 & 0 & N_3n_{my} \\ 0 & 0 & -N_2n_{mx} & 0 & 0 & -N_3n_{mx} \\ -N_3n_{my} & N_3n_{mx} & 0 & N_3n_{my} & -N_3n_{mx} & 0 \end{bmatrix}$$

6 Optimization – Hill Climbing Mutation Algorithm

The Hill Climbing Mutation Algorithm (HCMA) is a simple, stochastic search technique to maximize a defined function. This algorithm commences with a randomly generated set of design variables, and then in every iteration, mutates and updates the design vector only when the altered vector exhibits a better solution [27, 28]. The HCMA is used to alter the topology, shape and size design variables as follows (Table 3).

Table 3. Design variables from HCMA.

Parameter	Description	Range/Limits
Topology Optimization	1 = Element gets retained in the design 0 = Element gets removed	0 or 1
Width	Different widths used for each element in the study	[3, 5] mm
Displacement along x	Permitted displacement of each node along x direction	[0, 3] mm
Displacement along y	Permitted displacement of each node along y direction	[0, 3] mm
Magnitude of input force	Constant by which the input force/torque gets multiplied with	[0.5, 1]

Within each iteration, the HCMA generates a design variable vector that represents a candidate design, which is subjected to a test of deformations and contact forces between it and a surface. This test is based on geometrically nonlinear, large displacements with co-rotational beams capable of engaging in mutual contact, as shown in Sect. 5. After a chosen number of design iterations, the HCMA yields the best possible candidate model that maximizes the objective in Sect. 4. One such model is discussed next.

The Hill Climbing Mutation is used as a stochastic search because of the two following reasons: (a) members should be known to be completely present or absent for as-is and non-ambiguous realization of the topology and (b) a no convergence within Newton-Raphson or ARC length iterations can impede gradient search. For any intermediate design, successful completion of nonlinear analysis, especially with contact, cannot be guaranteed.

7 Optimization Results and Discussion

Following the formulation of the design presented in Sect. 4, using the theory of contact modeling in large deformations between co-rotational beams and external surfaces presented in Sect. 5, and using topology optimization with the Hill Climbing algorithm in Sect. 6; a series of candidate designs are obtained.

7.1 Initial Model and Inputs of the Design

The designs correspond to a base grid with dimensions of 150 mm long and 25 mm height, a total of 106 elements and 43 nodes, young Modulus of 3300 MPa, constant out of plane thickness of 4 mm, initial slopes of zero and an input force in $-x$ direction of 80N as seen in the Fig. 3.

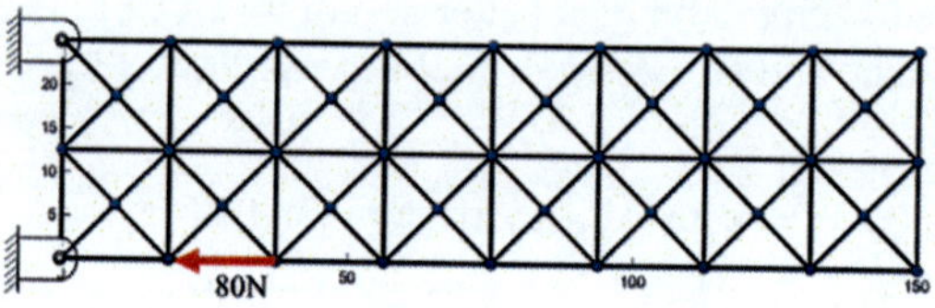

Fig. 3. Design specifications for a finger unit of a gripper.

A workflow of the process followed for the algorithm for each candidate design is presented in Fig. 4.

7.2 Results

A design of a compliant finger based on topology optimization was generated after 5500 iterations in a code developed in Matlab, as in Fig. 5.

The fingertip of the model traces a candidate path when subjected to an input force of 40 N, the free travel trajectory shows large deformation of the fingertip in absence of any external object, this curve has an arc of about 123 mm, as in Fig. 5a. A comparison between the desired path and the candidate path is shown in Fig. 5a, the average difference between the desired and the candidate path was 10.57 mm. This difference appears because the algorithm was considering the requirement of the path and the maximization of contact forces at the same time, as is explained in the workflow of the process in the previous Sect. 7.1.

In addition, contact appears at multiple points between the model and a circular external surface, which ensures the grasp stability, as in Fig. 5b. The force between the model and the surface was generated at two points, the total force at each point is 2.44 N and 1.6 N respectively.

The compliant finger presented has a total volume of 7069.8 mm^3 and a total of 45 elements, which is important in terms of optimization of the future manufacturability of the prototype.

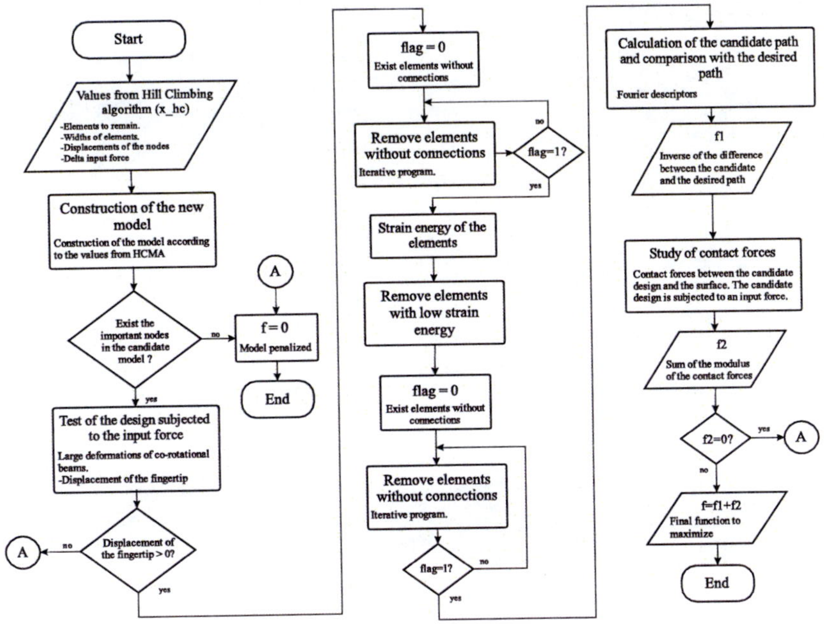

Fig. 4. Workflow for the design process.

With the thickness, nodal positions and topology data obtained in Matlab, a 3D model was created in Inventor and subsequently this model was subjected to a finite element analysis in Abaqus. Results of the FEA are shown in Fig. 5c and a comparison graph in Fig. 5d shows the difference of tip displacement of the contact model obtained in Matlab (co-rotational beams model) against the model in Abaqus (FEA). Paths of the tip node were modelled with a 5th degree polynomial equation Fig. 5d. The error between both data is 2,19%, which may be due to the difference in the type of modeling between Abaqus and Matlab. In the former, corner smoothing is performed and 3D elements are used while in the latter, co-rotational beams are employed. Nevertheless, the agreement is reasonable.

7.3 Discussion

This paper presents the design of a compliant finger based on a desired path and maximum possible contact forces. To get an optimal design that fulfils all the requirements made, several large deformation models were created by the algorithm, few of which are shown in Fig. 6. These previous designs have the following characteristics (Table 4).

The previous designs shown above are not considered, since it is observed that, in spite of being able to follow the desired trajectory, the contact points between the unit and the external surface are not enough to guarantee a good grip between them. Some (Fig. 6a and b) even show a sliding behavior between the workpiece and the grip region of the design. Other important aspect is the volume and number of elements, the chosen model has 45 elements and better gripping behavior in contrast to the 52 and 60 elements of designs 2 and 3 in Fig. 6.

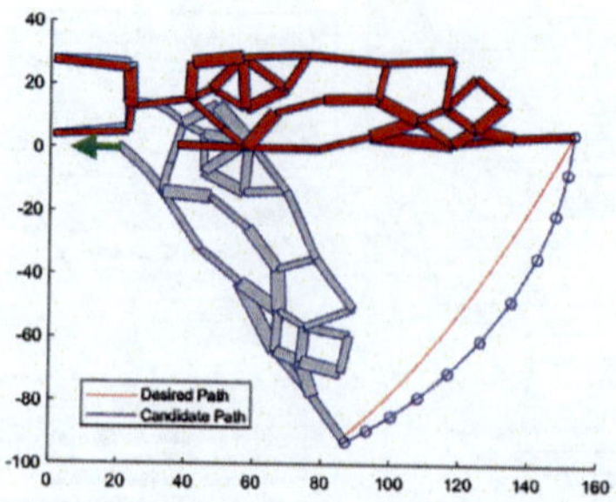

(a) Displacement of the design subjected to an input load of 40N. Green arrow: Input force.

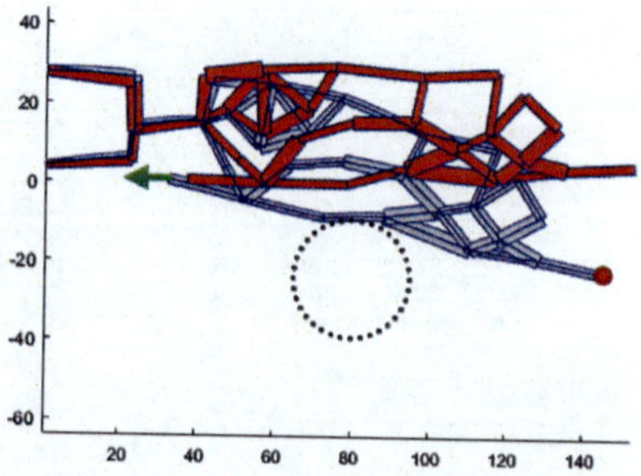

(b) Contact forces generated between a circular surface and the design with an input load of 40 N in Matlab. Green arrow: Input force. Red circle: Tip node.

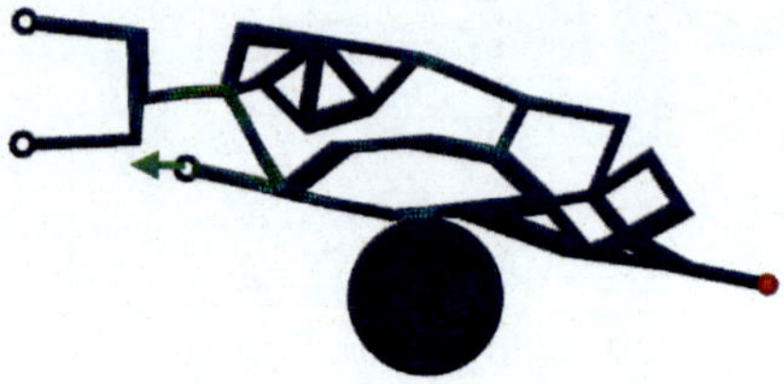

(c) Contact forces generated between a circular surface and the design with an input load of 40 N in Abaqus. Green arrow: Input force. Red circle: Tip node.

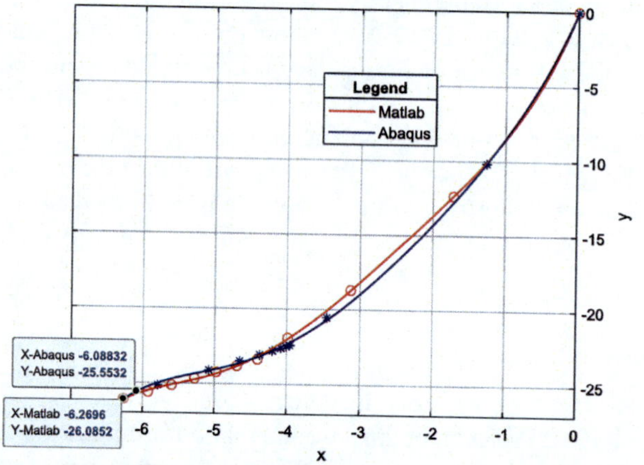

(d) Graph comparison of tip's finger displacement between Abaqus and Matlab.

Fig. 5. Design of a candidate compliant finger.

Another important point to discuss is the difference that appears between the desired and the candidate path (Fig. 5a and b), this difference may be the result of applying the algorithm considering the requirements of path following and maximization of contact

Table 4. Characteristics of previous attempts.

Number of Designs	Number of iterations	Volume	Total number of elements
Design 1	10000	6743.5 mm^3	38
Design 2	10000	9100 mm^3	52
Design 3	3000	9733.9 mm^3	60

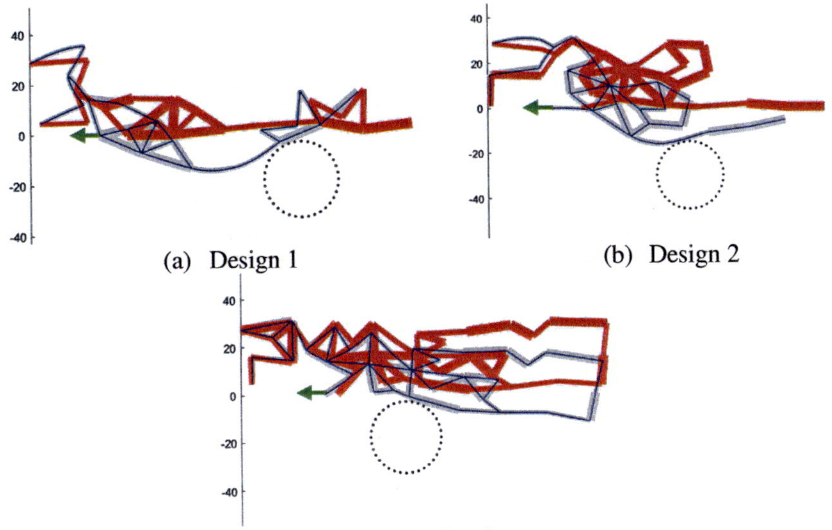

(a) Design 1 (b) Design 2

(c) Design 3

Fig. 6. Other (previous) candidate designs.

forces at the same time. These two parameters are coupled requirements (because we are working with compliant mechanisms), which means that while one may be maximizing, the other is also affected, it is not a design with decoupled requirements, like rigid-body mechanisms.

8 Conclusions

In this paper, large deformation compliant finger design based on topology optimization is presented, which satisfies the requirements of having a displacement along a given path and maximization of contact forces with a chosen external surface. A feasible starting grid is modelled according to the dimensions desired for the compliant finger, the desired minimum thickness and the knowledge in additive manufacturing (to be used to manufacture a prototype of the proposed design) was used. The base grid is composed of 106 elements and has a dimension of 150 mm long and 25 mm height.

The shape, position and size of the contact fixed surface is selected according to the grid modelled, meaning that this surface must not be in initial contact with the design in its initial position. The contact surface is a circle of 15 mm radius. Although a circular object was chosen to demonstrate the methodology, a finger design in the future could be accomplished with generically shaped or multiple objects placed at different positions relative to the finger. The chosen design was obtained after 5500 iterations and has a total volume of 7069.8 mm^3 and a total of 45 elements, which is important in terms of optimization of the future manufacturability of the prototype. The force between the model and the surface was generated at two points, the total force in each point is 2.44 N and 1.6 N respectively. The theory of co-rotational beams with contact applied in Matlab was verified through a comparison with a finite element analysis to be in agreement with the real behavior of deformations in flexible mechanisms. The design presented, being monolithic, can be miniaturized and actuated in small scales without losing precision; moreover, the mechanism can be used in assembly of small parts, sample collection, automation of some process that involves grasping of delicate objects, pick-and-place, and others tasks that involves high precision.

For future work, it is intended to test the prototype for different situations and make a physical model of it for experimental verification. It is also possible to change the order of the workflow process to optimize it or add different considerations, e.g., stress constraints into the algorithm, initial values of slopes of the elements or improve the criteria for the elimination of elements that have no function and protrude from the overall structure, to, possibly, obtain a better design for the finger unit.

Acknowledgements. The KAAD Scholarship and the Alexander von Humboldt (AvH) Fellowship to Hermoza, Estefania and Saxena, Anupam respectively are graciously acknowledged.

References

1. Howell, L., Magleby, S., Olsen, B.: Handbook of Compliant Mechanisms. Wiley, United Kingdom (2013)
2. Balasubramanian, R., Santos, V.J. (eds.): The Human Hand as an Inspiration for Robot Hand Development. STAR, vol. 95. Springer, Cham (2014). https://doi.org/10.1007/978-3-319-03017-3
3. Mogeeb, A.E.: Fibonacci-Compliant Finger Design. Wichtig Publishing, p. 6 (2016)
4. Renghao, L., Guanghua, X., Bo, H., Min, L., Zhicheng, T., Sicong, Z.: Developing of a rigid-compliant finger joint exoskeleton using topology optimization method. In: IEEE International Conference on Robotics and Automation 2021, Xi'an, China (2021)
5. Saurabh, S., Sachin, S., Sarang, P., Vishal, P.: Design of prosthetic finger using topology optimization. In: International Conference on Computational Intelligence and Knowledge Economy, Dubai (2021)
6. Hongying, Z., Kumar, S., Chen, F., Fuh, JYH, Yu Wang, M.: Topology optimized multimaterial soft fingers for applications on gripper, rehabilitation, and artificial hands. IEEE/ASME Trans. Mechatron. **2019** (2019)
7. Liu, C., Chen, Y., Yang, S.: Topology Optimization and Prototype of a Multimaterial-Like Compliant Finger by varying the Infill Density in 3D Printing. Mary Ann Liebert, Inc, Taiwan (2021)
8. Chen, Z., Rahimi Nohooji, H., Chew, C.: Development of topology optimized bending-twisting Sof finger. J. Mech. Robot. (2022)

9. Liu, C., Chiu, C., Chen, T., Pai, T., Hsu, M., Chen, Y.: Topology optimization and prototype of a three-dimensional printed compliant finger for grasping vulnerable objects with size and shape variations. J. Mech. Robot. **2018** (2018)
10. Petkovic, D., Pavlovic, N.: Compliant Multi-Fingered Adaptative Robotic Gripper. in Kolloquium Getriebetechnik - Ilmenau 2013. Ilmenau, Germany (2013)
11. Dechev, H., Cleghorn, W., Naumann, S.: Multiple finger, passive adaptive grap prosthetic hand. Mech. Mach. Theory **36**, 1157–1173 (2001)
12. Yang, J., Pitarch, E., Abdel-Malek, K., Patrick, A., Lindkvist, L.: A multi-fingered hand prosthesis. Mech. Mach. Theory **39**, 555–581 (2004)
13. Chen, F., Cao, J., Zhang, L., Zhang, H., Yu Wang, M., Zhu, J., Zhan, Y.: Networked soft actuators with large deformations. In: IEEE International Conference on Robotics and Automation (ICRA), pp. 4332–4337 (2017)
14. Li, C, Gu, X., Ren, H.: A cable driven flexible robotic grasper with lego-like modular and reconfigurable joints. IEEE/ASME Trans. Mechatron. 2757–2767 (2017)
15. Popov, D., Ganopov, I., Ryu, J.-H.: Portable exoskeleton glove with soft structure for hand assistance in activities of daily living. IEEE/ASME Trans. Mechatron. 865–875 (2017)
16. Manti, M., Hassan, T., Passetti, G., D'elia, N., Laschi, C., Cianchetti, M.: A bioinspired soft robotic gripper for adaptable and effective grasping. Soft Robotics SoRo, pp. 107–116 (2015)
17. Wehner, M, Truby, R., Fitzgerald, D., Mosadegh, B., Whitesides, G., Lewis, J., Wood, R.: An integrated design and fabrication strategy for entirely soft, autonomous robots. Nature 451–455(2016)
18. Kim, S., Laschi, C., Trimmer, B.: Soft robotics: a bioinspired evolution. Trends Biotechnol 287–294 (2013)
19. Drotman, D., Jadhav, S., Karimi, M., Tolley, M: 3D printed soft actuators for a legged robot capable of navigating unstructured terrain. In: IEEE International Conference on Robotics and Automation (ICRA), Singapore (2017)
20. Çulha, U., Iida, F.: Enhancement of fingermotion range with compliant anthropomorphic joint design. Bioinspired Biomimethic (2016). Art. no. 026001
21. Deimel, R., Brock, O.: A novel type of compliant and underactuated robotic hand for dexterous grasping. Int. J. Robot. Res. 161–185 (2016)
22. Polygerinos, P., Lyne, S., Wang, Z., Nicolini, L., Mosadegh, B., Whitesides, G., Walsh, C.: Towards a soft pneumatic glove for hand rehabilitation. In: IEEE/RSJ International Conference on Intelligent Robots and Systems (IROS), pp. 1512–1517, 3–7 November 2013
23. Mankame, N.D.: Investigations on contact-aided compliant mechanisms. University of Pennsylvania, ProQuest (2004)
24. Crisfield, M.: Nonlinear Finite Element Analysis of Solids and Structures. Wiley, New York (1991)
25. Belytschko, T., Glaum, L.: Applications of Higher Order Corotational Stretch Theories to Nonlinear Finite Element Analysis, pp. 175–182 (1979)
26. Belytschko, T., Hsieh, B.: Non-Linear Transient Finite Element Analysis with Convected Coordinates, pp. 255–271 (1973)
27. Liu, J., Perez-Liebana, D., Lucas, S.: Bandit-Based Random Mutation Hill-Climbing. United Kingdom (2016)
28. Morgan, B.: Towards data science. 23 September 2021. [Online]. Available: https://towardsdatascience.com/unit-1-hill-climber-optimization-985d5b79bd5

Development of a Database to Simulate and Adapt Compliant Mechanisms to a Given Characteristic for Improving Energy Efficiency of a Walking Robot

Marten Zirkel[1](✉), Yinnan Luo[2], Ulrich J. Römer[2], Alexander Fidlin[2], and Lena Zentner[1]

[1] Technische Universität Ilmenau, Compliant Systems, Ilmenau, Germany
{marten.zirkel,lena.zentner}@tu-ilmenau.de
https://www.tu-ilmenau.de/nsys

[2] Institute of Engineering Mechanics, Karlsruhe Institute of Technology, Karlsruhe, Germany
{yinnan.luo,ulrich.roemer}@kit.edu
https://www.itm.kit.edu/dynamik.php

Abstract. Compliant mechanisms with variable geometric parameters are investigated for the application in a bipedal robot to improve its walking efficiency. These mechanisms have nonlinear torque-angle characteristics and act like torsion springs to change the systems free oscillation frequency. High energy efficiency is achieved if the free oscillation frequency matches the step frequency, meaning the that the robot walks in resonance. For this purpose, the desired characteristic of the optimized elastic coupling is identified via optimization. Then, a database is developed, which consists of boundary conditions and beam elements. In this paper, there are three boundary conditions and three beam elements for demonstration purpose. To simulate a large number of compliant mechanisms with different characteristics, two boundary conditions and a beam element can be combined. The boundary conditions serve as bearing types to connect the beam element to the thighs of a robot. The large deformation behavior is assumed to be simulated by the Euler-Bernoulli beam theory, which is validated by FEM models. Thus, the desired characteristic from the proceeding optimization process is realized by a specific compliant mechanism.

Keywords: Mechanisms · Compliance · Design · Energy · Efficiency

Supported by the German Research Foundation (DFG) grant number FI 1761/4-1 and ZE 714/16-1

A. K. Pandey et al. (Eds.): MAMM 2022, 126, pp. 46–60, 2023.
https://doi.org/10.1007/978-3-031-20353-4_4

1 Introduction

Compliant mechanisms (CM) can be used for creating different nonlinear torque-angle characteristics depending on their geometric shapes. Using numerical simulation, the optimum deformation behavior in particular surroundings such as balancing problems [31] and systems with repeatable movement patterns [18,26,32] can be found. According to [23], the walking movement of a bipedal robot is predominantly periodic with alternating single support and double support phases of the system which, in this case, consists of five rigid bodies. The research focuses on improving the energy efficiency of a robot with stable walking gaits generated by a nonlinear controller. The energy efficiency depends on the system design and walking gaits. Minimizing the energy consumption is formulated as an optimization process in order to find a optimum match between the system and the gait parameters. This process shows great potential for improving the efficiency by adding elastic couplings between the robot's segments. An optimal design is achieved by adding an optimized elastic coupling between the thighs, thus reducing its energy consumption by up to 70% in the walking speed range from 0.2 to 1.4 $\frac{m}{s}$. The efficiency improvement can be also shown in different environments: ground inclinations in the range of$-10°$ to $10°$ are considered in [22]. In order to validate the theoretical improvements that are achieved by the simulated elastic coupling, a specific CM with similar characteristics we want to identify via simulations and subsequently apply it as a part of a robot prototype. Figure 1 depicts the development process of determining the system design of the robot using compliant mechanisms. Compliant mechanisms can be modeled by a number of different theories. such as the FACT/Screw theory [27,28,33–35,40], building blocks [9,12,13,19,29,36,43] and topology optimization [4,6–8,15,21,25,44,45]. As a further method, rigid body replacement [1,16,17,20,37,41] is considered which is mentioned in the overviews in [11] and [14]. Bilancia et al. [2] give an overview of procedures and tools to design beam-based compliant mechanisms. Howell et al. [14] present an approach which is similar to a database. The analysis of compliant mechanisms can be analytical [3,5,10,30,38] or based on finite element analysis [12,35,45]. The torque-angle-characteristic depends on the design of the mechanism and also influences the optimal gaits of the robot. Since a fully coupled optimization of the robot's gaits and a detailed model of

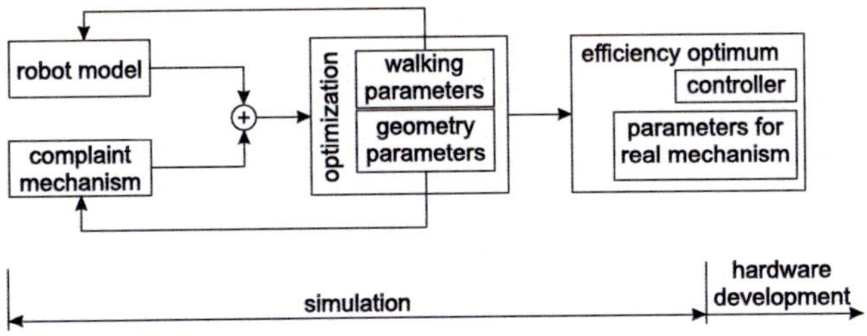

Fig. 1. Development process

a CM yields a large optimization problem that is computationally expensive and slow to solve, various other strategies have been investigated. In [47] the torque-angle-characteristic of a CM and the dependency on its geometry parameters is first fitted by suitably selected ansatz functions, making its transfer and application to other mechanisms tedious. Approaches without this limitation rely on a two stage procedure: first, a generic parametrization of the torque-angle-characteristic via splines is used in the optimization to identify the desired nonlinear characteristic. In the second stage, CMs with matching characteristics are synthesized. A method based on this approach is used in [46] where a CM is synthesized via genetic algorithms to produce the previously identified desired characteristic. However, a large number of simulations is needed to match the required behavior.

The present paper introduces the concept of a database to create different kinds of compliant mechanisms. The database elements are modeled by beam elements and corresponding boundary conditions. By changing the geometric configuration of the beam element or its boundary conditions, different characteristic curves under deformation can be generated. These are used to design feasible compliant mechanisms that meet the optimal characteristic curve for the application in the robot prototype. As a result, the database can also easily be integrated into the optimization process of the energy efficiency study [47]. In Sect. 2, the robot model, the elastic coupling and its desired characteristic are introduced; in Sect. 3 the database and its usage are explained; in Sect. 3.1 the mathematical model of describing the default beam element and three examples of beam elements and boundary conditions are introduced; in Sect. 4 the simulated deformation behavior is validated via FEA methods; in Sect. 5, the compliant mechanism is chosen to meet the requirements for the application in the robot system and in the last section we discuss the results and give an outlook.

2 Desired Characteristic

The investigated robot model is depicted in Fig. 2 and walks in the sagittal plane. It consists of five rigid body segments: one upper body, two thighs and two shanks. The segments are connected by four revolute joints, which are actuated by electric motors. A nonlinear controller synchronizes the joint angles to their reference trajectories and creates stable periodic walking gaits. According to the study in [23], connecting the thighs by a nonlinear spring reduces the energy consumption by up to 70% compared to the robot without elastic couplings. The simulated elastic coupling characteristic and the gaits are simultaneously optimized to minimize the cost of transport, which is a measure for the energy consumption of locomotion. As a result of the optimization, the leg's free oscillation frequency is adjusted to match with the current step frequency, which means the robot walks close to the system's resonance in order to achieve a high efficiency. As depicted in Fig. 2, the assumed optimum nonlinear torque-angle characteristic is derived from simulations of this system. This is regarded as the

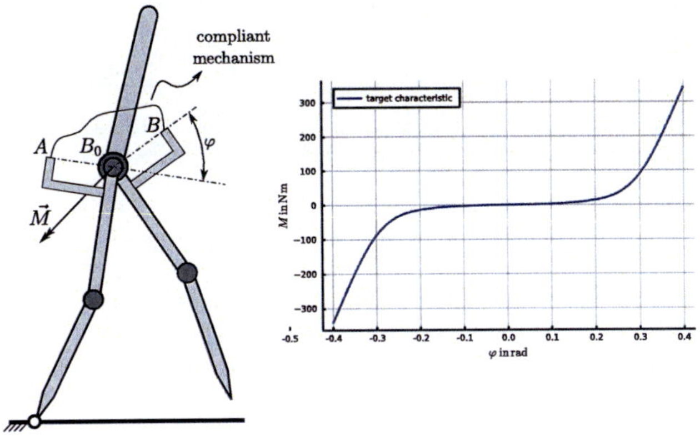

Fig. 2. The walker model with a compliant mechanism between the thighs (left) and the desired characteristic for this compliant mechanism (right)

required curve that should be created by a specific CM. The future goal is to realize the optimum nonlinear characteristic in praxis, which will be tested on a robot prototype to validate the optimization results. To this end, we consider compliant mechanisms based on Euler-Bernoulli beams with large displacements, which are modeled in the next section.

3 Database Design

As described above, the aim is to identify a compliant mechanism with a specific torque-angle-characteristic. For this purpose a database is introduced. The database holds different boundary conditions and beam elements. It provides the included objects for investigation and optimization purpose as shown in Fig. 3. Elements can be added by describing their geometric shape. The boundary conditions define how and where either end of a beam element is connected to its environment, in our specific example to the robot segments with the hip as the coordinates origin B_0, cf. Fig. 2. The location of the boundary conditions is independent of the beam elements. If the boundary conditions location don't match the location of the undeformed beam element ends a preload will be provoked. In Fig. 5 we are using this to fit the characteristic of the mechanisms to the desired characteristic from the previous section.

Three example beam elements based on the shapes in Fig. 4 and three example boundary conditions as depicted in Fig. 5 are used to demonstrate and validate the database design. These examples are explained in the next section.

3.1 Mathematical Model

Compliant mechanisms are regarded as deformable beam elements, whose deformations are described by the Euler-Bernoulli beam theory with large displacements [42]. The system of ordinary differential (1)–(4) describes the deformation

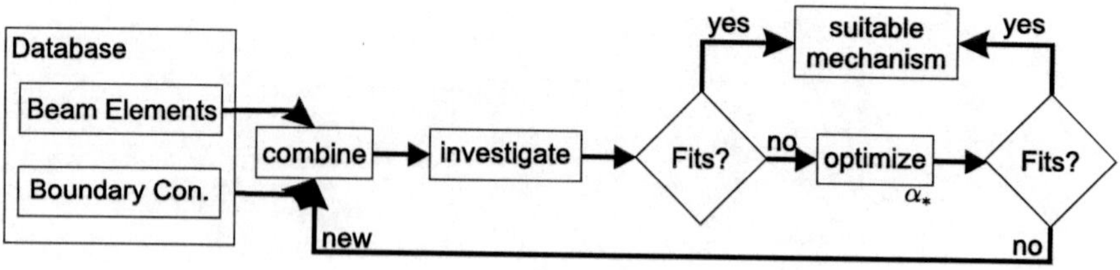

Fig. 3. The flowchart using the database

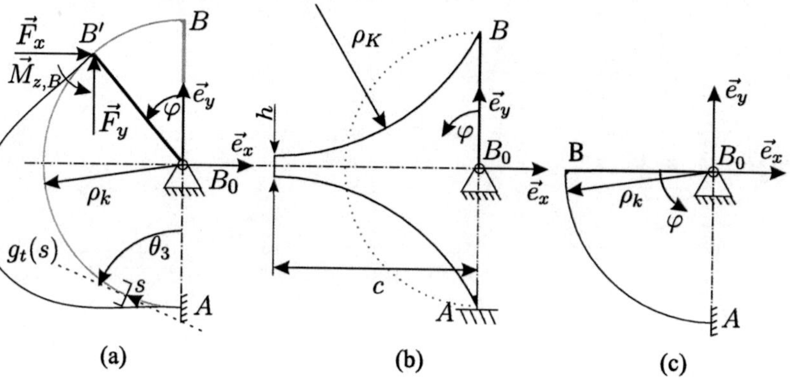

Fig. 4. Three different CMs: (a) a semi-circle, (b) a semi-rhombus and (c) a quarter-circle. The rotation of the movable segment and the displacement of B are determined by φ. Note that (b) and (c) are depicted in their undeformed configuration

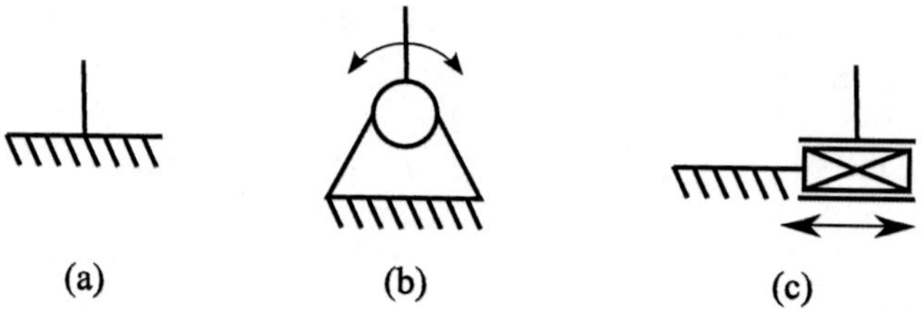

Fig. 5. Three different boundary conditions: (a) clamp, (b) joint and (c) slider

in the xy-plane (cf. Fig. 6) under the load by the forces F_x, F_y and the moment M_z at the end point. The geometry parameter of the beam element include the width w, the height h, the length L, the undeformed curvature κ and Young's modulus E. The variable s denotes the coordinate along the neutral axis of the beam element and the variable M is the deformation torque as a function of s. The angle θ describes the angle between the abscissa and the tangent $g_t(s)$ on the fiber at s, cf. Fig. 6.

$$\frac{dM}{ds} = F_x \sin(\theta) - F_y \cos(\theta), \tag{1}$$

$$\frac{d\theta}{ds} = \frac{M}{E I_z} + \kappa \quad \text{with} \quad I_z = \frac{w h^3}{12}, \tag{2}$$

$$\frac{dx}{ds} = \cos(\theta), \tag{3}$$

$$\frac{dy}{ds} = \sin(\theta). \tag{4}$$

Discontinuities are implemented via event-based interruptions of the integration process. During this interruption the integration variables are modified to fulfill the physical conditions. An example is shown in the next section. We suppose that the end points of the beam are connected to both thighs of the robot at points A and B. The rotational movements of the thighs around point B_0 cause the deformation of the entire beam element, as depicted in Fig. 4a. Three boundary conditions each are specified at points A and B (e.g. x_0, y_0 and θ_0 for a clamp at point A). To solve this boundary value problem for (1) to (4), we use a shooting method based on the trust region dogleg method, cf [24]. Given initial conditions for x_0, y_0, θ_0, M_0, F_x and F_y, the initial value problem for the ODE is solved by a Runge-Kutta method [39]. The initial conditions are then optimized by minimizing the residuals which are defined in the next section until a solution for the original boundary value problem is found. Based on this result, another integration of the ODE yields the corresponding reaction forces and torques at A and B.

Figure 4a is a beam element in the shape of a semi-circle. The length L and its undeformed curvature κ depend for this example on the radius ρ_k. The second shape is a semi-rhombus, a combination of two curved beam elements as shown in fig. 4b. The length of these beam elements is $L = 2\,\rho_k\, \arcsin(\frac{c}{\rho_k})$, c being the expansion in x direction. For this beam element we use a callback to interrupt

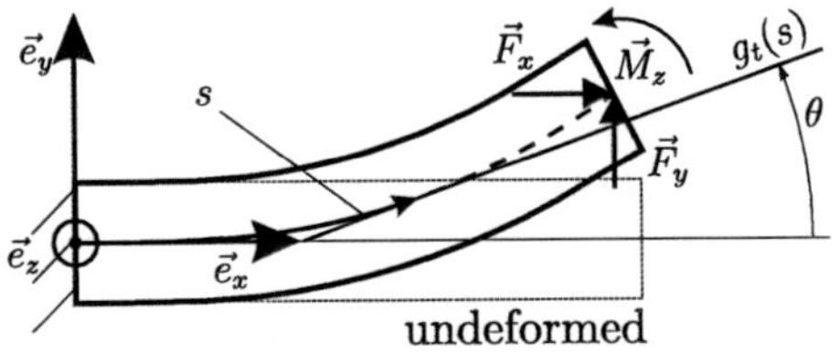

Fig. 6. A cantilever beam with loading forces

the integration at the discontinuous point and apply the modifications:

$$\theta^+\left(\frac{L}{2}\right) = \theta^-\left(\frac{L}{2}\right) - \pi$$
$$x^+\left(\frac{L}{2}\right) = x^-\left(\frac{L}{2}-\right) + h\,\cos\left(\theta^-\left(\frac{L}{2}\right) - \frac{\pi}{2}\right)$$
$$y^+\left(\frac{L}{2}\right) = y^-\left(\frac{L}{2}-\right) + h\,\sin\left(\theta^-\left(\frac{L}{2}\right) - \frac{\pi}{2}\right)$$

where "−" and "+" indicate the values before and after the modification, respectively. This simulates a rigid body between the circle segments in Fig. 4 (b). The third shape is a beam element in the form of a quarter-circle, see Fig. 4c. The length of this beam element is $L = \frac{\pi\,\rho_k}{2}$.

Besides the geometrical shape of the beam element, three different boundary conditions are considered. These boundary conditions describe the method of attaching both end points A and B of the beam element to the robot segments, as depicted in Fig. 5. The boundary condition clamp (a) restricts translational and rotational displacements about the fix point. The second one is a joint (b), which only allows rotational movement. The last one is the slider allowing movement in the radial direction without rotations. For each boundary condition a residual is defined to quantify the differences between the result of the Runge-Kutta integration (index c) and the required values due to the boundary condition (index bc). The residuals r_i with $i = 1, 2, 3$ are minimized by the shooting method as described in the previous section.

The residual function for the clamp is

$$\hat{r} = \sqrt{x_{bc}^2 + y_{bc}^2}$$
$$\hat{\varphi} = \arctan\left(\frac{y_{bc}}{x_{bc}}\right) + \varphi,$$
$$r_1 = x_c - \hat{r}\,\sin(\hat{\varphi}),$$
$$r_2 = y_c - \hat{r}\,\cos(\hat{\varphi}),$$
$$r_3 = \theta_c - \theta_{bc} - \hat{\varphi},$$

the residual function for the joint is

$$\hat{r} = \sqrt{x_{bc}^2 + y_{bc}^2}$$
$$\hat{\varphi} = \arctan\left(\frac{y_{bc}}{x_{bc}}\right) + \varphi,$$
$$r_1 = x_c - \hat{r}\,\sin(\hat{\varphi}),$$
$$r_2 = y_c - \hat{r}\,\cos(\hat{\varphi}),$$
$$r_3 = M_c$$

and the residual function for the slider is

$$
\begin{aligned}
\hat{\varphi} &= \arctan\left(\frac{y_{bc}}{x_{bc}}\right) + \varphi, \\
r_1 &= -\arctan(\frac{y_c}{x_c}) - \hat{\varphi}, \\
r_2 &= F_{x,c}\sin(\hat{\varphi}) - F_{y,c}\cos(\hat{\varphi}), \\
r_3 &= \theta_c - \theta_{bc} - \hat{\varphi}.
\end{aligned}
$$

The variables $\hat{\varphi}$ and $\hat{r}$ representing the polar coordinates of the particular boundary condition and are used to add the displacement φ. For each boundary condition there has to be an initialization method which defines the fixed initialization values and which values have to be optimized by the shooting method for the Runge-Kutta integration to fulfill the residual functions. For the clamp the torque at A and the forces at B have to be optimized by the shooting method and the position in x and y as well as the exit angle for the beam are set by the definition of the clamp. For the joint the exit angle in A and the forces in B have to be optimized. For the slider the direction of the summed forces in B has to be orthogonal to the free translation direction in A, so the shooting method optimizes the tangential force, the torque and the radial distance of A to the center B_0.

It is now possible to investigate up to 12 mechanisms by combining the beam elements with the boundary conditions in the manner of <boundary condition>-<beam element>-<boundary condition>. Mechanisms that are identical if they are kinematical inverted are only included once. The parameter set $\rho_k = 50\,\text{mm}, h = 1\,\text{mm}, E = 72,000\,\text{MPa}$ and $w = 10\,\text{mm}$ are used as initial values for creating different deformation behaviors. In order to validate the mathematical model, three example mechanisms with different beam elements and boundary conditions are modeled in FEM and investigated, as discussed in the next section.

4 Validation

In this section, the results for three example mechanisms from the above described method are compared to FEM simulations with Ansys. We use the standard settings in Ansys to simulate a static structural design of each mechanism. The option large deflection is set and the mesh size is set to 0.25 mm. All models have the same geometric parameters as their analytical models. The clamp is realized by a fixed connection. To implement the rotating clamp (B around B_0) we use a remote displacement with all degrees of freedom set to zero except the rotation angle is set to φ. Tables 1, 2 and 3 summarize the resulting moment M in B_0 and the displacements in $x\left(\frac{L}{2}\right)$ and $y\left(\frac{L}{2}\right)$ for $\varphi \in [-0.5, -0.4, \ldots, 0.5]$ rad.

Tables 1, 2 and 3 show the results of the FEM analysis compared to the analytical approach from Sect. 3.1, with an relative error of maximal $\approx 7.5\%$ for the torque and maximal $\approx 6.5\%$ for the displacement. Table 4 summarizes the results with the mean value of the relative error and the corresponding standard deviations.

Table 1. Comparison of the analytical and Ansys models of the clamp-semi-circle-clamp mechanism

Deflection	$x\left(\frac{L}{2}\right)$ in mm			$y\left(\frac{L}{2}\right)$ in mm			M in Nm		
φ in rad	Ansys	analytical	%	Ansys	analytical	%	Ansys	analytical	%
−0.5	15.081	15.049	0.21	8.917	8.925	0.09	−620.351	−600.002	3.39
−0.4	12.169	12.143	0.21	7.669	7.674	0.07	−545.443	−539.415	1.12
−0.3	9.181	9.163	0.20	6.169	6.172	0.05	−462.593	−451.845	2.38
−0.2	6.14	6.128	0.20	4.401	4.402	0.02	−341.204	−334.247	2.08
−0.1	3.069	3.064	0.16	2.349	2.349	0.00	−190.185	−184.198	3.25
0.1	−3.036	−3.031	0.16	−2.654	−2.654	0.00	225.987	219.259	2.98
0.2	−6.001	−5.993	0.13	−5.619	−5.618	0.02	470.742	473.721	0.63
0.3	−8.859	−8.475	4.53	−8.896	−8.894	0.02	789.981	762.819	3.44
0.4	−11.567	−11.555	0.10	−12.481	−12.478	0.02	1114.774	1085.398	2.64
0.5	−14.086	−14.074	0.09	−16.365	−16.361	0.02	1429.71	1439.848	0.70

Table 2. Comparison of the analytical and Ansys models of slider-quarter-circle-clamp mechanism

Deflection	$x\left(\frac{L}{2}\right)$ in mm			$y\left(\frac{L}{2}\right)$ in mm			M in Nm		
φ in rad	Ansys	analytical	%	Ansys	analytical	%	Ansys	analytical	%
−0.5	0.577	0.561	0.09	23.697	23.674	0.10	−648.509	−626.706	0.55
−0.4	0.735	0.721	0.07	19.376	19.358	0.09	−650.524	−656.097	0.60
−0.3	0.768	0.755	0.05	14.833	14.82	0.09	−657.907	−637.109	0.66
−0.2	0.666	0.657	0.02	10.072	10.063	0.09	−558.345	−539.416	0.71
−0.1	0.417	0.412	0.00	5.1131	5.109	0.08	−345.362	−334.248	0.76
0.1	−0.608	−0.603	0.00	−5.205	−5.202	0.06	487.415	473.721	0.89
0.2	−1.433	−1.422	0.02	−10.422	−10.419	0.03	1174.168	1085.398	0.95
0.3	−2.496	−2.48	0.02	−15.564	−15.564	0.00	1875.514	1824.254	1.02
0.4	−3.817	−3.795	0.02	−20.543	−20.547	0.02	2763.165	2674.598	1.11
0.5	−5.406	−5.379	0.02	−25.272	−25.285	0.05	3728.312	3619.943	1.19

Table 3. Comparison of the analytical and Ansys models of joint-semi-rhombus-clamp mechanism

Deflection	$x\left(\frac{L}{2}\right)$ in mm			$y\left(\frac{L}{2}\right)$ in mm			M in Nm		
φ in rad	Ansys	analytical	%	Ansys	analytical	%	Ansys	analytical	%
−0.5	5.341	5.26	3.28	7.931	7.975	3.36	−483.389	−458.841	5.08
−0.4	4.187	4.122	1.11	6.482	6.521	0.85	−411.656	−390.859	5.05
−0.3	3.071	3.022	2.32	4.964	4.997	3.16	−328.431	−312.315	4.91
−0.2	1.997	1.964	2.04	3.3768	3.401	3.39	−228.964	−221.906	3.08
−0.1	0.971	0.955	3.15	1.7218	1.735	3.22	−124.752	−118.274	5.19
0.1	−0.876	−0.894	2.98	−1.786	−1.802	2.81	141.974	134.392	5.34
0.2	−1.686	−1.721	0.63	−3.634	−3.669	7.56	306.636	286.452	6.58
0.3	−2.424	−2.474	3.44	−5.539	−5.596	2.73	484.827	457.822	5.57
0.4	−3.083	−3.146	2.64	−7.496	−7.58	3.21	689.595	650.270	5.70
0.5	−3.658	−3.731	0.70	−9.501	−9.615	2.91	920.543	865.738	5.95

Table 4. The relative mean error with the standard deviation for all three mechanisms over the range of $\varphi \in [-0.5, -0.4, \dots, 0.5]$ rad

	clamp-semi-circle-clamp	joint-semi-rhombus-clamp	slider-quarter-circle-clamp
displacement x	$0.58\% \pm 1.25\%$	$1.8\% \pm 0.21\%$	$1.22\% \pm 0.69\%$
displacement y	$0.03\% \pm 0.027\%$	$0.84\% \pm 0.21\%$	$0.06\% \pm 0.03\%$
Moment M_{B_0}	$2.23\% \pm 1.02\%$	$5.25\% \pm 0.86\%$	$3.32\% \pm 1.58\%$

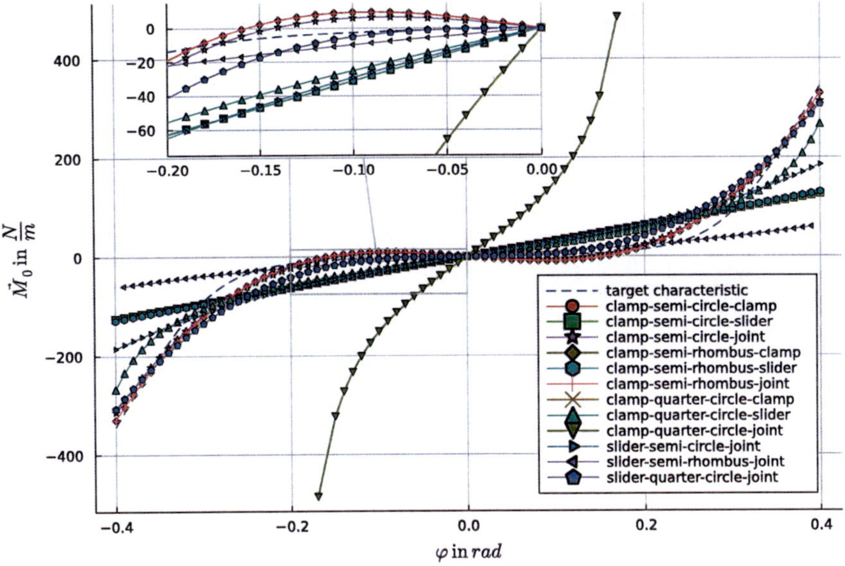

Fig. 7. The optimized characteristics of 12 symmetrical mechanisms compared to the target characteristic from Fig. 2

5 Choosing a Compliant Mechanism

In order to meet the requirement, the symmetry of the elastic coupling as depicted in Fig. 2, a further modification of the introduced mechanisms is needed. Two identical mechanisms are combined in the following way:

$$M_{B_0,new}(\varphi) = M_1(\varphi) - M_2(-\varphi).$$

Here, M_1 is the torque created by the first mechanism, M_2 is the torque of the mirrored mechanism. This combination results in $M_{B_0,new}$ with symmetrical characteristics as shown in Fig. 7. These characteristics are also optimized.

To achieve the characteristics in Fig. 7 the least squares error at $\phi = 0.1, 0.2, 0.3$ and 0.4 rad between the target curve and the characteristic curve of the mechanisms are minimized by a Nelder Mead algorithm. The optimized parameters are displayed in Table 5. The optimization algorithm can vary the

Table 5. The parameters of the optimized symmetrical mechanisms

symmetrical mechanism	α_{ρ_k}	$\alpha_{\overline{B\,B_0}}$
clamp-semi-circle-clamp	0.404	1.602
clamp-semi-circle-slider	3.066	1.566
clamp-semi-circle-joint	0.200	1.617
clamp-semi-rhombus-clamp	0.508	1.193
clamp-semi-rhombus-slider	12.315	1.599
clamp-semi-rhombus-joint	0.342	1.059
clamp-quarter-circle-clamp	50.0	1.015
clamp-quarter-circle-slider	32.920	1.601
clamp-quarter-circle-joint	49.997	1.6
slider-semi-circle-joint	16.169	3.0
slider-semi-rhombus-joint	3.245	0.434
slider-quarter-circle-joint	0.289	1.701

parameters in the range of $\alpha_{\rho_k} \in [0.2, 50]$ and $\alpha_{\overline{B\,B_0}} \in [0.2, 3]$, where the first factor scales the whole mechanism (not h and w) and the second one scales only the lever $\overline{B\,B_0}$. The limits are chosen since smaller values make it difficult to attach the mechanism on the robot and bigger values can cause undesired dynamic behaviors of the beam element in the walking robot. There is no mechanism in between the investigated ones which fits perfectly. But there are mechanisms with a qualitative characteristic similar to the desired characteristic of the elastic coupling.

6 Discussion and Outlook

In this manuscript we present a database which includes different types of the beam elements as well as attachment methods described as boundary conditions. Through variating the geometrical parameter and the boundary conditions, a large number of different elastic characteristics with large displacements are created using Euler-Bernoulli beam theory. The purpose is to find one compliant mechanism that matches with the required nonlinear characteristic, which is the result of the optimization process aiming on improving the energy efficiency of a bipedal robot through smart system designs. Three example beam elements and three boundary conditions are displayed in this approach to demonstrate the process of finding a viable mechanism. The beam element is described by five parameters: length, height, width, undeformed curvature and Youngs modulus; and each boundary condition at the end of the beam element is defined by rotational and translational displacements. Twelve different compliant mechanisms can be achieved through different combinations of the beam elements and two boundary conditions. The size of the beam and the length of the lever can be optimized to match the characteristics of the mechanisms to the desired characteristic of the elastic coupling. With the introduced beam elements and boundary conditions a perfect fit of characteristic of a mechanism could not be found, but there are mechanisms with a qualitative similar characteristic which are candidates for further investigations. These mechanisms can replace the elastic coupling in the optimization process of the robot. In addition we will add more beam elements and boundary conditions to the database. It is also planned to add branching.

References

1. Alfattani, R., Yunus, M., Alamro, T., Alnaser, I.: Linkage factors optimization of multi-outputs of compliant mechanism using response surface. Int. J. Nonlinear Anal. Appl. **12**(1), 59–74 (2021). https://doi.org/10.22075/ijnaa.2021.4658
2. Bilancia, P., Berselli, G.: An overview of procedures and tools for designing non-standard beam-based compliant mechanisms. Comput.-Aided Des. **134**, 103001 (2021). ISSN: 0010-4485. https://doi.org/10.1016/j.cad.2021.103001
3. Bodkhe, S., Ananthasuresh, G.K.: Multi-patch isogeometric analysis of planar compliant mechanisms. In: Mechanism and Machine Science, pp. 671–690. Springer, Singapore (2021). https://doi.org/10.1007/978-981-15-4477-4
4. Cao, L., Dolovich, A.T., Chen, A., Zhang, W.: Topology optimization of efficient and strong hybrid compliant mechanisms using a mixed mesh of beams and flexure hinges with strength control. Mech. Mach. Theory **121**, 213–227 (2018). ISSN: 0094-114X. https://doi.org/10.1016/j.mechmachtheory.2017.10.022
5. Choi, M.-J., Cho, S.: Isogeometric optimal design of compliant mechanisms using finite deformation curved beam built-up structures. J. Mech. Des. **142**(8) (2020). ISSN: 1050-0472. https://doi.org/10.1115/1.4043585
6. Chu, S., Gao, L., Xiao, M., Luo, Z., Li, H.: Stress-based multi-material topology optimization of compliant mechanisms. Int. J. Numer. Methods Eng. **113**(7), 1021–1044 (2018). ISSN: 00295981. https://doi.org/10.1002/nme.5697
7. Da Silva, G. A., Beck, A.T., Sigmund, O.: Topology optimization of compliant mechanisms considering stress constraints, manufacturing uncertainty and geometric nonlinearity. Comput. Methods Appl. Mech. Eng. **365**, 112972 (2020). ISSN: 0045-7825. https://doi.org/10.1016/j.cma.2020.112972
8. Da Silva, G. A., Beck, A.T., Sigmund, O.: Topology optimization of compliant mechanisms with stress constraints and manufacturing error robustness. Comput. Methods Appl. Mech. Eng. **354**, 397–421 (2019). ISSN: 0045-7825. https://doi.org/10.1016/j.cma.2019.05.046
9. Danun, A.N., Palma, P.D., Klahn, C., Meboldt, M.: Building block synthesis of self-supported three-dimensional compliant elements for metallic additive manufacturing. J. Mech. Des. **143**(5) (2021). ISSN: 1050-0472. https://doi.org/10.1115/1.4048220
10. Delissen, A.A., Radaelli, G., Herder, J.L.: Design and optimization of a general planar zero free length spring. Mech. Mach. Theory **117**, 56–77 (2017). ISSN: 0094-114X. https://doi.org/10.1016/j.mechmachtheory.2017.07.002
11. Gallego, J.A., Herder, J.: Synthesis methods in compliant mechanisms: an overview. In: American Society of Mechanical Engineers Digital Collection, pp. 193–214 (2010). https://doi.org/10.1115/DETC2009-86845

12. Gandhi, I., Zhou, H.: Synthesizing constant torque compliant mechanisms using precompressed beams. J. Mech. Des. **141**(1) (2019). ISSN: 1050-0472. https://doi.org/10.1115/1.4041330
13. Hampali, S., Pai, S.A., Ananthasuresh, G.K.: A tunable variable-torque compliant hinge using open-section shells. J. Mech. Robot. **12**(6) (2020). ISSN: 1942-4302. https://doi.org/10.1115/1.4047440
14. Howell, L.L., Magleby, S.P., Olsen, B.M. (eds.): Handbook of Compliant Mechanisms. Wiley, Chichester (2013). ISBN: 9781119953456. https://doi.org/10.1002/9781118516485
15. Ibrahim, A., Warsame, A.A., Pervaiz, S.: Finite element (FE) assisted investigation of a compliant mechanism made of various polymeric materials. Mater. Today: Proc. **28**, 1181–1187 (2020). ISSN: 2214-7853. https://doi.org/10.1016/j.matpr.2020.01.105
16. Kalpathy Venkiteswaran, V., Su, H.-J.: Pseudo-rigid-body models of initially-curved and straight beams for designing compliant mechanisms. In: American Society of Mechanical Engineers Digital Collection (2017). https://doi.org/10.1115/DETC2017-67431
17. Kalpathy Venkiteswaran, V.K., Su, H.-J.: A versatile 3R pseudo-rigid-body model for initially curved and straight compliant beams of uniform cross section. J. Mech. Des. **140**(9) (2018). ISSN: 1050-0472. https://doi.org/10.1115/1.4040628
18. Khoramshahi, M., Nasiri, R., Shushtari, M., Ijspeert, A.J., Ahmadabadi, M.N.: Adaptive natural oscillator to exploit natural dynamics for energy efficiency. Robot. Auton. Syst. **97**, 51–60 (2017). https://doi.org/10.1016/j.robot.2017.07.017
19. Kuppens, P.R., Bessa, M.R., Herder, J.L., Hopkins, J.B.: Monolithic binary stiffness building blocks for mechanical digital machines. Extreme Mech. Lett. **42**, 101120 (2021). ISSN: 2352-4316. https://doi.org/10.1016/j.eml.2020.101120
20. Liang, R., Xu, G., Li, M., Bo, H., Khalique, U.: Fusing topology optimization and pseudo-rigid-body method for the development of a finger exoskeleton. IEEE Robot. Autom. Lett. 1 (2021). ISSN: 2377-3766. https://doi.org/10.1109/lra.2021.3114418
21. Lieu, Q.X., Lee, J.: Multiresolution topology optimization using isogeometric analysis. Int. J. Numer. Methods Eng. **112**(13), 2025–2047 (2017). ISSN: 00295981. https://doi.org/10.1002/nme.5593
22. Luo, Y., Romer, U.J., Fidlin, A.: The influence of ground inclination on the energy efficiency of a bipedal walking robot. PAMM **20**(1), e202000142 (2021)
23. Luo, Y., Romer, U.J., Zentner, L., Fidlin, A.: Improving energy efficiency of a Bipedal Walker with optimized nonlinear elastic coupling. In: Advances in Nonlinear Dynamics. Springer Nature, Switzerland (2021). https://doi.org/10.1007/978-3-030-81166-2_23
24. Mogensen, P.K., et al.: JuliaNLSolvers/NLsolve.jl: v4.5.1. Version v4.5.1. (2020). https://doi.org/10.5281/zenodo.4404703
25. Murphy, M.D., Midha, A., Howell, L.L.: The topological synthesis of compliant mechanisms. Mech. Mach. Theory **31**(2), 185–199 (1996). ISSN: 0094-114X
26. Nasiri, R., Khoramshahi, M., Shushtari, M., Ahmadabadi, M.N.: Adaptation in variable parallel compliance: towards energy efficiency in cyclic tasks. IEEE/ASME Trans. Mechatron. **22**(2), 1059–1070 (2016)
27. Pan, J., Zhang, Y., Fu, Z., Shen, L.: Stiffness modelling of flexible support module for large-aperture laser transmission unit. J. Phys.: Conf. Ser. **2113**(1), 012069 (2021). ISSN: 1742-6596. https://doi.org/10.1088/1742-6596/2113/1/012069

28. Qiu, C., Dai, J.S.: Large deformation analysis of compliant parallel mechanisms. In: Analysis and Synthesis of Compliant Parallel Mechanisms—Screw Theory Approach, pp. 121–145. Springer, Cham (2021). https://doi.org/10.1007/978-3-030-48313-5_8
29. Racu, C.M., Doroftei, I.: Compliant mechanism for ankle rehabilitation device; Part I: modelling and design. IOP Conf. Ser. Mater. Sci. Eng. **444**(8), 052014 (2018). ISSN: 1757-899X. https://doi.org/10.1088/1757-899x/444/5/052014
30. Radaelli, G.: Synthesis of Mechanisms with Prescribed Elastic Load-Displacement Characteristics. Delft University of Technology (2017). https://doi.org/10.4233/uuid:d518b379-462a-448f-83ef-5ba0e761c578
31. Richiedei, D., Trevisani, A.: Optimization of the energy consumption through spring balancing of servo-actuated mechanisms. J. Mech. Des. **142**(1) (2020)
32. Scalera, L., Palomba, I., Wehrle, E., Gasparetto, A., Vidoni, R.: Natural motion for energy saving in robotic and mechatronic systems. Appl. Sci. **9**(17) (2019). https://doi.org/10.3390/app9173516
33. Soek, T.: Spatial Flexure Elements in a Compliant Motion Stage (2020)
34. Sun, P., Li, Y., Chen, K., Zhu, W., Zhong, Q., Chen, B.: Generalized kinematics analysis of hybrid mechanisms based on screw theory and Lie Groups Lie Algebras. Chin. J. Mech. Eng. **34**(1), 1–14 (2021). https://doi.org/10.1186/s10033-021-00610-2
35. Synthesis method of two translational compliant mechanisms with redundant actuation. Mech. Sci. 12 (2), 983–995 (2021). ISSN: 2191-9151. https://doi.org/10.5194/ms-12-983-2021
36. Thanaki, M., Zhou, H.: Synthesizing bidirectional constant torque compliant mechanisms using precompressed beams. In: American Society of Mechanical Engineers Digital Collection (2019). https://doi.org/10.1115/IMECE2018-86469
37. Tran, N.T., Dao, T.-P.: Statics analysis and optimization design for a fixed-guided beam flexure. J. Adv. Eng. Comput. **4**(2), 125 (2020). ISSN: 1859-2244. https://doi.org/10.25073/jaec.202042.276
38. Tschiersky, M., Hekman, E.E.G., Brouwer, D.M., Herder, J.L.: Gravity balancing flexure springs for an assistive elbow orthosis. IEEE Trans. Med. Robot. Bionics **1**(3), 177–188 (2019). ISSN: 2576-3202. https://doi.org/10.1109/tmrb.2019.2930341
39. Tsitouras, C.: Runge-Kutta pairs of order 5(4) satisfying only the first column simplifying assumption. Comput. Math. Appl. **62**(2), 770–775 (2011)
40. Wu, Z., Bandara, D., Kiguchi, K., Arata, J.: Design strategy for a surgical manipulator based on a compliant mechanism: rigidity and range of motion: finding the optimized balance. In: 2019 IEEE International Conference on Robotics and Biomimetics (ROBIO). IEEE (2019). https://doi.org/10.1109/robio49542.2019.8961727
41. Xu, H., Gan, J., Zhang, X.: A generalized pseudo-rigid-body PPRR model for both straight and circular beams in compliant mechanisms. Mech. Mach. Theory **154**, 104054 (2020). ISSN: 0094-114X. https://doi.org/10.1016/j.mechmachtheory.2020.104054
42. Zentner, L., Lins, S.: Compliant Systems: Mechanics of Elastically Deformable Mechanisms, Actuators and Sensors. Walter de Gruyter GmbH & Co KG, Berlin (2019)
43. Zhang, J., Guo, H.-W., Wu, J., Kou, Z.-M., Eriksson, A.: Design of flexure revolute joint based on compliance and stiffness ellipsoids. Proc. Inst. Mech. Eng. Part G: J. Aerosp. Eng. 095441002110169 (2021). https://doi.org/10.1177/09544100211016978

44. Zhang, X.: Topology Optimization of Compliant Mechanisms. Springer Nature, Singapore (2018). ISBN: 9789811304323
45. Zhu, B., Zhang, X., Zhang, H., Liang, J., Zang, H., Li, H., Wang, R.: Design of compliant mechanisms using continuum topology optimization: a review. Mech. Mach. Theory **143**, 103622 (2020). ISSN: 0094-114X. https://doi.org/10.1016/j.mechmachtheory.2019.103622
46. Zirkel, M., Luo, Y., Fidlin, A., Zentner, L.: Synthese eines Nachgiebigen Systems mit gegebener Kennlinie. In: Siebte IFToMM D-A-CH Konferenz 2021: 18./19. February 2021, Online-Konferenz (2021). https://doi.org/10.17185/duepublico/74063
47. Zirkel, M., Luo, Y., Romer, U.J., Fidlin, A., Zentner, L.: Parameter study of compliant elements for a Bipedal Robot to increase its walking efficiency. In: Conference on Microactuators and Micromechanisms, pp. 58–75. Springer, Cham (2020)

Analytical Characterization of Spatial Compliant Mechanisms Using Beam Theory

Stefan Henning(✉) and Lena Zentner

Compliant Systems Group, Technische Universität Ilmenau, 98693 Ilmenau, Germany
{stefan.henning,lena.zentner}@tu-ilmenau.de
https://www.tu-ilmenau.de/nsys

Abstract. Due to the advantages of compliant mechanisms, they are widely used in technical fields such as precision engineering, measurement and medical technology. The movement of the mostly monolithically designed mechanisms is primarily achieved by bending the individual structural sections. Due to geometric nonlinearities caused by large deflections, the analytical description of the deformation behavior under the influence of external loads is not trivial. While plane compliant mechanisms can be reliably calculated with different models, the analysis of spatial compliant mechanisms is accordingly more difficult. Due to the complexity of the motion, simplifications are often made in the model assumptions, which have a detrimental effect on the precision of the results. Therefore, this paper presents a method to apply nonlinear beam theory to complex spatial structures that can include both curved sections and sections with non-constant cross-sections. Using selected examples of varying complexity, the model is validated using FEM simulations. In addition, a procedure to characterize branched spatial compliant mechanisms with the proposed analytical model and to calculate their deformation is explained. By efficiently solving the analytical model numerically, the deformed state of the mechanism can be obtained in seconds. Therefore, the model is particularly suitable for early stages of development or iterative synthesis tasks.

Keywords: Compliant mechanism · Flexure hinge · Nonlinear · Beam theory · 3D · Analytical model · FEM

1 Introduction

Due to the advantages of compliant mechanisms, such as backlash- and frictionless motion as well as high reproducibility, they are often used in robotics, precision engineering, measurement technology or medical technology [3,5,6,11,14].

A. K. Pandey et al. (Eds.): MAMM 2022, 126, pp. 61–76, 2023.
https://doi.org/10.1007/978-3-031-20353-4_5

Typically, these are monolithic bodies that deform elastically under the influence of external loads. The deformation behavior is influenced by the geometry of the structure. The compliance can be concentrated to individual sections or distributed over the entire mechanism. Especially in precision engineering, it is therefore of particular importance to develop models for describing the deformation behavior under the influence of external loads that take into account the deformation of the entire mechanism.

Nevertheless, due to the geometric nonlinearity resulting from large deflections, the analysis of the motion behavior is not trivial. It is known that, with the help of the nonlinear beam theory, 2D compliant mechanisms can be analyzed reliably [8,10].

However, e.g. in precision engineering, further tasks of compliant mechanisms can also represent the positioning or manipulation of objects in space. While in the case of guiding tasks the guiding can usually be reduced to movements within single planes, positioning and manipulation usually take place spatially. The type of motion can be, for example, spherical around a kinematic center (e.g. spherical crank-and-rocker mechanism [15,17,24]) or multi-axial (e.g. toggle mechanism [19], micro-manipulator [12], parallel robot [18]).

Several models exist in the literature for the purpose of calculating the deformation behavior of 3D compliant mechanisms like the pseudo-rigid-body-model [15,16], the chained-beam-constrained-model [4] or finite element method (FEM) (e.g. [2,13]), each offering specific advantages and disadvantages. Additionally, some approaches for the calculation of spatial compliant mechanisms using beam theory exist in the literature. However, these are mostly limited to beams with constant cross-sections and small deformations [7,22].

According to the state of the art, the main difficulty in analytically modeling spatial, but also planar compliant mechanisms using beam theory is the simultaneous consideration of the following:

- all sections of a compliant mechanism as compliant (continuum),
- sections with constant or non-constant curvature,
- branched mechanisms,
- non-constant cross-sections,
- large deflections.

In this paper, the analytical model based on nonlinear beam theory is given. Rather than modeling individual beam structures as in [21] and [27], the given equations are applied to arbitrarily shaped spacial compliant mechanisms under external loads or displacements. In the model, the aforementioned points are all considered. The model is then applied to different examples of varying complexity and the results are compared to nonlinear FEM results.

2 Analytical Model

Bending theories are typically classified according to their order. A classification is made between three orders, where a distinction is made about whether the equilibrium conditions are derived in the undeformed or in the deformed state and whether large deflections are taken into account. Analytical approaches to modeling compliant mechanisms from the literature are predominantly based on 3rd order theory [25,27]. This means that the equilibrium conditions are established in the deformed state and large deflections are considered. This approach leads directly to a geometrically nonlinear character. This is justified by the fact that large deflections may well occur in compliant mechanisms. In the case of concentrated compliance, this can represent joint angles of more than ten degrees [9]. Significantly larger deflections can still be expected for elastic deformation of mechanisms with distributed compliance.

The term "theory for large deflections of (curved) rod-like structures" has been established for referring to this analytical model [21,27].

2.1 Mathematical Description of the Deformation State

In the following, a complete derivation will be omitted. Instead, the assumptions and equations necessary for the compliant mechanisms considered in the context of this work will be discussed in more detail. Detailed derivations can be taken from [21] and [27]. For the analytical model, the following assumptions are made:

- the equilibrium conditions are formulated on the motionless body (static problem),
- any boundary conditions can be defined at the beginning and end of the beam, whereby a static problem must be present,
- Hooke's law applies, linear material behavior is assumed and a plane stress state results,
- only solutions are valid where the maximum occurring strain or stress does not cause plastic deformations in the given material,
- only directional forces are considered,
- the principle of St.-Venant is valid, i.e. the stresses at a force application point are distributed as if the force would act on the entire cross-section,
- the beam axis s passes through the centroid of the cross-section and is continuous,
- geometrical nonlinearity is assumed (large deflections are taken into account),
- the length of the beam axis does not change (inextensible),
- the slenderness condition is satisfied (cross-sectional dimensions at least ten times smaller than the length and the curvature radius of the beam),
- the Bernoulli hypothesis is valid: the cross-sections before and during deformation are always perpendicular to the beam axis,
- the beam axis can be oriented arbitrarily in space,
- the surface of the beam is load-free,
- only solid cross-sections are considered,
- the warping due to torsion is constant over the length of the beam,
- small torsion angles are assumed.

A sketch of the problem is shown in Fig. 1. In [27], the differential equations for describing the spatial deformation state are given in vectorial form for a

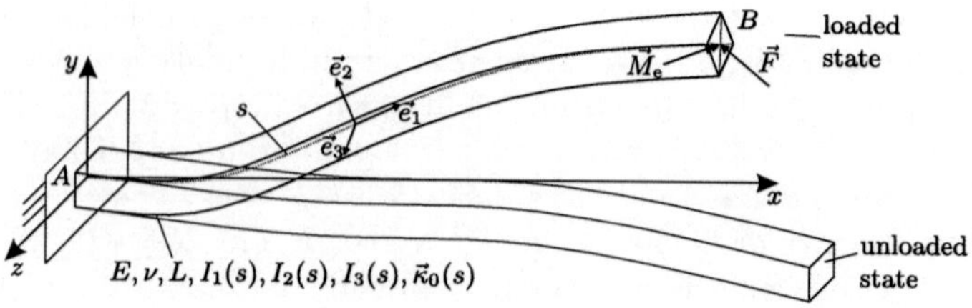

Fig. 1. Deformed state of a beam spatially loaded with the force $\vec{F}$ and the moment $\vec{M}_e$

Cartesian coordinate system (matrices are shown in bold, A matrix with superscript "T" represents a transposed matrix and with superscript "−1" represents an inverse matrix): Based on the given assumptions, the following system of differential equations results for describing the deflected state of the beam in space:

$$\frac{\mathrm{d}\vec{M}}{\mathrm{d}s} = -(\mathbf{T}^{\mathrm{T}}\vec{e}_x \times \vec{F}), \tag{1}$$

$$\frac{\mathrm{d}\vec{\theta}}{\mathrm{d}s} = \mathbf{T}_{\boldsymbol{\theta}}^{-1}\vec{\kappa}, \tag{2}$$

$$\frac{\mathrm{d}\vec{r}}{\mathrm{d}s} = \mathbf{T}^{\mathrm{T}}\vec{e}_x, \tag{3}$$

$$\text{with } \vec{\kappa} = \mathbf{S}^{-1}\mathbf{T}\vec{M} + \vec{\kappa}_0. \tag{4}$$

For a beam clamped on one side in point A with a free end in point B according to Fig. 1, the boundary conditions result:

$$\vec{M}(L) = \vec{M}_e,\ \vec{\theta}(0) = \vec{0},\ \vec{r}(0) = \vec{0}. \tag{5}$$

In the differential equations 1–4, the individual vectors represent the following parameters in Eq. 6 with respect to the Cartesian inertial system (in addition, $\vec{e}_x$ is the unit vector in the x-direction).

$$\vec{M} = \begin{pmatrix} M_x \\ M_y \\ M_z \end{pmatrix},\ \vec{F} = \begin{pmatrix} F_x \\ F_y \\ F_z \end{pmatrix},\ \vec{r} = \begin{pmatrix} x \\ y \\ z \end{pmatrix},\ \vec{\theta} = \begin{pmatrix} \theta_1 \\ \theta_2 \\ \theta_3 \end{pmatrix},\ \vec{\kappa} = \begin{pmatrix} \kappa_1 \\ \kappa_2 \\ \kappa_3 \end{pmatrix},\ \vec{\kappa}_0 = \begin{pmatrix} \kappa_{10} \\ \kappa_{20} \\ \kappa_{30} \end{pmatrix} \tag{6}$$

The vector $\vec{\theta}$ contains the relative angles of the beam-fixed coordinate system $\vec{e}_1$, $\vec{e}_2$, $\vec{e}_3$ to the stationary coordinate system $\vec{e}_x$, $\vec{e}_y$, $\vec{e}_z$ in the deformed state. Here, the angle θ_1 represents rotation about the $\vec{e}_1$ axis, the angle θ_2 analogously about the $\vec{e}_2$ axis, and the angle θ_3 about the $\vec{e}_3$ axis, see Fig. 2.

To describe a location vector $\vec{r}$ whose components are defined in the basis $\vec{e}_{x,\dots,z}$ to the basis $\vec{e}_{1,\dots,3}$, transformations are necessary. Transformations with the angles θ_1, θ_2 or θ_3 can be performed with the following transformation matrices:

$$\mathbf{T_1} = \begin{pmatrix} 1 & 0 & 0 \\ 0 & \cos\theta_1 & \sin\theta_1 \\ 0 & -\sin\theta_1 & \cos\theta_1 \end{pmatrix}, \quad \mathbf{T_2} = \begin{pmatrix} \cos\theta_2 & 0 & -\sin\theta_2 \\ 0 & 1 & 0 \\ \sin\theta_2 & 0 & \cos\theta_2 \end{pmatrix}, \quad \mathbf{T_3} = \begin{pmatrix} \cos\theta_3 & \sin\theta_3 & 0 \\ -\sin\theta_3 & \cos\theta_3 & 0 \\ 0 & 0 & 1 \end{pmatrix}. \tag{7}$$

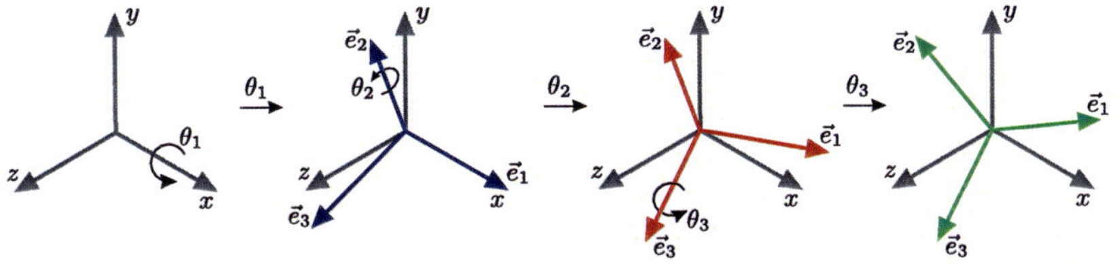

Fig. 2. Illustration of the transformation from x, y, z to $\vec{e}_1$, $\vec{e}_2$, $\vec{e}_3$ with the rotation angles θ_1, θ_2 and θ_3 in the given order

If the three individual rotations are combined, the transformation matrix $\mathbf{T} = \mathbf{T_1 T_2 T_3}$ in Eq. 8 is obtained. Here the transformation matrix depends on the transformation order used, cf. Fig. 2. The transformation matrix in Eq. 8 arises for the transformation order with θ_1, θ_2 and then with θ_3 for intrinsic rotations.

$$\mathbf{T} = \begin{pmatrix} \cos\theta_2\cos\theta_3 & \cos\theta_2\sin\theta_3 & -\sin\theta_2 \\ \sin\theta_1\sin\theta_2\cos\theta_3 - \cos\theta_1\sin\theta_3 & \cos\theta_1\cos\theta_3 + \sin\theta_1\sin\theta_2\sin\theta_3 & \sin\theta_1\cos\theta_2 \\ \sin\theta_1\sin\theta_3 + \cos\theta_1\sin\theta_2\cos\theta_3 & -\sin\theta_1\cos\theta_3 + \cos\theta_1\sin\theta_2\sin\theta_3 & \cos\theta_1\cos\theta_2 \end{pmatrix} \tag{8}$$

To characterize the shape of the beam, the vector $\vec{\kappa}$ is considered and related to the angles $\vec{\theta}$. This relation can be described by a matrix $\mathbf{T}_{\boldsymbol{\theta}}$ according to Eq. 9:

$$\vec{\kappa} = \mathbf{T}_{\boldsymbol{\theta}} \frac{\mathrm{d}\vec{\theta}}{\mathrm{d}s}. \tag{9}$$

Transformed Eq. 2 follows. The matrix $\mathbf{T}_{\boldsymbol{\theta}}^{-1}$ also depends on the transformation order and becomes for the chosen order:

$$\mathbf{T}_{\boldsymbol{\theta}}^{-1} = \begin{pmatrix} 1 & \sin\theta_1\tan\theta_2 & \cos\theta_1\tan\theta_2 \\ 0 & \cos\theta_1 & -\sin\theta_1 \\ 0 & \sin\theta_1\sec\theta_2 & \cos\theta_1\sec\theta_2 \end{pmatrix}. \tag{10}$$

The material properties and loads of the beam are linked by Eq. 4. In it, the material properties of the beam are implemented by the stiffness matrix $\mathbf{S}$:

$$\mathbf{S} = \begin{pmatrix} GI_1 & 0 & 0 \\ 0 & EI_2 & 0 \\ 0 & 0 & EI_3 \end{pmatrix}. \tag{11}$$

In addition to the material properties (shear modulus G and Young's modulus E), it includes I_1 the torsional moment of inertia of the cross-section, I_2 the equatorial area moment of inertia about the $\vec{e}_2$ axis, and I_3 the equatorial area moment of inertia about the $\vec{e}_3$ axis. While I_2 and I_3 are known, for example, for beams with rectangular cross-sections with height H and width w:

$$I_2 = \frac{Hw^3}{12}, \quad I_3 = \frac{wH^3}{12}, \tag{12}$$

the torsional moment of inertia I_1 is non-trivial for solid non-circular cross-sections. Its derivation is discussed in detail e.g. in [20] and [26]. I_1 for rectengular cross-sections ($I_{1\text{rect.}}$) can be expressed by:

$$I_{1\text{rect.}} = \frac{1}{3} Hw^3 \left(1 - \frac{192w}{\pi^5 H} \sum_{i=1}^{i} \frac{1}{i^5} \tanh \frac{i\pi H}{2w}\right). \tag{13}$$

In this work, a value of $i = 100$ is used in Eq. 13.

2.2 Solution Process

As can be seen in the differential equations 1–4, the given analytical model represents a boundary value problem, whereby the boundary conditions in Eq. 5 can be adapted depending on the specific application. Due to the non-constant cross-sections within the compliant mechanisms considered, for example in the form of flexure hinge contours, a heteronomous system of differential equations is present. Therefore, deriving an explicit analytical solution to the equations is not possible. However, numerical methods are suitable to approximate a solution for such nontrivial problems [23].

A numerical solution is possible when the number of equations matches the number of boundary conditions. A conventional solution approach for boundary value problems is a combination of a Runge-Kutta method with the so-called shooting method (see e.g. [1]). The solution process is shown in Fig. 3.

After defining the material, geometry, loads, and boundary conditions, unknown system parameters are determined. The unknown system parameters can be the reaction forces and moments of the supports or, in the case of displacement specification, unknown deflection forces (more detailed information follows with concrete examples in Sect. 3).

As part of the solution process, an initial estimate for the unknown system parameters is first required. These can have a significant influence on the computation time. The closer the estimates are to the actual result, the faster the

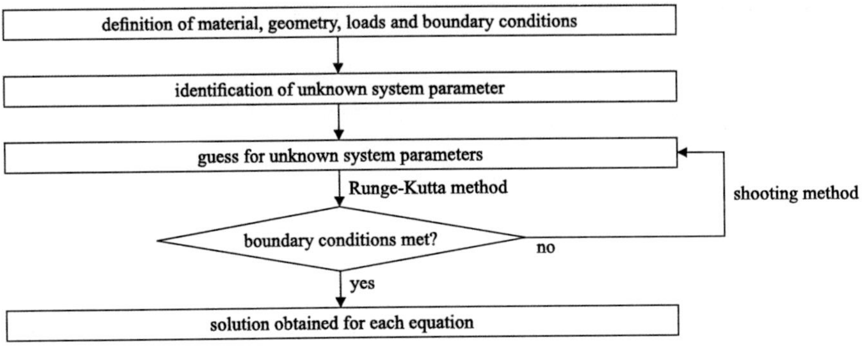

Fig. 3. Sequence of the calculation for solving the boundary value problem

solution takes place. In the context of the investigations in this work, it has been shown that for a single analysis an initial estimate of zero for each unknown parameter is sufficient to obtain a solution within a few seconds.

However, parametric studies or multiple load steps can be optimized by taking the results of the previous individual analysis into account for the next estimation in each case. This procedure is applied for the example in Fig. 4. After an initial estimate is made, the boundary value problem is treated as an initial value problem. Due to the modeling of the mechanisms as continua, the system of differential equations can thereby be integrated along the path of the beam axis s using a four-step Runge-Kutta method. Subsequently, the results of this calculation are compared with the boundary conditions. If the boundary conditions are not satisfied, the initial estimate must be modified. This change is performed by the algorithm implemented in the shooting method. After that, another calculation loop is made to solve the differential equations. This process continues until the boundary conditions are satisfied within a certain tolerance. If this is true, the calculation is terminated and the results for all differential equations are available. Thereupon, the deformed state of the mechanism can finally be displayed and further parameters like the elastic strain can be evaluated.

The following calculations are performed in MATLAB®. The four-step Runge-Kutta method is implemented by the function "ode45", while the shooting method is implemented by the function "fsolve".

3 Verification and Application

In this section, the analytical model for spatial considerations will be verified utilizing 3D-FEM simulations. The considerations are carried out based on selected examples of different complexity. First follows, the calculation of a simple beam under spatial loading. This is followed by the calculation of the deformation of a mechanism with flexure hinges based on the rigid-body mechanism of a spherical crank-and-rocker mechanism. Lastly, the procedure for the calculation of branched mechanisms is shown.

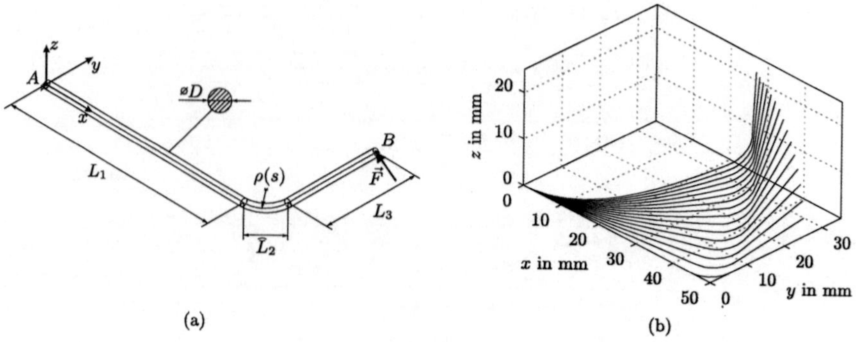

Fig. 4. Spatially loaded beam with circular cross-section: (a) geometry, loading and boundary conditions, (b) deformed state of beam axis

Verification on the example of angled beams under spacial loading First, the angled beam shown in Fig. 4a is considered, which is fixed at point A and loaded at the free end B by a force $\vec{F}$. The example represents a plane beam since the course of the beam axis lies within the xy-plane, but the beam is loaded spatially so that a spatial deformation results.
The force direction vector is chosen as follows:

$$\vec{F} = F \cdot \begin{pmatrix} -1 \\ 1 \\ 1 \end{pmatrix}. \tag{14}$$

To describe the shape of the beam, its curvature in the unloaded state is characterized by $\vec{\kappa}_0(s)$, in the given example specifically by κ_{30}. For this purpose, the beam can be mentally divided into three sections with the lengths L_1, L_2 and L_3. From $0 \leq s \leq L_1$ the curvature is 0. Then in the second section from $L_1 < s \leq L_1 + L_2$ the curvature is $\kappa_{30} = \frac{1}{\rho}$. In the third section, it is zero again. In this way, any beam shape can be characterized using the given analytical model. The same applies to changes in the cross-section or even the material parameters. The given example represents a beam with a constant circular cross-section. The selected geometrical parameters are:

$$L_1 = 45\,\text{mm},\ \rho = 5\,\text{mm},\ L_2 = \frac{\pi}{2}\rho,\ L_3 = 20\,\text{mm},\ D = 1.5\,\text{mm}. \tag{15}$$

Therein, D represents the diameter of the circular beam. The torsional moment of inertia for the circular cross-section is:

$$I_1 = \frac{\pi D^4}{32}. \tag{16}$$

In addition, the same beam equally loaded and with a square cross-section ($H = w = 1.5\,\text{mm}$) is considered. Its torsional moment of inertia is implemented with Eq. 13. Both beams are calculated numerically as described using the given analytical model. The displacements of the beam end in the three spatial directions are compared with the 3D-FEM simulation results.

The FEM model is meshed using *Solid186* elements. Before that, a mesh study is performed, which results in an element size of 0.35 mm for both beams. Large deflections are considered and the load vector is applied to the free end face.

For the comparison, the force F is iterated from 0 N to 7.5 N in steps of 0.5 N with the given direction vector from Eq. 14. The deformed state of the beam axis of the beam with circular cross-section calculated with the use of MATLAB® is shown for all load steps in Fig. 4b. From the deformed beam axes, it can be seen that there are large deflections in all three spatial directions. The beams are both bent and twisted. In addition, the deflection results u_x, u_y, u_z of the calculations for the maximum load of $F = 7.5$ N are given in Table 1.

Based on the results, despite the large deflections, with a maximum deviation of −0.12% at u_x, a very good agreement between the analytical and the FEM simulation results is evident. Thus, the analytical model can be considered verified for spatial beams with circular and square cross-sections.

Table 1. Numerical results of the analytical model (ana.) of the displacements of the beam ends and the relative deviation to the FEM-based results of the two beams at $F = 7.5$ N

Beam	u_x in mm			u_y in mm			u_z in mm		
	FEM	ana.	Δ_{FEM}	FEM	ana.	Δ_{FEM}	FEM	ana.	Δ_{FEM}
Circular	−24.70	−24.67	−0.12%	11.71	11.72	0.09%	22.40	22.39	−0.04%
Square	−16.47	−16.46	−0.06%	9.43	9.43	–	17.14	17.13	−0.06%

Verification on the example of a spacial compliant mechanism In this paragraph, the analytical model is applied to a spatial compliant mechanism. The rigid-body mechanism of a spherical crank-and-rocker mechanism shown in Fig. 5 (a)

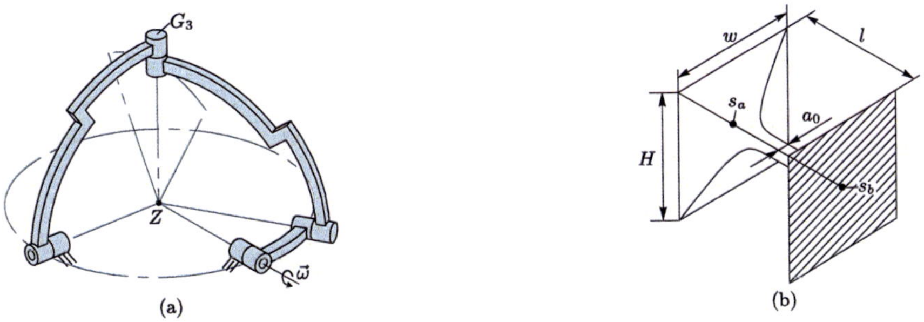

Fig. 5. Design of the spatial compliant mechanism: (a) rigid body mechanism, (b) implemented flexure hinge contours

serves as a template for the realization of the compliant mechanism in Fig. 6. This specific mechanism is characterized by the special arrangement of the four individual pivot joints. In this arrangement, all four axes of rotation intersect at a central point Z, causing the pivot G_3 (Fig. 5 (a)) to oscillate on a circular arc. While the rigid-body mechanism has a rotatable input link (input labeled $\vec{\omega}$), only a small range of possible motion of the rigid-body mechanism can be realized due to material strain in the compliant mechanism. For this purpose, the specific position shown in Fig. 5 (a) is chosen for the design of the compliant mechanism in Fig. 6. In the design of the compliant mechanism, flexure hinges are used at the points where conventional pivot joints are located in the rigid-body mechanism. Since torsion is to be expected in the hinges due to the spatial loading, only square instead of rectangular cross-sections are utilized in the mechanism here. The hinges used in this process are shown in Fig. 5 (b). The hinge contour shown therein is described by the power function given in Eq. 17:

$$a(s) = a_0 + \frac{2^n (H - a_0)}{l^n} \left| s - \frac{l}{2} \right|^n . \tag{17}$$

The cross-sectional dimension $a(s)$ is calculated from the beginning of the contour (s_a) and the end of the contour (s_b) using the minimum hinge height a_0, the contour length l, the maximum height H and the exponent of the power function n, see Fig. 5.

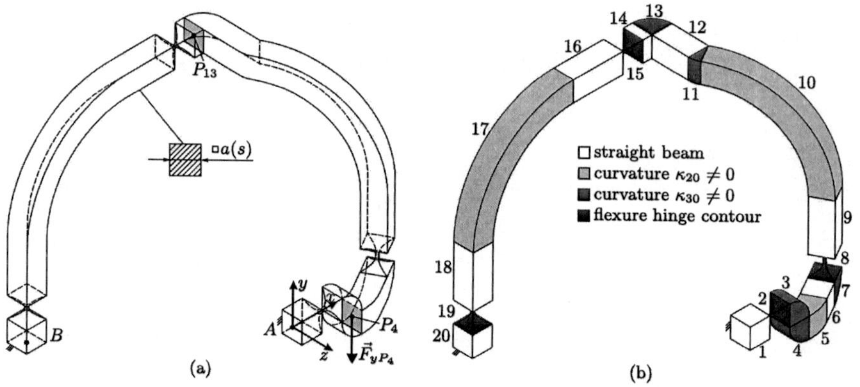

Fig. 6. Model of the compliant mechanism of the spherical crank-and-rocker mechanism: (a) chosen geometric parameters, (b) division into individual sections

To characterize the compliant mechanism as a whole using the given differential equations, it is possible to regard the mechanism as a single beam that is fixed at both ends A and B. The shape and geometry of the mechanism can be described in the same way as explained in the example of the beam in Fig. 4. It is possible to divide the mechanism up into a total of j several sections. Sectioning is performed, when changes in the beam's cross-section, changes in curvature, or points with external loading occur.

The considered complaint mechanism is split up into 20 different sections as shown in Fig. 6 (b). In the given example, the parameters $a_0 = 1$mm, $H = w = l = 10$mm, $n_2 = n_8 = 3$ and $n_{15} = n_{19} = 2$ are considered with n_j being the exponents of the according flexure hinge sections in Fig. 6 (b). The lengths and curvatures of each individual section of the considered mechanism are given in the appendix in Table 4.

In the performed investigations, the loading of the mechanism, as shown in Fig. 6 (a), is carried out by the force F_{yP_4} on the end face of the fourth section at the depicted point P_4 in y-direction. The deformed state is calculated using the analytical model from Sect. 2 and, to compare the results, 3D-FEM simulations. The meshing of the FEM models is again done using *Solid186* elements. A mesh study has been carried out and refinement is done at the hinges. It is specified that the force is varied until a displacement at the end of the fourth section of $u_{yP_4} = -2$ mm results. The force required for this is evaluated in the respective model. In addition, the displacement of the points P_4 and P_{13} in all three spatial directions is evaluated.

In Table 2 the results for the compliant mechanism of the spherical crank-and-rocker mechanism are listed. Based on the results from Table 2, it is clear that the deviations of the numerically determined results of the analytical model from the FEM results are small. The maximum deviation from FEM occurs at the displacement $u_{zP_{13}}$ and is −5.12% which results because of the small absolute value of this particular displacement. The force F_{yP_4} deviates by 4.69%. Thus, even for more complex beam structures in the form of compliant mechanisms, the analytical model is considered verified.

Table 2. Comparison of the compliant spherical crank-and-rocker mechanism at a given displacement of $u_{yP_4} = -2$ mm

Result	FEM	ana.	Δ_{FEM}
u_{xP_4} in mm	−1.5204	−1.5267	0.41%
u_{zP_4} in mm	−0.1661	−0.1696	2.11%
$u_{xP_{13}}$ in mm	3.1808	3.1704	−0.33%
$u_{yP_{13}}$ in mm	−3.9634	−3.9491	−0.36%
$u_{zP_{13}}$ in mm	−0.5701	−0.5409	−5.12%
F_{yP_4} in N	−31.3450	−32.8154	4.69%

In particular, the computational times to obtain the results using both models should be emphasized. While the numerical solution of the differential equation system is obtained within one second, the FEM calculation with the required mesh fineness takes significantly longer with one and a half hours on the same computer using a single core. Therefore, the described method for the computation of spatial compliant mechanisms is particularly suitable for early development phases in which frequent adjustments of the mechanism geometry have to be carried out to obtain quickly adapted results.

Consideration of branched beam structures Lastly, branched beam structures will be considered. For this purpose, the branched beam shown in Fig. 7 under spatial loading is considered as an example. In the initial position, the mechanism is located in the xy-plane and is spatially deformed by two force vectors.

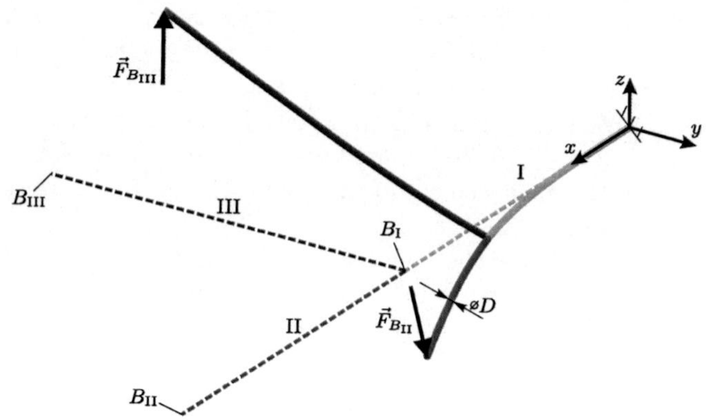

Fig. 7. Undeformed (dashed) and deformed state of the selected branched beam under spatial loading.

First, the procedure for analytically modeling branching points is needed to be investigated in more detail. Due to the branching point, the beam axis s is not continuous anymore. Because of this, one of the assumptions in Sect. 2.1 is not respected. Nevertheless, to describe branched beam structures using the given analytical model, it is possible to divide the structure into three individual beams I, II and III as shown in Fig. 7. Each of the three resulting beams thus has a continuous beam axis s_{I}, s_{II} and s_{III} and fulfills the assumptions made.

Then, to evaluate the deformation of the whole structure under spacial loading, transition conditions are necessary at the branching point. This is because the introduction of a branching point results in further unknown system parameters, which are compensated by the respective boundary conditions as well as by transition conditions at the branching point. The transition conditions for the location vector $\vec{r}$ are:

$$\vec{r}(s_{\mathrm{II}} = 0) = \vec{r}(s_{\mathrm{I}} = L_{\mathrm{I}}),\ \vec{r}(s_{\mathrm{III}} = 0) = \vec{r}(s_{\mathrm{I}} = L_{\mathrm{I}}). \tag{18}$$

The same can be expressed for the angles $\vec{\theta}$ depending on the orientation of each beam in the unloaded state. For the bending moment, the following transition conditions apply:

$$\vec{M}_{\mathrm{I}}(s_{\mathrm{I}} = L_{\mathrm{I}}) - \vec{M}_{\mathrm{II}}(s_{\mathrm{II}} = 0) - \vec{M}_{\mathrm{III}}(s_{\mathrm{III}} = 0) = \vec{0} \tag{19}$$

With the method of splitting up the structure into individual beams and the formulation of the transition conditions, the whole branched mechanism can be calculated numerically with the given differential equations.

For the given example in Fig. 7, the chosen force vectors are given in Eq. 20:

$$\vec{F}_{B_{\mathrm{II}}} = \begin{pmatrix} 0.5\,\mathrm{N} \\ 0.5\,\mathrm{N} \\ 0\,\mathrm{N} \end{pmatrix}, \ \vec{F}_{B_{\mathrm{III}}} = \begin{pmatrix} 0\,\mathrm{N} \\ 0\,\mathrm{N} \\ 0.5\,\mathrm{N} \end{pmatrix}. \tag{20}$$

These loads are applied to the according beams individually. For example, beam II is loaded with $\vec{F}_{B_{\mathrm{II}}}$ and beam III is loaded with $\vec{F}_{B_{\mathrm{III}}}$. The first beam (I) however, is loaded by the sum of both forces.

An FEM simulation will again serve as a comparison. In the FEM model, *Solid186* elements with an element size of 0.15 mm are used. For the selected load, large deflections result, which are shown to scale in Fig. 7. As a result, the displacements in x-, y- and z-direction at the end of the three single beams are evaluated for the analytical and the FEM model. The resulting displacements and the relative deviations of the numerically determined results of the analytical model from the FEM results are given in Table 3.

Table 3. Numerical results of the analytical model (ana.) of the displacements of the beam ends and the relative deviation to the FEM-based results of the branched beams

Point	u_x in mm			u_y in mm			u_z in mm		
	FEM	ana.	Δ_{FEM}	FEM	ana.	Δ_{FEM}	FEM	ana.	Δ_{FEM}
B_{I}	−1.444	−1.444	–	9.410	9.407	−0.03%	5.010	5.013	0.06%
B_{II}	−5.784	−5.807	0.40%	28.936	28.982	0.16%	10.674	10.693	0.18%
B_{III}	10.651	10.650	−0.01%	17.987	18.038	0.28%	30.166	30.261	0.31%

Based on the results from Table 3, it becomes clear that both models agree very well with deviations of less than 0.5%. Thus, it can be confirmed that the presented approach to modeling branched spatial mechanisms is valid.

4 Conclusions

In this paper, it was shown that nonlinear beam theory is suitable for modeling spatial compliant mechanisms. First, the analytical model was described for given assumptions. Based on selected examples of varying complexity, the analytical model could be validated with the use of FEM simulations. Even for large deflections, the numerically derived solution of the analytical model and the FEM-based results are in good correlation with deviations of under 0.2% for spatially loaded beams and under 5.2% for a complex mechanism.

In addition, a possibility to characterize branched mechanisms under spatial loading was presented and deviations of the analytical results from the FEM-based results of less than 0.5% were obtained.

A major advantage of the presented section-by-section modeling approach of spatial compliant mechanisms is that curvature or cross-section changes can be implemented into the model rather conveniently. Thus, respecting the assumptions of the model, arbitrary structures with concentrated or distributed compliance can be considered.

Especially in the early development phases, the deformed state can be calculated quickly with the help of a numerical solution to the resulting boundary value problem. The efficient solution thus allows adjustments to be made to the mechanism geometry and recalculations to be performed. Furthermore, it may be considered for synthesis, too.

Appendix

Table 4. Length and curvatures of each section for the analytical model of the compliant mechanism of the spherical crank-and-rocker mechanism ($l = 10\,\text{mm}$)

Section j	L_j	κ_{10j}	κ_{20j}	κ_{30j}
1	10	0	0	0
2	l	0	0	0
3	7.854	0	0	−0.200
4	7.854	0	0	0.200
5	9.958	0	0.053	0
6	10	0	0	0
7	7.854	0	0	0.200
8	l	0	0	0
9	25	0	0	0
10	70.686	0	0.022	0
11	2.618	0	0	−0.200
12	18.481	0	0	0
13	7.854	0	0	0.200
14	4.330	0	0	0
15	l	0	0	0
16	25	0	0	0
17	70.686	0	0.022	0
18	25	0	0	0
19	l	0	0	0
20	10	0	0	0

Acknowledgments. Supported by the German Research Foundation (DFG) under grant number ZE714/22-1.

References

1. Ascher, U.M., Petzold, L.R.: Computer Methods for Ordinary Differential Equations and Differential-Algebraic Equations. Society for Industrial and Applied Mathematics, Philadelphia (1998)
2. Bilancia, P., Berselli, G., Bruzzone, L., Fanghella, P.: A practical method for determining the pseudo-rigid-body parameters of spatial compliant mechanisms via CAE tools. Procedia Manuf. **11**, 1709–1717 (2017). https://doi.org/10.1016/j.promfg.2017.07.374
3. Böhm, V., Schorr, P., Feldmeier, T., Chavez-Vega, J.H., Henning, S., Zimmermann, K., Zentner, L.: An approach to robotic end effectors based on multistable tensegrity structures. In: Pisla, D., Corves, B., Vaida, C. (eds.) New Trends in Mechanism and Machine Science, Mechanisms and Machine Science, vol. 89, pp. 470–478. Springer International Publishing, Cham (2020). https://doi.org/10.1007/978-3-030-55061-5_53
4. Chen, G., Bai, R.: Modeling large spatial deflections of slender bisymmetric beams in compliant mechanisms using chained spatial-beam constraint model. J. Mech. Robot. **8**(4) (2016). https://doi.org/10.1115/1.4032632
5. Cosandier, F., Henein, S., Richard, M., Rubbert, L. (eds.): The Art of Flexure Mechanism Design, 1st edn. EFPL Press and CRC Press Taylor & Francis Group, Lausanne and Boca Raton, FL (2017)
6. Griebel, A., Henning, S., Schale, F., Griebel, S., Zentner, L.: Modellbasierte Untersuchungen zur Kraftüberwachung anhand des Verformungsverhaltens einer Matratzenfeder. In: Gössner, S. (ed.) Tagungsband 13. Kolloquium Getriebetechnik, Fachhochschule Dortmund, 18. - 20. September 2019, pp. 191–200. Logos Berlin, Berlin (2019)
7. Hao, G., Kong, X., Reuben, R.L.: A nonlinear analysis of spatial compliant parallel modules: multi-beam modules. Mech. Mach. Theory **46**(5), 680–706 (2011). https://doi.org/10.1016/j.mechmachtheory.2010.12.007
8. Henning, S., Linß, S., Gräser, P., Theska, R., Zentner, L.: Non-linear analytical modeling of planar compliant mechanisms. Mech. Mach. Theory **155**, 104067 (2020). https://doi.org/10.1016/j.mechmachtheory.2020.104067
9. Henning, S., Linß, S., Zentner, L.: detasFLEX—a computational design tool for the analysis of various notch flexure hinges based on non-linear modeling. Mech. Sci. **9**(2), 389–404 (2018). https://doi.org/10.5194/ms-9-389-2018
10. Henning, S., Zentner, L.: Analysis of planar compliant mechanisms based on non-linear analytical modeling including shear and lateral contraction. Mech. Mach. Theory **164**, 104397 (2021). https://doi.org/10.1016/j.mechmachtheory.2021.104397
11. Howell, L.L., Magleby, S.P., Olsen, B.M. (eds.): Handbook of Compliant Mechanisms. Wiley, Chichester (2013). https://doi.org/10.1002/9781118516485
12. Ivanov, I.: Methodical development of a parallel kinematic positioning system based on monolithic structures with flexure hinges. Dissertation, RWTH Aachen, Aachen (2016)
13. Lin, S., Wang, J., Xiong, W., Hu, Q., Liu, H., Wang, Q.: Design and modeling of a curved beam compliant mechanism with six degrees of freedom. Micromachines **13**(2), 208 (2022). https://doi.org/10.3390/mi13020208

14. Marangoni, R.R., Rahneberg, I., Hilbrunner, F., Theska, R., Fröhlich, T.: Analysis of weighing cells based on the principle of electromagnetic force compensation. Meas. Sci. Technol. **28**(7), 075101 (2017). https://doi.org/10.1088/1361-6501/aa6bcd
15. Parlaktaş, V., Tanık, E., Tanık, Ç.M.: On the design of a novel fully compliant spherical four-bar mechanism. Adv. Mech. Eng. **11**(9) (2019). https://doi.org/10.1177/1687814019879548
16. Parvari Rad, F.: Design and characterization of curved a spherical flexure hinges for planar and spatial compliant mechanisms. Dissertation, Alma Mater Studiorum Università di Bologna, Bologna (2014). https://doi.org/10.6092/unibo/amsdottorato/6544
17. Parvari Rad, F., Berselli, G., Vertechy, R., Parenti-Castelli, V.: Design and stiffness analysis of a compliant spherical chain with three degrees of freedom. Precis. Eng. **47**, 1–9 (2017). https://doi.org/10.1016/j.precisioneng.2016.06.011
18. Raatz, A.: Stoffschlüssige Gelenke aus pseudo-elastischen Formgedächtnislegierungen in Pararellrobotern. Dissertation, Technische Universität Braunschweig, Braunschweig (2006)
19. Rösner, M., Lammering, R.: Effiziente räumliche Modelle komplexer nachgiebiger Mechanismen auf Basis elastischer Festkörpergelenke: Hamburg, Helmut-Schmidt-Univ., Diss., 2015, /Bericht aus dem Institut für Mechanik, vol. Heft Dezember 2015. Helmut-Schmidt-Universität Bibliothek, Hamburg and Hamburg (2016). www.nbn-resolving.de/urn:nbn:de:gbv:705-opus-31154
20. Selvadurai, A.P.S.: Partial Differential Equations in Mechanics 2: The Biharmonic Equation, Poisson's Equation. Springer, Berlin and Heidelberg (2000). https://doi.org/10.1007/978-3-662-09205-7
21. Svetlitsky, V.A.: Statics of Rods. Foundations of Engineering Mechanics. Springer, Berlin Heidelberg (2000)
22. Tian, Y., Zhou, C., Wang, F., Lu, K., Zhang, D.: A novel compliant mechanism based system to calibrate spring constant of AFM cantilevers. Sens. Actuators A: Phys. **309**, 112027 (2020). https://doi.org/10.1016/j.sna.2020.112027
23. Venkiteswaran, V.K., Su, H.J.: Extension effects in compliant joints and pseudo-rigid-body models. J. Mech. Des. **138**(9), 535 (2016). https://doi.org/10.1115/1.4034111
24. Wilding, S.E., Howell, L.L., Magleby, S.P.: Spherical lamina emergent mechanisms. Mech. Mach. Theory **49**, 187–197 (2012). https://doi.org/10.1016/j.mechmachtheory.2011.10.009
25. Will, C.: Continuum models for biologically inspired tactile sensors: theory, numerics and experiments. Dissertation, Technische Universität Ilmenau, Ilmenau (2018)
26. Wunderlich, W., Kiener, G.: Statik der Stabtragwerke. Vieweg+Teubner Verlag, Wiesbaden (2004). https://doi.org/10.1007/978-3-322-80128-9
27. Zentner, L., Linß, S.: Compliant Systems: Mechanics of Elastically Deformable Mechanisms, Actuators and Sensors. De Gruyter, Berlin (2019)

Study of Curved Beam Based Displacement Amplifying Compliant Mechanism for Accelerometer Design

Mithlesh Kumar[1](✉), P. Krishna Menon[2], and Ashok Kumar Pandey[1]

[1] Mechanical and Aerospace Engineering, Indian Institute of Technology Hyderabad, Kandi, Sangareddy, Telangana, India 502285
mithlesh.ece@gmail.com ,
ashok@mae.iith.ac.in
[2] Research Centre Imarat, Hyderabad 500069, India
https://people.iith.ac.in/ashok/

Abstract. The performance of an inertial micro-device largely depends on its displacement sensitivity. Different techniques like mechanical amplification using compliant mechanisms, geometric anti-spring design, and mode-localized can be employed for improving it. In this paper, a compliant mechanism-based accelerometer is studied and the design is optimized by incorporating a curved shaped beam. A detailed analysis presents interesting insight about the operation of the curved-shaped displacement amplifying compliant mechanism (DaCM). Study suggest that a larger sense mass displacement could also be achieved by DaCM with smaller geometrical amplification. The design simulations are carried out in finite element analysis (FEA) based CoventorWare and COMSOL Multiphysics simulator.

Keywords: Accelerometer · Displacement amplification · Compliant mechanism · Geometric anti-spring · MEMS

1 Introduction

Microelectromechanical systems (MEMS) have been of significant interest for design of inertial sensors because of their low cost, small size, and ruggedness. Devices like MEMS accelerometer and gyroscope find applications in consumer electronics, automobiles, missiles, aircraft navigation [1]. Applications like inertial navigation system for ships, aircraft, submarines, guided missiles, and spacecraft requires very high sensitivity and resolution (of an order of 10−20 µg

Supported by DRDO, New Delhi, India.

A. K. Pandey et al. (Eds.): MAMM 2022, 126, pp. 77–93, 2023.
https://doi.org/10.1007/978-3-031-20353-4_6

for accelerometer and 1−5°/h for gyroscopes) [2]. Resolution of inertial MEMS devices is mainly limited by Brownian noise in micro-mechanical structures and interface electronics circuit noise [3]. A low mechanical sensitivity of MEMS devices may cause the electronics noise to dominate and limit the resolution [4]. Thus, there has been a significant interest to increase the sensitivity of MEMS. Different techniques like mechanical amplification using compliant mechanism, geometric anti-spring design, mode-localized design are being explored to improve the performance [5–7].

A compliant mechanism consists of flexible members that undergo deformation for applied force [8]. The deformation stores elastic energy in the flexible member, similar to strain energy in a deflected spring, which can be transferred or transformed to obtain a desired force-deflection relationship. In contrast, a rigid body mechanism consists of rigid link connected at movable joints to achieve a required force displacement behaviour. Thus, compared to a rigid body mechanism, a compliant structure has significantly lesser number of parts and movable joints to accomplish a specified task. Thus, it increases the precision of the compliant structures. Other advantages include miniaturization and reduction in weight of the compliant structure [8]. Further, compliant mechanism based design also suits the micro-fabrication process. It has allowed implementation of various compliant mechanism based MEMS devices like microgripper, crimping mechanism, compliant pliers [8]. The feasibility of micro-fabricated compliant structure has attracted research interest toward improving the displacement sensitivity of MEMS accelerometer; which can significantly improve their performance. In [9], a micro-lever based displacement amplifying mechanism was proposed for MEMS accelerometer. The lever mechanism works on the principle of displacement amplification at the free end of a beam for an applied force applied near to the fixed end of a beam as shown in Fig. 1(a). It shows a significant amplification of the output displacement and sensitivity. A simplest compliant structure for displacement amplification is an angled shaped beam shown in Fig. 1(b) [10]. The folded shape allow realization of amplification mechanism in smaller area. Different designs of displacement amplifying compliant mechanism have been studied in literature for accelerometer design [5,11,12].

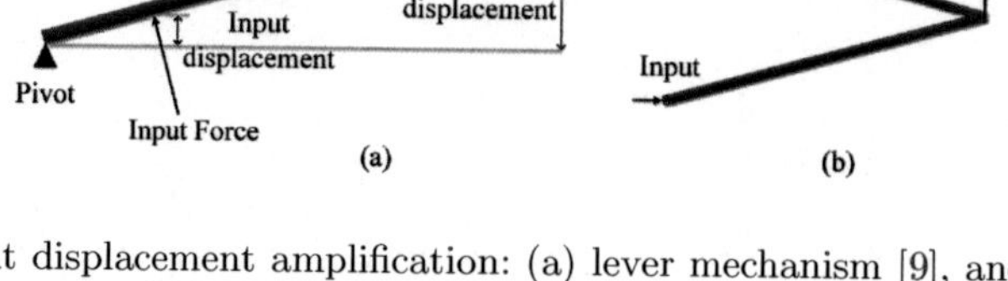

Fig. 1. Output displacement amplification: (a) lever mechanism [9], and (b) a simple compliant mechanism [10]

Applications like gravimeter for distributed gravity monitoring requires ultra-sensitive seismic accelerometer [6,13–15]. These accelerometers employ a curved shaped geometric anti-spring having reduced stiffness and very high sensitivity. In the present paper, the geometric anti-spring is incorporated in the design of displacement amplifying compliant mechanism and studied for its performance improvement. The designed accelerometer is also compared with the conventional accelerometer.

This article is organized as follows: A brief discussion on the design of conventional and compliant-mechanism accelerometer is presented in Sect. 2. Next, the optimization of displacement amplifying compliant mechanism is presented in Sect. 3. This is followed by the design and study of accelerometer with geometric anti-spring based compliant mechanism in Sect. 4. In the end, result and discussion is presented.

2 Accelerometer Designs

In this section, discussion on conventional (without mechanical amplification), and DaCM-based accelerometers are presented. In the following section, a novel design of DaCM is explored towards its performance improvement. The performance improvement by employing a DaCM in a design must not be at a cost of device footprint area. Thus, all designs considered in the manuscript have the same footprint area of 2 mm × 3 mm.

2.1 Conventional Accelerometer

A conventional accelerometer consist of a proof mass suspended for four springs at the corner such that the mass undergoes displacement in response to external vibrations. This allows displacement of combs attached to the proof mass and result in sense gap variations with respect to fixed combs attached to the substrate. A measure of this sense gap variations determine the acceleration of the external vibration. A conventional accelerometer is schematically illustrated in Fig. 2. The resonant frequency of the structure is 5.76 kHz and has a static displacement sensitivity of 6.76 nm/g. This results in a nominal capacitance of 9.0169 pF/g and differential capacitance change of 6.095 fF/g. The detailed dimensions of the accelerometer is shown in Table 1.

2.2 Accelerometer Employing Compliant Mechanism

Displacement amplifying mechanism provides an opportunity to improve the sensitivity of an accelerometer without increasing the device footprint. The layout of conventional and DaCM-based accelerometer is compared in Fig. 3. The DaCM-based accelerometer comprised of proof-mass assembly, DaCM, and sense combs. The proof-mass assembly drives the DaCM resulting in an amplified displacement of the sense combs. DaCM is characterized by inherent geometrical

Fig. 2. A conventional accelerometer designed for a resonant frequency of 5.76 kHz with a footprint area of 2 mm × 3 mm

Table 1. Dimensions of conventional accelerometer

Dimensions	
Proof mass	3000 µm × 480 µm
Spring	319 µm × 7 µm
Combs	760 µm × 7 µm
Sense gap	10 µm
Comb pairs	67 × 2
Thickness	100 µm

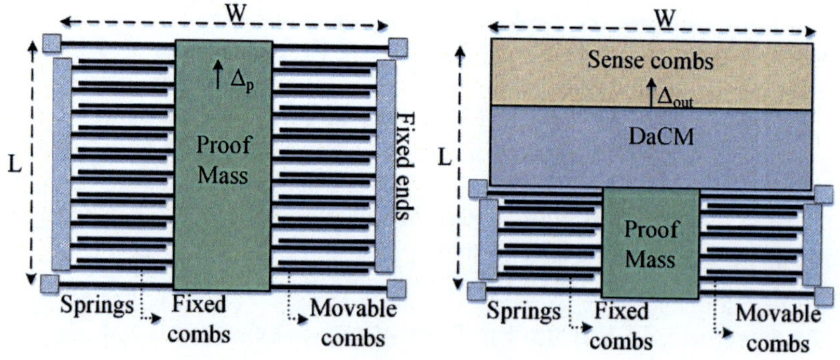

Fig. 3. Illustration of DaCM-based accelerometer and a conventional accelerometer having the same footprint [16]

amplification $n = \frac{\Delta_{out}}{\Delta_{in}}$ and net amplification $NA = \frac{\Delta_{out}}{\Delta_p}$, where, Δ_p is displacement of the conventional accelerometer, Δ_{in} and Δ_{out} are the input and output displacement of DaCM respectively.

In literature, different compliant mechanisms have been reported. They have been implemented for technologies requiring ultra-high precision motion generation, such as scanning probe microscopes, mechanisms for nano-imprint lithography, precision manufacturing, cell manipulation, micro- grippers and optical steering mechanisms [12]. For applications like accelerometer, a higher net displacement amplification is desired. In [11], Krishnan et al. have presented a comparative study of the displacement amplification of different DaCM's for accelerometer application. The study shows that DaCM shown in Fig. 4 has relatively simple construction and a higher amplification [16]; and thus it is considered for analysis in this article. The two arms of the DaCM are fixed and force F_{in} is applied at the input node results in an input displacement of Δ_{in} and output displacement of Δ_{out}.

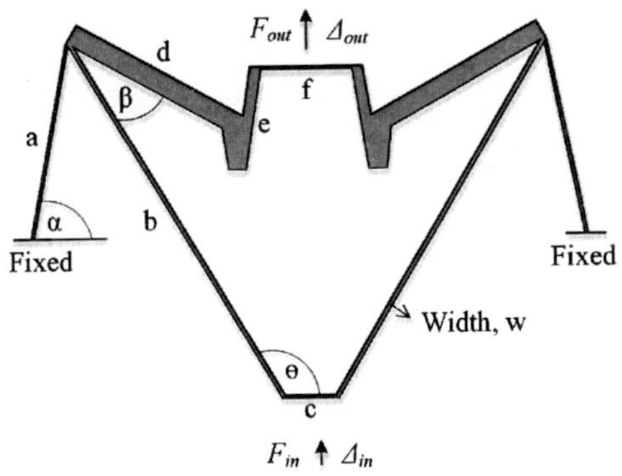

Fig. 4. Schematic of displacement amplifying compliant mechanism (DaCM). Here, a, b, c, d, e, f, w, α, β, and θ are the dimensional parameters.

Analysis of DaCM: The kinematics and the elastic deformation of the single-input-single-output compliant mechanisms can be expressed by the spring-leverage (SL) model, shown in Fig. 5, to account for its displacement-amplification [11]. In this model, five parameters are defined to fully characterize a compliant mechanism, i.e., input stiffness k_{ci}, output stiffness k_{co}, input mass m_{ci}, output mass m_{co}, and inherent geometrical amplification n. Analytical solutions to these parameters are challenging. Usually, finite element analysis (FEA) is used to estimate these parameters. In the first FEA run, a known force F_{in} is applied, keeping $F_{out} = 0$, to compute Δ_{in1}, and Δ_{out1}. The input stiffness k_{ci} and inherent geometrical amplification n is given as [11]

$$k_{ci} = \frac{F_{in}}{\Delta_{in1}} \quad (1)$$

$$n = \frac{\Delta_{out1}}{\Delta_{in1}} \quad (2)$$

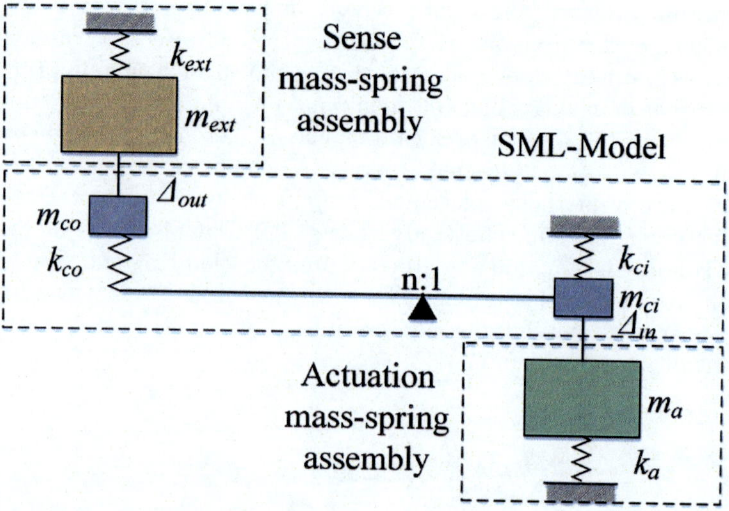

Fig. 5. Spring-leverage model of compliant mechanism connected to actuator and sense-mass-spring assembly. Here, k_a and m_a are stiffness and mass of actuator, whereas k_{ext} and m_{ext} are stiffness and mass of the sense side. The DaCM parameters corresponding to stiffness (k_{ci}, k_{co}), inertial masses (m_{ci}, m_{co}), and inherent geometrical amplification n are also illustrated

In the next FEA run, a known force F_{out} is applied, keeping $F_{in} = 0$, to compute Δ_{in2} and Δ_{out2}. The output stiffness k_{co} can be given as [11]

$$k_{co} = \frac{F_{out}}{\Delta_{out2} - n\Delta_{in2}} \tag{3}$$

The estimation of input and out inertial masses of DaCM requires another FEA run where the first and the second modal frequencies are computed. Referring to Fig. 5, equations for modal frequencies are derived by first writing the Lagrangian for the system and then applying the Euler-Lagrange (E-L) equations to get [17]

$$\omega^2 = \left(\frac{k_{ci} + k_a + n^2 k_{co}}{2(m_{ci} + m_a)} + \frac{k_{ext} + k_{co}}{2(m_{co} + m_{ext})}\right) \pm \sqrt{\left(\frac{k_{ci} + k_a + n^2 k_{co}}{2(m_{ci} + m_a)} - \frac{k_{ext} + k_{co}}{2(m_{co} + m_{ext})}\right)^2 + \frac{(nk_{co})^2}{(m_{co} + m_{ext})(m_{ci} + m_a)}} \tag{4}$$

The input-output inertial masses of DaCM are estimated by solving Eq. (4) for $k_a = m_a = k_{ext} = m_{ext} = 0$. The above discussed method is followed to estimate different parameters, i.e., n, k_{ci}, k_{co}, m_{ci}, and m_{co}, for the considered DaCM (shown in Fig. 4) and is presented in Table 2.

Table 2. Estimated parameters of the DaCM in Fig. 4

Parameters	FEA run 1	FEA run 2
Δ_{in}	1.2757 nm	1.1937 nm
Δ_{out}	9.9732 nm	11.994 nm
F_{in}	$9.81 \times 4.188 \times 10^{-7}$ N	$9.81 \times 5.025 \times 10^{-8}$ N
f_1	8.437 kHz	
f_2	10.567 kHz	
n	7.81	
k_{ci}	3217.24 N/m	
k_{co}	184.35 N/m	
m_{ci}	7.65×10^{-5} kg	
m_{co}	9.75×10^{-7} kg	

3 Optimization of DaCM

The design of displacement amplifying mechanism needs to be optimized for performance and footprint area. Further, feasibility of its fabrication using the existing technology also needs to be ensured. The optimization function for maximization of Δ_{out} can be defined as as a function of the geometrical parameters as

$$max(\Delta_{out}) = fun(a, b, c, d, e, f, \alpha, \beta, \theta, w) \tag{5}$$

$$Constraints : w \geq w_{min} \tag{6}$$

Equation (5) is constrained by permissible beam width w_{min}. The parametric optimization function available with FEA software, i.e., Coventor, COMSOL, IntelliSuite, can be used to study the variations of these parameters on Δ_{out}.

Typically, DaCM is classified as inverting and non-inverting type. In non-inverting type DaCM, Δ_{in} and Δ_{out} have displacements in the same direction, whereas it is in opposite directions for inverting DaCM. The DaCM in Fig. 4 is of non-inverting type. As shown in Fig. 7, increasing the length of beam b results in significant increase in Δ_{in} and Δ_{out}. However, for shorter beam length, displacement of Δ_{out} is reversed and DaCM becomes inverting. Thus, there is a trade-off between the DaCM displacements, and footprint area for its realization. The optimum dimensions of the different design parameters of DaCM estimated using FEA analysis are $a = 546\,\mu m$, $b = 1148\,\mu m$, $c = 160\,\mu m$, $d = 500\,\mu m$, $e = 115\,\mu m$, $w = 7\,\mu m$, $\alpha = 78.9°$, $\beta = 30°$, and $\theta = 121°$.

The optimized dimensions, presented above, are considered as reference for exploring the variations in design of DaCM as discussed below:

Curved beam: A curved microbeam with load across one end and fixed at other acts as a geometric anti-spring (GAS) structure, i.e., spring with significantly reduced stiffness. GAS has been incorporated in MEMS design to achieve higher sensitivity for MEMS accelerometer and gravimeter [6,15]. In this section, analysis is presented to optimize the considered DaCM by exploiting curved beam based anti-spring geometry for beam a and b (see Fig. 6(a)). A Bezier polygon with quadratic function available in COMSOl Multiphysics is used for designing the curved beam. The curvature of beam a and b are varied by varying the parameter a_h and b_h, respectively, as shown in Table 3.

Table 3. Curvature of beam a and b with variations of a_h and b_h

a_h, b_h (μm)	0	10	20	30	40	50	60	70	80	90	100	110	120	130	140	150	160	170	180	190
Curvature of beam a ($\tan^{-1}\left(\frac{a_h}{0.5a}\right)$, in deg)	0	2.1	4.2	6.3	8.3	10.4	12.4	14.4	16.3	18.2	20.1	21.9	23.7	25.5	27.2	28.8	30.4	31.9	33.4	34.8
Curvature of beam b ($\tan^{-1}\left(\frac{b_h}{0.5b}\right)$, in deg)	0	1	2	2.99	3.99	4.98	5.97	6.95	7.93	8.91	9.88	10.85	11.81	12.76	13.71	14.65	15.58	16.5	17.41	18.32

The variations of input and output displacements, i.e., Δ_{in}, Δ_{out}, of DaCM with curved beams for an applied input boundary load is shown in Fig. 6. As shown in Fig. 6(b), Δ_{in} increases with increasing curvature of beam b, however, it decreases with the curvature of beam a. This is because the anti-spring design of beam b reduces its stiffness and allows larger displacement of Δ_{in}. Thus, the combination of straight beam a and curved beam b have maximum Δ_{in} displacement. Further, it is also observed that Δ_{out} increases significantly with the curvature of beam b and remains nearly independent to the curvature of beam a (see Fig. 6c)). Thus, it can be argued replacing straight beam b with a curved beam in the design of DaCM, shown in Fig. 4, increases the input and output displacements. In addition to the input and output displacements, an important parameter defined to characterize a DaCM is inherent geometrical amplification n. It is defined as the ratio of Δ_{out} by Δ_{in}. Interestingly, a higher value of n is observed for straight beams based design of DaCM, i.e., $a_h = b_h = 0$ (see Fig. 6(d))). However, Δ_{in} and Δ_{out} are relatively less for straight beam based design compared to the curved beam DaCM design. This is further evident from the total displacement plot shown in Fig. 6(e)). The curved beam based design has higher total displacement, which is the sum of Δ_{in} and Δ_{out}. Thus, it can be concluded that a higher inherent geometrical amplification should not be the main criteria in the choice of DaCM as it does not guarantee a larger input-output displacements.

Variations of curved beam design: Different variations of the curved beam based DaCM have also been explored, i.e., *inverted curved beam*, and *double curved beam design*. The inverted curved beam DaCM, shown in Fig. 8(a) has curvature of beam b in opposite direction to the curved beam DaCM. The double curved beam design of DaCM is shown in Fig. 8(c). This design is the combination of the curved and inverted curved DaCM design. For brevity, a detailed analysis of inverted curved and double curved beam DaCM designs is not presented. However, in the next section, a comparative analysis of the designed accelerometer using these designs are included.

In the next section, the design of accelerometer using compliant mechanism is presented.

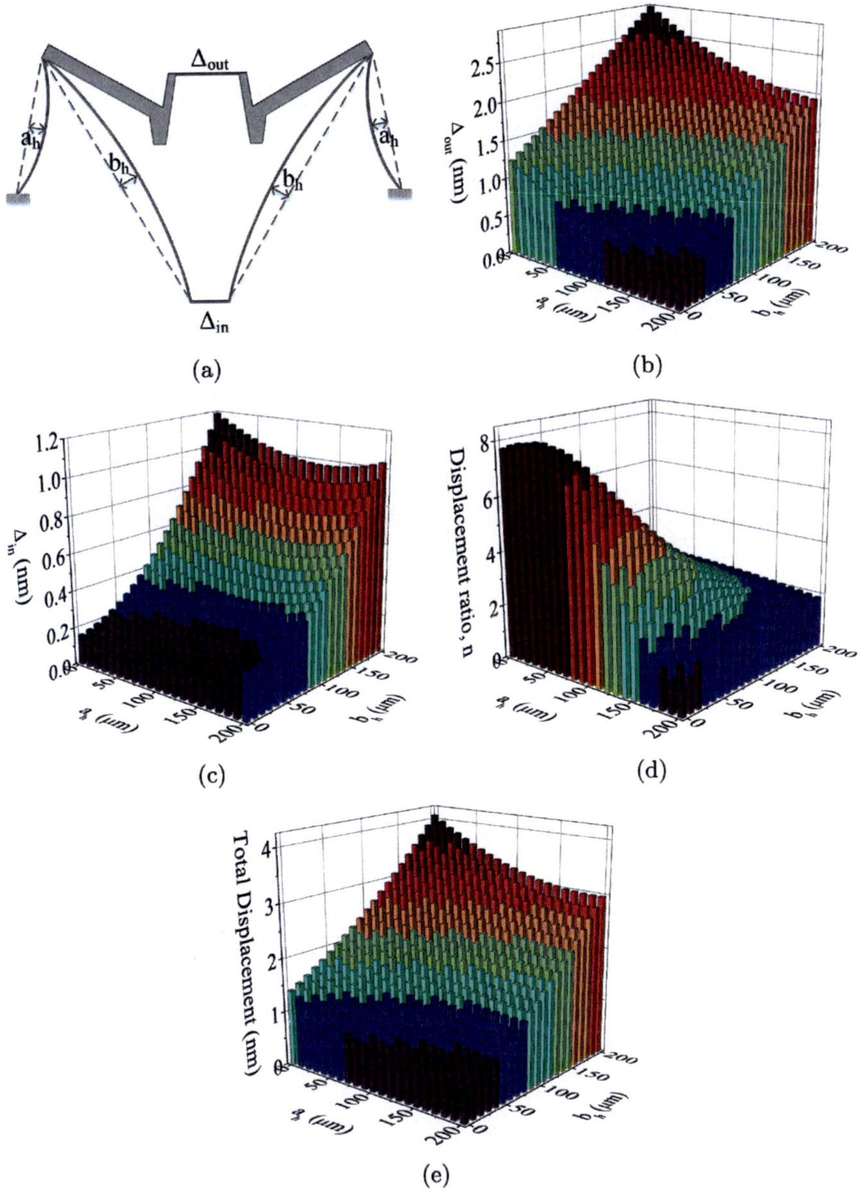

Fig. 6. Analysis of curved beam based DaCM design: (a) Schematic of curved beam DaCM; displacement of Δ_{out} and Δ_{in} are shown in (b) and (c), respectively, for different curvature of beam a and b. Displacement ratio, i.e., $n = \frac{\Delta_{out}}{\Delta_{in}}$ and total displacement, i.e., $(\Delta_{in} + \Delta_{out})$, are presented in (d) and (e) respectively

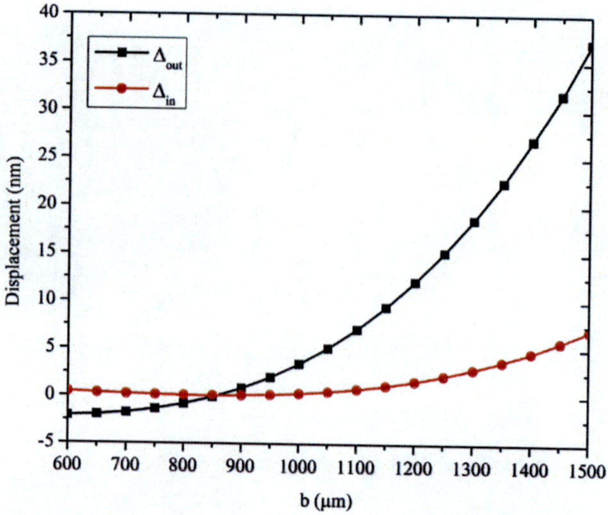

Fig. 7. Displacements of Δ_{out} and Δ_{in} with parametric variation of beam length b

(a) (b) (c)

Fig. 8. Schematic of different DaCM designs: (a) Curved beam, (b) Inverted curved beam and (c) Double curved beam design

4 Accelerometer Using Compliant Mechanism

The schematic of accelerometers with straight and curved beam based DaCM is shown in Fig. 9. The accelerometer consists of a proof-mass of 1500 μm × 1000 μm suspended by springs of 700 μm × 7 μm for driving the compliant structure. The amplified displacement Δ_{out} drives the sense comb assembly resulting in larger sense gap variations and sensitivity improvement. The sense gap of the combs is 10 μm. The sense mass has a dimension of 1000 μm × 50 μm and is supported by springs of 500 μm × 7 μm. The proof mass and the sense mass have 68 and 38 pairs of electrodes, respectively, of size 600 μm × 7 μm. The thickness of the structure considered for the analysis is 100 μm and the device footprint area is 2 mm × 3 mm. The analysis considers polycrystalline silicon as structural material with Young's modulus of 160 GPa, density of 2330 kg/m^3.

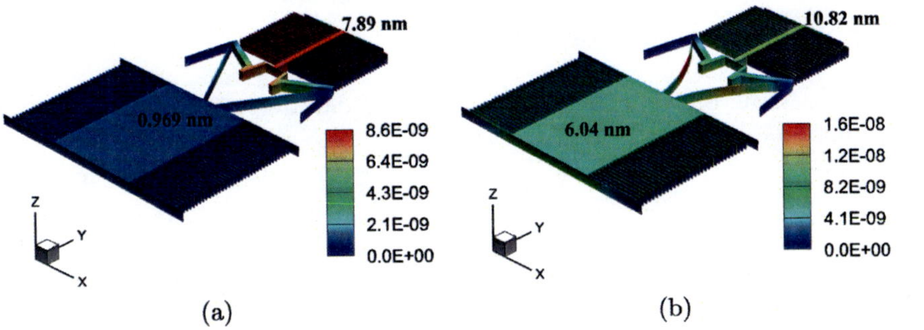

Fig. 9. Displacement sensitivity of (a) Straight beam DaCM, and (b) Curved beam DaCM based accelerometer design for applied acceleration of $1g$ along Y-axis. Accelerometer in (a) and (b) have resonant frequencies of 7.09 kHz and 5.875 kHz, respectively, for the same dimensions

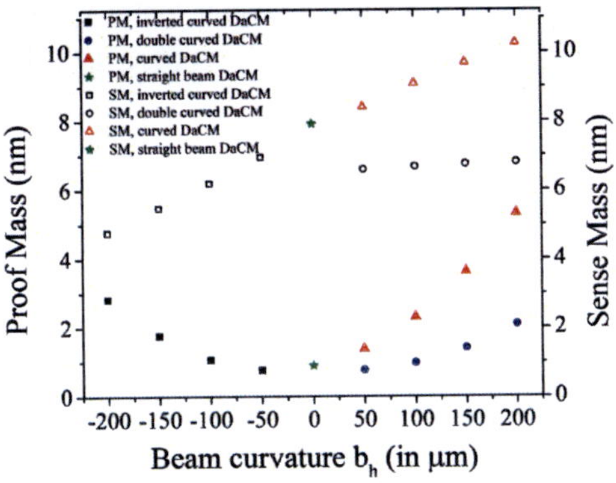

Fig. 10. Displacement sensitivity of accelerometer with different DaCM, i.e., straight beam, curved beam, inverted curved beam, and double curved beam, are compared for and applied load of $1g$

Figure 10 shows the comparison of the proof mass and the sense mass displacement for the different compliant mechanism based accelerometers. These displacements are estimated for an applied acceleration of $1g$. Analysis shows the straight beam DaCM has a sense mass displacement of 7.89 nm. The amplified sense mass displacement decreases with the curvature of the inverted curved beam DaCM (Fig. 8(b)). However, for a curved beam design, the sense mass displacement increases with the curvature of beam b. In case of the double curved beam design, the sense mass displacement remains same with increasing the curvature of the beams. Thus, it can be concluded that the curved beam DaCM outperforms the other DaCM variants in terms of the amplified sense mass displacement. Hence, it is preferred and considered for further analysis in the next section.

5 Results and Discussion

In this section, different parameters, i.e., modal frequencies, displacements for applied acceleration, stress, non-linearity, and capacitance charges, of the designed straight and curved beam DaCM based accelerometer are compared.

5.1 Modal Frequencies

The modal analysis of the accelerometer is carried out in CoventorWare. The design with straight and curves beam DaCM have in-plane displacement with the first modal frequency at 7.09 kHz and 5.875 kHz respectively. As the same dimensions are considered for both designs, it is apparent that the curved beam based DaCM reduces the overall stiffness of the design and thus results in resonant frequency reduction. The displacement sensitivity and modal frequencies of the straight and curved beam DaCM based accelerometer is shown in Fig. 9.

5.2 Cross-Axis Sensitivity

Cross-axis sensitivity is a measure of output observed across one axis when the acceleration is imposed on a different axis. The proposed accelerometer is exposed to an acceleration of $1g$, $10g$, and $30g$ along Y-axis and the cross axis sensitivity along X-axis and Z-axis are determined. The estimated cross-axis sensitivity along X-axis and Y-axis are $\sim 0.003\%$ and $\sim 0.005\%$, respectively, for the sense mass displacement of both types of accelerometer. This suggests that the curved beam based compliant design of does not contribute to the cross-axis sensitivity.

5.3 Stress

The stress gradient in the accelerometer must be less than the yield strength of the silicon. Typically, the yield strength of silicon is around 165 MPa [18]. The stress gradients shown in Fig. 11 is calculated for an applied body load of $100g$. The straight beam and curved beam DaCM based accelerometer have a maximum stress concentration of 11 MPa and 15 MPa respectively, which is less compared to the yield stress of silicon. This ensures the reliable operation of the designed accelerometer.

5.4 Non-linearity

The design of an accelerometer must ensure that the non-linearity in the displacement of the movable masses are minimum over the range of their operation. A compliant mechanism based accelerometer must ensure that the both the proof mass displacement and the amplified sense mass displacement are liner the over desired range. The non-linearity is studied by estimating the deviations in the displacement of the masses with reference to the extrapolated linear displacement for varying acceleration. As shown in Fig. 12, the simulated and the

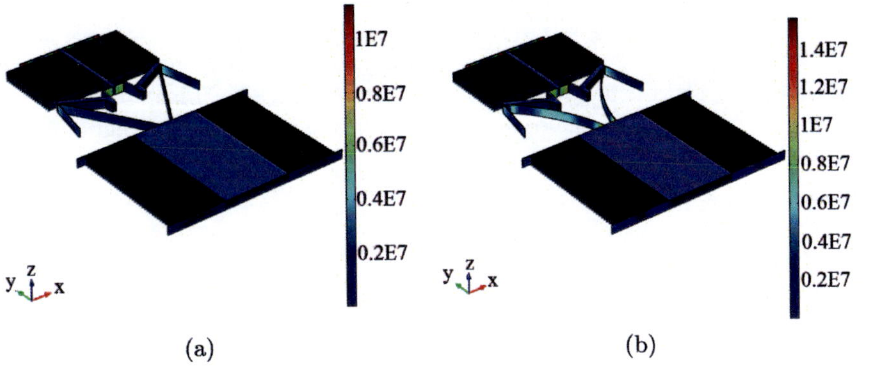

Fig. 11. Stress analysis for an applied body load of 100g: (a) Straight beam, and (b) Curved beam DaCM based accelerometer

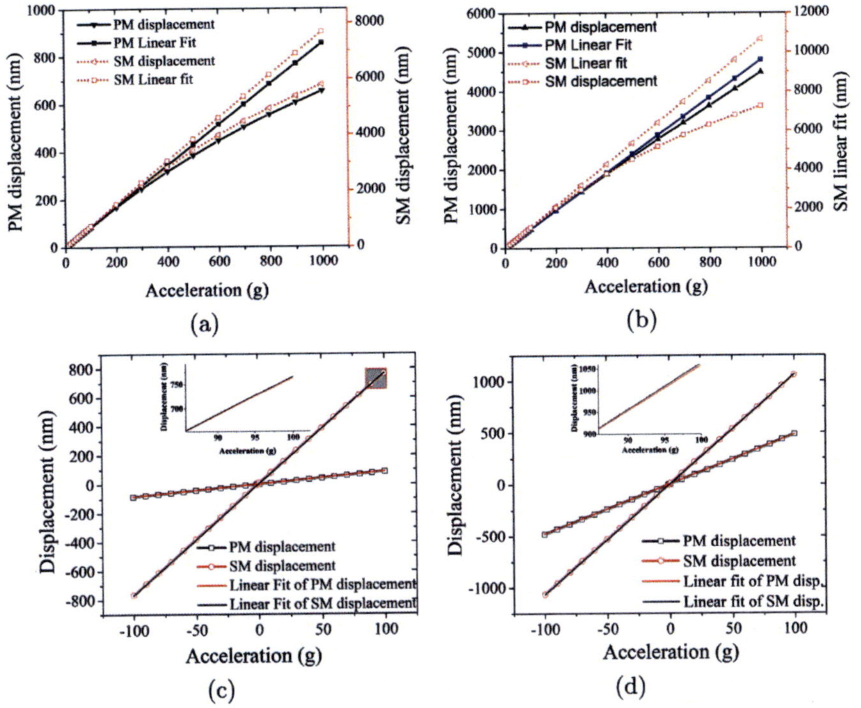

Fig. 12. Non-linearity in displacement of proof mass (PM, before amplification) and sense mass (SM, after amplification) over applied acceleration: (a), (c) Straight beam, and (b), (d) Curved beam based DaCM. The non-linearity is 1.345% and 2.739% for straight and curved beam based accelerometer over a range of 100g

linearly fitted displacement of the sense masses has a non-linearity of 1.345% and 2.739% for straight and curved beam based accelerometer over a range of 100g. An increase in applied acceleration increases the displacement deviation from the linear fit line and thus the non-linearity increases. This observed trend is same for both straight beam and curved beam DaCM based accelerometer.

5.5 Capacitance

The proof mass has a smaller displacement, however it can accommodate a larger comb pairs across its length. The proof mass has 68 comb pairs and can be used for actuation purpose. The capacitive sensitivity of the proof mass comb pairs are 1.2 fF/g and 4.1 fF/g respectively for the accelerometer with straight beam and curved beam DaCM. Incorporating DaCM in the design amplifies the output displacement of the sense mass. This allows larger capacitive sensitivity for sense mass comb pairs. The considered design has a sense mass electrodes count of 38. This gives the output capacitive sensitivity of 3.4 fF/g and 4.4 fF/g respectively for the sense mass comb pairs of the accelerometer with straight beam and curved beam DaCM.

Table 4. Simulated parameters of accelerometers in Fig. 9

Parameters	Design in Fig. 9(a)	Design in Fig. 9(b)
Resonant frequency, kHz	7.09	5.875
Input displacement sensitivity, nm/g	0.969	6.04
Output displacement sensitivity, nm/g	7.89	10.82
Geometrical amplification	8.14	1.79
Cross-axis sensitivity	S_{YX}=0.003%, S_{YZ}=0.005%	
Stress at 100g, MPa	11	15
Non-linearity (%, for ±100g)	1.345	2.739
Proof mass electrode nominal capacitance, pF	7.4383	
Sense mass electrode nominal capacitance, pF	4.7694	

6 Analysis of Fabrication Induced Variations

Silicon-on-insulator (SOI) based micro-fabrication is a four mask process, i.e., PAD METAL, SOI, TRENCH, and BLANKET METAL. The PAD METAL mask is used to deposit a layer of 20 nm chrome and 500 nm gold on the silicon and is used for interconnects. The patterning of silicon is carried out using SOI mask layer. Accordingly, the silicon layer is etched out using DRIE (Dry reactive ion etch) to realize the required geometry. Next, it is ensured that the designed structure is properly released. This is ensured by removing the substrate and oxide layer beneath the structure using the TRENCH mask. In the end, bond pads are realized using the BLANKET METAL mask. These bond pads are used for electrical connections during the packaging of the fabricated devices.

Ideally, the etched silicon layer should have a steep vertical (orthogonal) faces. Practically, the etched silicon faces are slanted. The slanting depends on the thickness of the silicon layer. Typically, a 100 μm thick silicon layer has a sidewall slanting angle of an order of 0.8° [19]. This results in non-uniform width variations across the thickness of the silicon layer. As a result, the stiffness of the springs used to suspend the proof mass is changed. The variation in stiffness of the spring, induced due to the fabrication process, has a significant effect the resonant frequencies and other performance parameters of the devices.

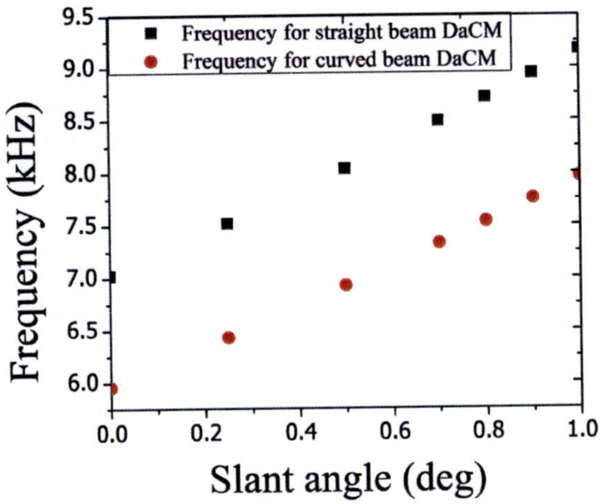

Fig. 13. Effect of sidewall angle due to fabrication process on resonant frequency

7 Conclusion

Curved beam based displacement amplifying compliant mechanism, inspired from the geometrical anti-spring design, has been explored in this article. A detailed analysis of the curved beam DaCM is presented and compared to the conventional straight beam DaCM. Despite having a lesser geometrical amplification ratio, the curved beam DaCM has a larger sense mass displacement. This is because of the lower input stiffness of the curved beam DaCM which allows a lager proof mass displacement. Other variations of the curved beam DaCM, i.e., inverted curved beam, and double curve beam DaCM, are also studied and compared. The curved beam DaCM design outperforms the other DaCM designs and has a larger proof mass and the sense mass displacements. Further, a detailed analysis of the complete accelerometer design using the curved beam DaCM is also presented and different performance parameters like modal frequencies, cross-axis sensitivity, stress, and non-linearity are estimated. The effect of the fabrication tolerances by considering the side-wall angle during dry reactive ion etching is also studied. Analysis suggests that an increase in side-wall angle increases the resonant frequency of the designed accelerometer. In future, we plan to fabricate the designed accelerometer and characterize.

Acknowledgments. The authors would like to acknowledge DRDO, New Delhi, India for funding the research work through the grant number DRDO/./IITHRC-011. We would also like to acknowledge useful suggestions and discussion with Prof. Ananthsuresh, IISc Bangalore, and Dr. Safvan, IIT Hyderabad on Mechanical Amplifiers.

References

1. Tanaka, M.: An industrial and applied review of new mems devices features. Microelectron. Eng. **84**(5–8), 1341–1344 (2007)
2. Barbour, N.M.: Inertial Navigation Sensors. Tech. Rep, Charles Stark Draper Laboratory, Cambridge MA, USA (2010)
3. Handtmann, M., Aigner, R., Meckes, A., Wachutka, G.K.: Sensitivity enhancement of mems inertial sensors using negative springs and active control. Sens. Actuators A Phys. **97**, 153–160 (2002)
4. Gabrielson, T.B.: Mechanical-thermal noise in micromachined acoustic and vibration sensors. IEEE Trans. Electron Dev. **40**(5), 903–909 (1993)
5. Khan, S., Ananthasuresh, G.: Improving the sensitivity and bandwidth of in-plane capacitive microaccelerometers using compliant mechanical amplifiers. J. Microelectromech. Syst. **23**(4), 871–887 (2014)
6. Zhang, H., Wei, X., Ding, Y., Jiang, Z., Ren, J.: A low noise capacitive mems accelerometer with anti-spring structure. Sens. Actuators A Phys. **296**, 79–86 (2019)
7. Zhang, H., Li, B., Yuan, W., Kraft, M., Chang, H.: An acceleration sensing method based on the mode localization of weakly coupled resonators. J. Microelectromech. Syst. **25**(2), 286–296 (2016)
8. L.L. Howell, Compliant mechanisms. In: 21st Century Kinematics, pp. 189–216. Springer (2013)
9. Zeimpekis, I., Sari, I., Kraft, M.: Characterization of a mechanical motion amplifier applied to a MEMS accelerometer. J. Microelectromech. Syst. **21**, 1032–1042 (2012)
10. Hetrick, J., Kota, S.: An energy formulation for parametric size and shape optimization of compliant mechanisms (1999)
11. Krishnan, G., Ananthasuresh, G.: Evaluation and design of displacement-amplifying compliant mechanisms for sensor applications. J. Mech. Des. Trans. ASME **130**(10), 1023041–1023049 (2008)
12. Iqbal, S., Malik, A.: A review on mems based micro displacement amplification mechanisms. Sens. Actuators A Phys. **300**, 111666 (2019)
13. Boom, B.A., Bertolini, A., Hennes, E., Brookhuis, R.A., Wiegerink, R.J., Brand, J.F.V.D., Beker, M.G., Oner, A., Wees, D.V.: Nano-*g* accelerometer using geometric anti-springs. In: Proceedings of the IEEE International Conference on Micro Electro Mechanical Systems (MEMS), vol. 2, pp. 33–36 (2017)
14. Mansouri, B.E., Middelburg, L.M., Poelma, R.H., Zhang, G.Q., van Zeijl, H.W., Wei, J., Jiang, H., Vogel, J.G., van Driel, W.D.: High-resolution MEMS inertial sensor combining large-displacement buckling behaviour with integrated capacitive

readout. Microsyst. Nanoeng. **5** (2019.) [Online]. Available: https://doi.org/10.1038/s41378-019-0105-y
15. Middlemiss, R., Samarelli, A., Paul, D., Hough, J., Rowan, S., Hammond, G.: Measurement of the earth tides with a mems gravimeter. Nature **531**(7596), 614–617 (2016)
16. Khan, S.: Development of micromachined and meso-scale multi-axis accelerometers with displacement-amplifying compliant mechanisms. Ph.D. dissertation, Citeseer (2013)
17. Hegde, S., Ananthasuresh, G.: A spring-mass-lever model, stiffness and inertia maps for single-input, single-output compliant mechanisms. Mech. Mach. Theory **58**, 101–119 (2012)
18. Silicon - strength - hardness - elasticity - crystal structure, Nov 2020. [Online]. Available: www.material-properties.org/silicon-mechanical-properties-strength-hardness-crystal-structure/
19. Biswas, A., Pawar, V.S., Menon, P.K., Pal, P., Pandey, A.K.: Influence of fabrication tolerances on performance characteristics of a mems gyroscope. Microsyst. Technol. **27**(7), 2679–2693 (2021)

Model of a Micromechanical Modal-Localized Accelerometer with an Initially Curvedmicrobeam as a Sensitive Element

Nadezhda Mozhgova(✉), Alexey Lukin, and Ivan Popov

Peter the Great St. Petersburg Polytechnic University, Saint-Petersburg, Russia
nmojgova@yandex.ru

Abstract. The present study is devoted to mathematical modeling of the proposed new architecture of a microelectromechanical mode-localized acceleration sensor (MEMS accelerometer/gravimeter) with a sensitive element in the form of a clamped-clamped microbeam with an initial curvature, made in the form of the first asymmetric mode of free oscillations. The paper shows that with an asymmetric form of the initial curvature in the region of positive axial forces, there are zones of proximity of the frequency branches corresponding to the second symmetric and the first asymmetric oscillation modes. The proposed configuration of the oscillation excitation and output signal pickup electrodes makes it possible, with the help of a feedback loop, to stabilize the oscillation amplitude at the required level according to the working (third) symmetric form and, at the same time, measure the oscillation amplitude associated with the change in the value of the measured component of the external acceleration according to the asymmetric form. The effect of energy exchange between the symmetric and asymmetric modes of the sensitive element can be used as the basis for the development of a new subclass of high-precision resonant sensors with amplitude pickup of the output signal. The parametric studies of the nonlinear statics and dynamics of free oscillations of the sensing element have shown a high level of potentially achievable accuracy of measurements of the external acceleration.

Keywords: MEMS · Initially curved beam · Bistability · Accelerometer · Electrically actuated · Veering phenomena · Modal localization

1 Introduction

Increasingly widespread in the modern industry of nano-microelectromechanical systems (N/MEMS) is the use of essentially nonlinear characteristics of designed mechanical architectures and the principles of oscillation generation, as well as

A. K. Pandey et al. (Eds.): MAMM 2022, 126, pp. 94–118, 2023.
https://doi.org/10.1007/978-3-031-20353-4_7

the collective dynamic properties of moving elements of devices, which makes it possible to increase the accuracy and stability of measurements by orders of magnitude for a wide class of sensors physical quantities [1]. Research and applied developments in the field of application of modal interaction features in nonlinear and weakly coupled oscillatory systems with two or more degrees of freedom are being intensively developed.

One of the priority tasks of the MEMS industry is the creation of vibration and temperature stable highly sensitive elements of micromechanical inertial navigation systems (accelerometers, gravimeters, gyroscopes). In terms of MEMS accelerometers and MEMS gravimeters, along with classical amplitude modulated circuits (see, for example, [2]), the development of resonant and mode-localized sensors is being intensively developed [3–6]. The main principle of operation of such systems is the influence of the measured component of the acceleration of a moving object on the operating frequency or the type of natural mode of oscillation of the micromechanical architecture. A large number of recent papers (see, for example, [7]) report that for mode-localized systems with an amplitude feedback loop, orders of magnitude higher sensitivity can be achieved than for sensors with a frequency output.

Of considerable scientific and practical interest is the study of the unique spectral (modal and resonant) properties of micromechanical systems with projected geometry perturbations (the initial curvature of beam, plate and other structures) and the use of these properties for high-precision measurements. There are a number of works that demonstrate the possibility of successful practical use of such mechanical architectures for high-precision measurement problems: [8,9] force or translational acceleration, [10] displacements, [11] velocity and fluid flow velocity gradient [12], gas sensors [13,14]. There are also works on the use of systems with initial curvature as actuators [15] and logical elements of advanced computing devices [16]. At the moment, in most of the devices created, a microbeam with an initial curvature is used as a sensitive element, which has the property of bistability under the action of a constant electric field in the interelectrode gap.

Mathematical modeling and design of mechanical structures of this class requires the use of quite complex geometrically and physically nonlinear models of elastic deformation of continuum systems. There is an extensive literature devoted to this group of issues (see, for example, [17–29]). A separate intensively developing area of research is the problems of nonlinear statics and dynamics of micromechanical structures with projected geometry perturbations acting in electric fields of various configurations. Namely, a significant number of works are devoted to the study of issues of branching, bifurcations, symmetry breaking of equilibrium positions and switching modes between them for microbeams with initial curvature [30–39]. Spectral properties and nonlinear dynamic characteristics of such structures are also subject to extensive research [40–53]. There are works devoted to the synthesis of algorithms for controlling oscillations of such systems and processing output signals [54–56]. The direction of micromechanical design with projected geometry perturbations of a more complex shape is actively developing: membranes and plates [57–64].

The present study is devoted to mathematical modeling of the proposed new architecture of a microelectromechanical mode-localized acceleration sensor with a sensitive element in the form of a clamped-clamped microbeam with an initial curvature, made in the form of the first asymmetric mode of free oscillations. To the best of the authors' knowledge, microbeams with an initial curvature have not yet been considered as elements of promising resonant (mode-localized) sensors of translational accelerations—accelerometers or gravimeters. The paper shows that under certain conditions in an oscillatory system, conditions can be provided for efficient energy exchange between the symmetric and asymmetric modes of the sensing element. This effect can be used as the basis for the development of a new subclass of high-precision resonant sensors with amplitude pickup of the output signal.

2 Proposed Accelerometer Model

The model of the accelerometer, the operating principle of which is based on the localization of oscillations, is shown in Fig. 1. An initially curved beam, clamped on both sides, which is in the field of action of one fixed vibration-excitation electrode and two fixed vibration-pickup electrodes, acts as a sensitive element. The electrical voltage imparted to the ends of the beam makes it possible to change the natural frequencies of the sensitive element and adjust to the operating mode. The moving mass M is elastically attached to the microbeam by means of a system of suspensions. External acceleration sets the mass in motion and there is a axial force N_m, acting on the microbeam. The operating mode is excited by the central electrode and represents vibrations along the third (second symmetrical) natural shape of the beam. In the presence of external acceleration, the axial force changes the values of the natural frequencies of the beam, and with the right choice of system parameters, the effect of frequency proximity between the second (first asymmetric) and the third (second symmetric) vibration modes of the sensing element is observed. There is an exchange of energy between the specified forms and oscillations in the second eigenform are detected by the side electrodes.

3 Mathematical Model

We consider an initially curved clamped-clamped microbeam of length L, having a rectangular cross section with thickness d and width b (Fig. 2). The beam is made of a homogeneous isotropic elastic material with Young's modulus E. The initial shape of the beam is described by the function $\hat{z}_0(\hat{x}) = \hat{h} \cdot w_0(\hat{x})$, where $\hat{h}$—is the initial deviation of the top point of the beam from the midline in the uncurved position, and $w_0(\hat{x})$ is a dimensionless function for which the condition $\max\limits_{\hat{x}\in[0,L]} |w_0(\hat{x})| = 1$. The beam is located in the field of one fixed electrode, located

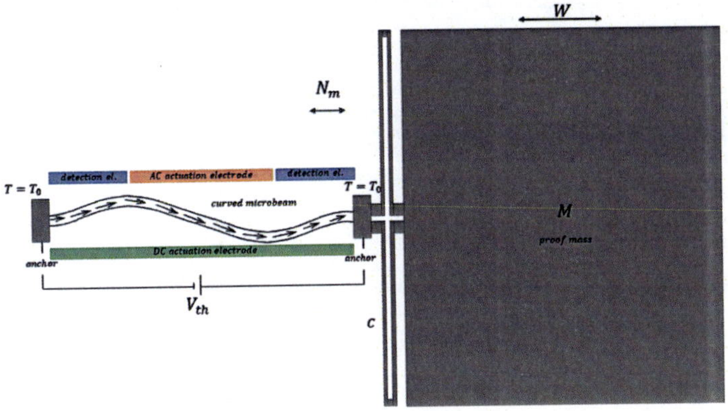

Fig. 1. Mode-localized accelerometer model

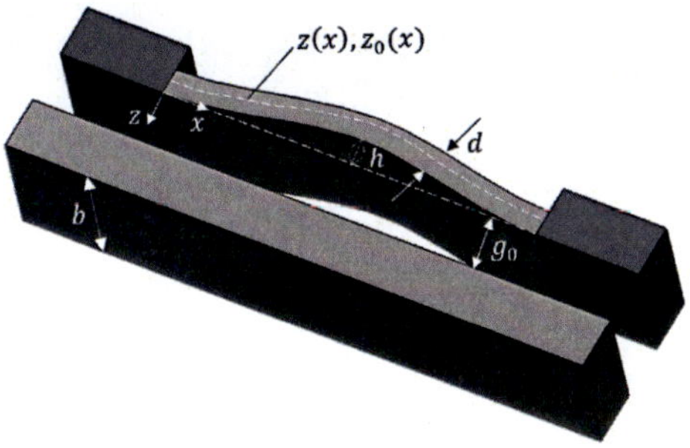

Fig. 2. Initially curved microbeam if field of one stationary electrode

at a distance g_0. It is assumed that $d \ll L$ and that the deflection is moderately large compared to the thickness of the beam, so there is tension in the midplane.

The equation of system dynamics has the form:

$$EI(\hat{z}'''' - \hat{z}_0'''') + \rho A \frac{\partial^2 \hat{z}}{\partial \hat{t}^2} + c\frac{\partial \hat{z}}{\partial \hat{t}} = \left(N + \frac{EA}{2L} \int_0^L (\hat{z}'^2 - \hat{z}_0'^2) d\hat{x} \right) \hat{z}'' + F, \quad (1)$$

where $\hat{z}(\hat{x})$—deflection function , $I = \frac{bd^3}{12}$—moment of inertia of the beam section, ρ—beam material density, c—coefficient of friction, N—axial force, $A = bd$—beam cross-sectional area , $()'$—derivative with respect to coordinate x $\frac{d}{d\hat{x}}$, F—electrostatic force, which is described by the following expression:

$$F = -\frac{1}{2} \frac{\varepsilon b V_{DC}^2}{(g_0 + \hat{z})^2}, \quad (2)$$

where ε—relative permittivity of the medium in the gap, b—beam width, V_{DC}—amplitude of the constant voltage component.

The axial force is caused by three factors:
1. (N_{th}) Joule effect due to the current flowing along the beam,
2. (N_m) by the inertial force caused by the acceleration of the external mass M,
3. (N_0) mechanical force caused by the prestressed state of the beam (can be either tensile or compressive):

$$N = N_{th} + N_m + N_0. \tag{3}$$

The inertial force is related to the mass acceleration as follows:

$$N_m = MW, \tag{4}$$

where W—body acceleration. Fourier's law for the temperature function:

$$-\frac{d}{d\hat{x}}\left(k\frac{dT}{d\hat{x}}\right) = J^2\rho_e, \tag{5}$$

where k—coefficient of thermal conductivity of the beam material, J—current density, ρ_e—specific resistance of the beam material. It is assumed that the properties of the beam material do not depend on temperature. The current density can be expressed in terms of the voltage applied at different ends of the beam V_{th}:

$$J = \frac{V_{th}}{\rho_e L}.$$

Then the Fourier law takes the form

$$-k\frac{d^2T}{d\hat{x}^2} = \frac{V_{th}^2}{\rho_e L^2}. \tag{6}$$

The solution to the Eq. (6) is

$$T(\hat{x}) = -\frac{V_{th}^2}{k\rho_e L^2}\left(x^2 + C_1 x + C_2\right), \tag{7}$$

where the integration constants C_1, C_2 are determined from the boundary conditions of the first kind at the ends of the beam, where the ambient temperature T_0 is maintained:

$$T(0) = T_0, \quad T(L) = T_0, \tag{8}$$

so

$$C_1 = -L, \quad C_2 = -\frac{T_0 k\rho_e L^2}{V_{th}^2}. \tag{9}$$

Thus, the temperature of each point of the beam is determined by the equation:

$$T(\hat{x}) = -\frac{V_{th}^2}{k\rho_e L^2}\left(x^2 - Lx\right) + T_0, \tag{10}$$

The axial force caused by the Joule effect depends on the temperature as follows [42]:

$$N_{th} = -\frac{EA}{L}\int_0^L \alpha(T[\hat{x}] - T_0)d\hat{x} = -\frac{EAV_{th}^2\alpha}{6k\rho_e}, \tag{11}$$

where α—coefficient of thermal expansion of the beam material, T_0—the initial temperature of the beam, the sign "-" means that the force is compressive. Thus, the equation of motion of an initially curved beam in the field of one fixed electrode with a current flowing through it has the form:

$$EI(\hat{z}'''' - \hat{z}_0'''') + \rho A \frac{\partial^2 \hat{z}}{\partial \hat{t}^2} + c\frac{\partial \hat{z}}{\partial \hat{t}} = \left(N_{th} + N_m + N_0 + \frac{EA}{2L}\int_0^L (\hat{z}'^2 - \hat{z}_0'^2) d\hat{x} \right) \hat{z}'' - \frac{1}{2}\frac{\varepsilon b V_{DC}^2}{(g_0 + \hat{z})^2}. \quad (12)$$

After introducing dimensionless parameters

$$z = \frac{\hat{z}}{g_0}, \quad x = \frac{\hat{x}}{L}, \quad t = \frac{\hat{t}}{T},$$

where

$$T = \sqrt{\frac{\rho b d L^4}{EI}}$$

– time scale,

$$\alpha_1 = 6\left(\frac{g_0}{d}\right)^2, \quad c_{non} = \frac{12cL^4}{ETbd^3}, \quad \alpha_2 = \frac{6\varepsilon L^4}{E g_0^3 d^3}, \quad \lambda = \alpha_2 V_{DC}^2$$

$$N_{non,th} = \frac{12L^2}{Ebd^3} N_{th}, \quad N_{non,m} = \frac{12L^2}{Ebd^3} N_m, \quad N_{non,0} = \frac{12L^2}{Ebd^3} N_0,$$

an equation in dimensionless form can be obtained:

$$\frac{\partial^4 z}{\partial x^4} - \frac{\partial^4 z_0}{\partial x^4} + \ddot{z} + c_{non}\dot{z} = \left[\alpha_1 \int_0^1 (z'^2 - z_0'^2) dx + N_{non,th} + N_{non,m} + N_{non,0}\right] \frac{\partial^2 z}{\partial x^2} - \frac{\alpha_2 V_{DC}^2}{(1+z)^2}, \quad (13)$$

where $\dot{()}$ is the dimensionless time derivative $\frac{\partial}{\partial t}$.

The paper will consider the problem of static equilibrium, as well as the spectral problem near the equilibrium positions found at the first stage.

3.1 Static Equilibrium Equations

Static equilibrium equation in dimensionless form:

$$z'''' - z_0'''' = \left[\alpha_1 \int_0^1 (z'^2 - z_0'^2) dx + N_{non,th} + N_{non,m} + N_{non,0}\right] z'' - \frac{\alpha_2 V_{DC}^2}{(1+z)^2}, \quad (14)$$

where $()'$ is the derivative with respect to the dimensionless coordinate $\frac{\partial}{\partial x}$.

To obtain a reduced order model, the Galerkin method is used, that is, an approximate solution of the (14) equation is sought in the form:

$$z(x) = \phi_0(x) + \sum_{i=1}^{n} C_i \phi_i \tag{15}$$

where $\phi_0(x) = 0$ since the boundary conditions of the problem are homogeneous. The decomposition (15) is carried out in orthogonal eigenforms of the straight beam, satisfying the Eq. (16):

$$\phi_i^4 = \omega_{non,i}^2 \phi_i, \quad i = 1, 2, ..., n, \tag{16}$$

where $\omega_{non,i}^2$ is the square of the dimensionless natural frequency. Taking into account the expansion into a series in terms of eigenforms and boundary conditions, the static equilibrium equation takes the form

$$\begin{aligned} &\int_0^1 \phi_j \left(1 - \sum_{l=1}^{n} C_l \phi_l\right)^2 \left(\sum_{i=1}^{n} C_i \omega_{non,i}^2 \phi_i\right) dx + \lambda \int_0^1 \phi_j dx - \\ &-\alpha_1 \int_0^1 \phi_j \left(1 - \sum_{l=1}^{n} C_l \phi_l\right)^2 \left[\sum_{i=1}^{n} C_i \phi_i'' \int_0^1 \left(\sum_{k=1}^{n} C_k \phi_k\right)^2\right] dx = 0, \end{aligned} \tag{17}$$

where ϕ_i and C_i are the eigenforms and their coefficients, respectively. The resulting algebraic equation with respect to the coefficients C_i of the expansion of the deflection function in terms of eigenforms is solved using the methods of the theory of bifurcations in the MATLAB MATCONT software package [65].

3.2 The Equation of Small Oscillations Around Static Equilibrium

Equation of system dynamics is:

$$\begin{aligned} &z^{IV} - z_0^{IV} + \ddot{z} + c_{non} \dot{z} \\ &= \left[\alpha_1 \int_0^1 (z'^2 - z_0'^2) dx + N_{non,th} + N_{non,m} + N_{non,0}\right] z'' - \frac{\lambda}{(1+z)^2}. \end{aligned} \tag{18}$$

The dimensionless deflection function $z(x,t)$ can be decomposed into the static part z_s (found from the static equilibrium equation (12)) and the time-dependent dynamic component z_d:

$$z(x,t) = z_s(x) + z_d(x,t). \tag{19}$$

Substituting this expansion into the equation of dynamics leads to an equation for the dynamic component of the deflection function:

$$\begin{aligned} &z_s^{IV} + z_d^{IV} - z_0^{IV} + \ddot{z}_d + c_{non} \dot{z}_d = -\frac{\lambda}{(1+z_s+z_d)^2} \\ &+ \left[\alpha_1 \int_0^1 ((z_s' + z_d')^2 - z_0'^2) dx + N_{non,th} + N_{non,m} + N_{non,0}\right] \frac{\partial^2 (z_s + z_d)}{\partial x^2}. \end{aligned} \tag{20}$$

After opening the brackets, expanding the term of the electrostatic force in a Taylor series in terms of the function z_d and neglecting the terms of the second and higher orders of smallness, the equation will take the following form:

$$z_s^{IV} + z_d^{IV} - z_0^{IV} + \ddot{z}_d + c_{non}\dot{z}_d = \alpha_1 \int_0^1 ({z_s'}^2 - {z_0'}^2)dx(z_s'' + z_d'')$$
$$+2\alpha_1 \int_0^1 z_s' z_d' dx z_s'' + (N_{non,th} + N_{non,m} + N_{non,0})(z_s'' + z_d'') \tag{21}$$
$$-\frac{\lambda}{(1+z_s)^2} + \frac{2\lambda z_d}{(1+z_s)^3}.$$

Some of the terms that make up the static equilibrium equation are mutually compensated. Assuming the absence of dissipation $c_{non} = 0$, small free oscillations around static equilibrium are determined by the equation

$$z_d^{IV} + \ddot{z}_d = 2\alpha_1 \int_0^1 z_s' z_d' dx z_s'' + \frac{2\lambda z_d}{(1+z_s)^3}$$
$$+\left[\alpha_1 \int_0^1 ({z_s'}^2 - {z_0'}^2)dx + N_{non,th} + N_{non,m} + N_{non,0}\right] z_d''. \tag{22}$$

Similar to the equation of statics, the dynamic addition to the deflection function can also be expanded into a series in terms of the eigenforms of a straight beam:

$$z_d(x,t) = \sum_{i=1}^{n} u_i(t)\phi_i(x). \tag{23}$$

Next, the expansion is substituted into the (22) equation, the fourth derivative is replaced by the corresponding term according to the (16) equation, and the entire equation is multiplied by its eigenform ϕ_j and integrated from 0 to 1. As a result of these actions, the following expression was obtained:

$$\ddot{u}_j + u_j\omega_{non,j}^2 = \left[\alpha_1 \int_0^1 ({z_s'}^2 - {z_0'}^2)dx + N_{non,th} + N_{non,m} + N_{non,0}\right]$$
$$\int_0^1 \phi_j \sum_{i=1}^{n} u_i \phi_i'' dx + 2\alpha_1 \int_0^1 \phi_j z_s'' \left(\int_0^1 z_s' \sum_{i=1}^{n} dx\right) dx \tag{24}$$
$$+\int_0^1 \phi_j \frac{2\lambda}{(1+z_s)^3} \sum_{i=1}^{n} u_i\phi_i dx, \quad j = 1, 2, ..., n.$$

Equation (24) is a system of linearly coupled ordinary differential equations with respect to the modal coordinates u_j. Eigenvalue problem for this equation gives the first n natural frequencies of the beam.

4 Parameters of the Structure Under Consideration

The paper will consider two types of the function of the initial curvature, that demonstrated on Table 1. Table 2 presents the geometrical parameters used in the calculations.
Beam material—isotropic silicon, gap material—air, material properties are presented in Table 3.

Table 1. Two types of the function of the initial curvature

The shape of the initial curvature corresponds to the first eigenform of the bending vibrations of the beam	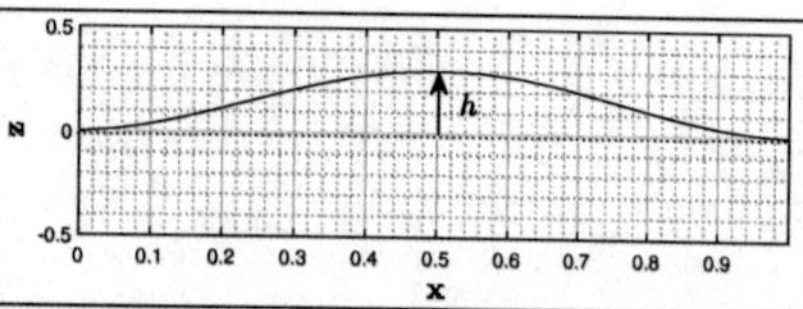
The shape of the initial curvature corresponds to the second eigenmode of the beam bending vibrations	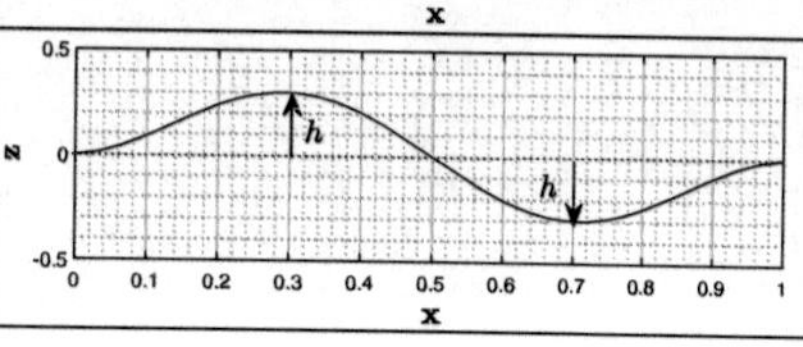

Table 2. Geometric dimensions of a beam with an initial curvature

Size	Value μm
Beam length	1000
Beam thickness	2
Beam width	25
Gap	8
Initial curvature amplitude	2.6μm

5 Diagrams of Static Equilibrium

In this paragraph, static diagrams of equilibrium positions are given, obtained by continuation along the static voltage parameter λ in the Eq. (17). The first step was to study the effect of the amplitude of the initial curvature on the shape of the equilibrium diagram (Fig. 3).
Figure 3 clearly demonstrates that not for all values of the the amplitude of the initial curvature for the first natural shape of the sensing element, the system will experience the effect of bistability, that is, the presence of two stable equilibrium

Table 3. Properties of the materials

Value	Silicon	Air
Young's modulus	170 GPa	–
Poisson's ratio	0.23	–
Density (kg/m^3)	2330	1.225
Conductivity (W/(m · K))	130	0.0259
Specific heat (J/(kg · K))	700	1030
Coefficient of linear thermal expansion	2.6×10^{-6} $1/K$	–
Relative permittivity	–	1

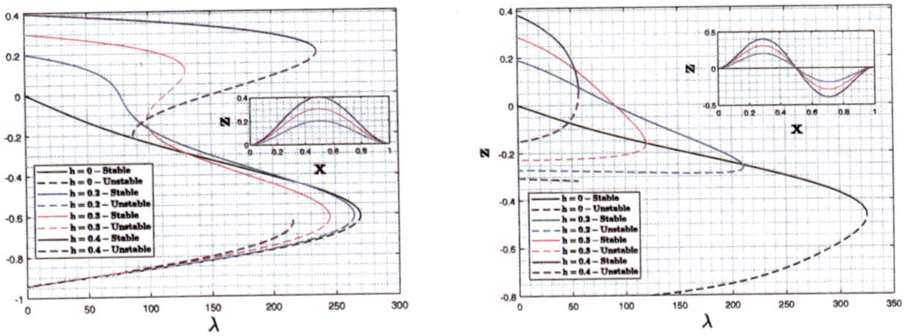

(a) first eigenform curvature, deflection in the center of the beam (b) second eigenform curvature, deflection in $\frac{1}{4}$ of the beam length

Fig. 3. Equilibrium diagram for different values of the initial curvature amplitude ($\alpha_1 = 96, N_{non} = 0$)

positions at the same value of the voltage parameter λ . With the function of the initial curvature corresponding to the second eigenmode of the sensitive element, no bistability effect is observed for the given values of the parameters, and a strong dependence of the critical voltage on the amplitude of the initial curvature is also observed.

Having chosen the value of the amplitude of the initial curvature equal to $h = 0.325$ (in the dimensional form $\hat{h} = 2.6\,\mu\text{m}$), then the parameter $\alpha_1 = 6(\frac{g_0}{d})^2$ is varied. Figure 4 shows that the parameter α_1 also significantly affects the shape of the bifurcation diagram, and its value must also be selected for the manifestation of system bistability in the case of first eigenform initial curvature. At the second eigenform initial curvature, the static equilibrium diagram has one stable branch and the value of the parameter α_1 along with the amplitude of the initial curvature determines the value of the critical voltage. It should be noted that the graph is given for the deflection value in $\frac{1}{4}$ of the beam length and, in case of buckling, the deflection value at this point is about 0.2 of the interelectrode gap, while the deflection in $\frac{3}{4}$ beam length already reaches critical levels of $0.8 - 0.9$ of the gap.

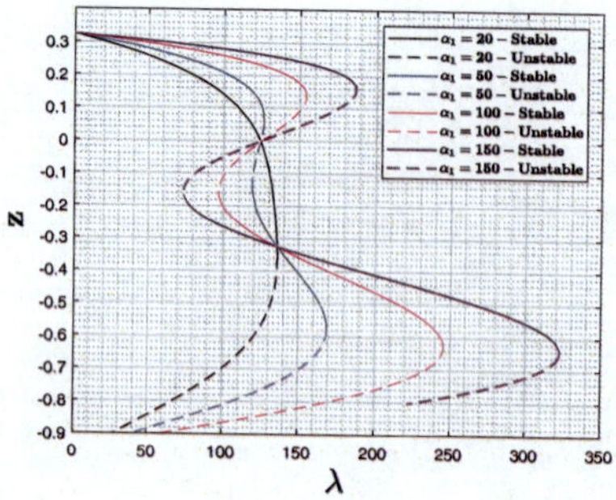

(a) first eigenform curvature, deflection in the center of the beam

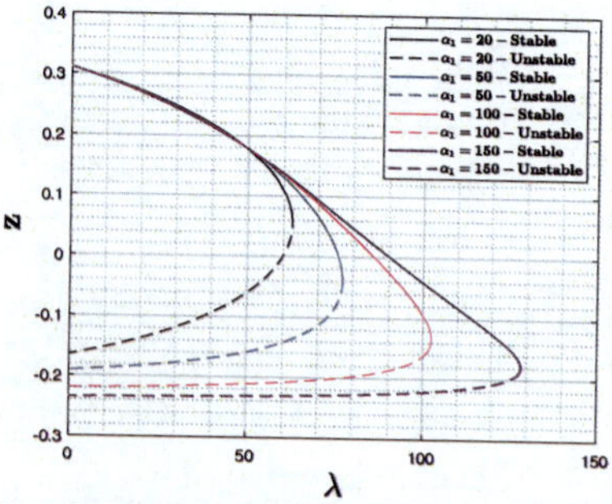

(b) second eigenform curvature, deflection in $\frac{1}{4}$ of the beam length

Fig. 4. Equilibrium diagram for different values of the parameter α_1 ($h = 0.325, N_{non} = 0$)

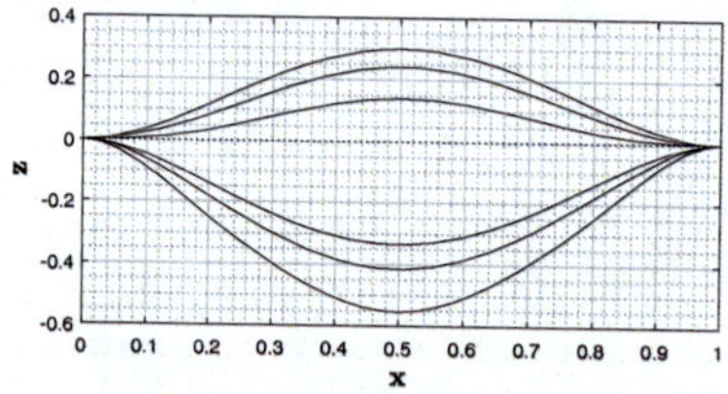

(a) Evolution of the deflection function when moving along the bifurcation diagram

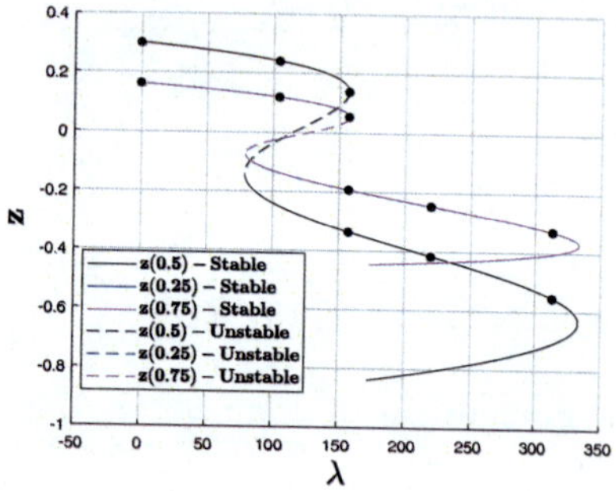

(b) Equilibrium diagram (first eigenform, $\alpha_1 = 150, h = 0.3, N_{non} = 0$)

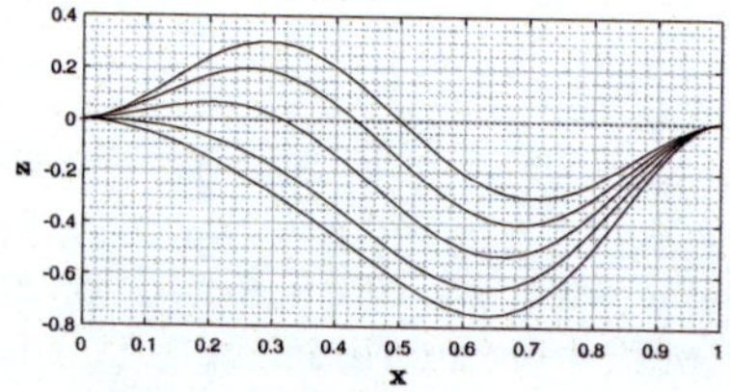

(c) Evolution of the deflection function when moving along the bifurcation diagram

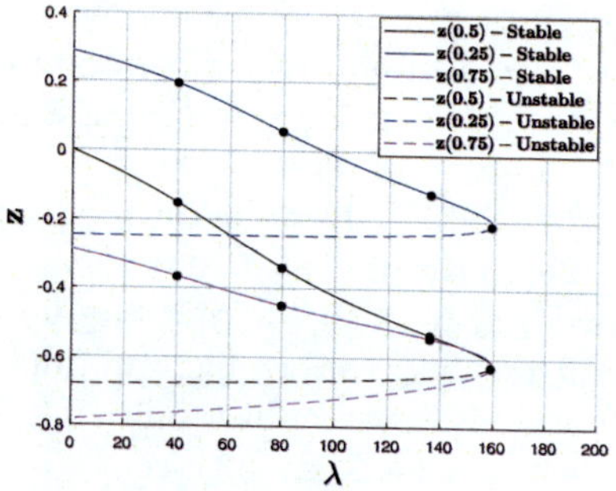

(d) Equilibrium diagram (second eigenform, $\alpha_1 = 150, h = 0.3, N_{non} = 0$)

Fig. 5. Demonstration of the influence of the eigenshape number on the form of the static equilibrium diagram and the evolution of the deflection function with an increase in the voltage parameter

Figure 5b, d show the dependences of the deflection at the center of the beam and at the points corresponding to $\frac{1}{4}$ and $\frac{3}{4}$ beam length versus static voltage parameter. The markers (black dots) in Fig. 5b, d illustrate the points at which Fig. 5a, c show the corresponding functions of the beam deflection (as if in the process of moving along the bifurcation diagram).

Further it was performed a study of the static behavior of the system depending on the magnitude of the axial force N. Figure 6 is a 3D static equilibrium diagram in the parameter space z, N_{non}, λ.

It can be noted that the influence of the axial force on the behavior of the system is significant. With an increase in the value of the axial force parameter, the critical value of the electrostatic force parameter also increases, at which the stability of the system is lost. And in states close to the unstable branch, the behavior of systems with an increase in axial force differs from other stable states—they pass into unstable equilibrium states. As the axial force decreases, the behavior of the system is similar.

Of interest is the graph of the dependence of fold-type bifurcation points ("limit point") in the parameter space axial force N—static voltage λ for a microbeam with first eigenform curvature, which is shown in Fig. 7.
The blue line reflects the dependence of the first bifurcation point, at which the microbeam exhibits the property of bistability and snaps-through towards the electrode to another stable equilibrium position. The black line reflects the dependence of the second bifurcation point, which is the beginning of the second stable branch on the static deflection-voltage diagram. The red line reflects the dependence of the third bifurcation point on the static diagram, after which there is a complete loss of stability and the microbeam "sticks" to the stationary electrode ("pull-in effect").

Also in Fig. 7, nested graphs illustrate the appearance of a static equilibrium diagram for a certain value of the axial force parameter N. For values of the dimensionless force $N > 30$ (blue graph), one bifurcation point will be observed on the static equilibrium diagram—the system will not have the property of bistability, since a positive value of the parameter N corresponds to the tensile force, which leads to straightening of the initially curved microbeam. The value of the dimensionless force $N = 30$ (burgundy graph) corresponds to the boundary, after which the system exhibits the property of bistability and has three bifurcation points on the static equilibrium diagram. The value of the dimensionless force $N = -30$ (purple graph) corresponds to the case when two critical voltages (snap-through voltage and pull-in voltage) have the same numerical value. In the range of values $N = [-70, -30]$ (grey graph), the critical pull-in voltage value is less than the critical snap-through voltage value. With values of the dimensionless force $N < -70$ (green graph), the system formally has two bifurcation points, however, the value of the voltage of the first point is small and in reality the system will lose stability when the larger value of the voltage of the two, which is determined by the black curve in Fig. 4, is reached. It is also worth noting that this range corresponds to a large compressive axial force, which bulges the microbeam, increasing the curvature amplitude, which leads

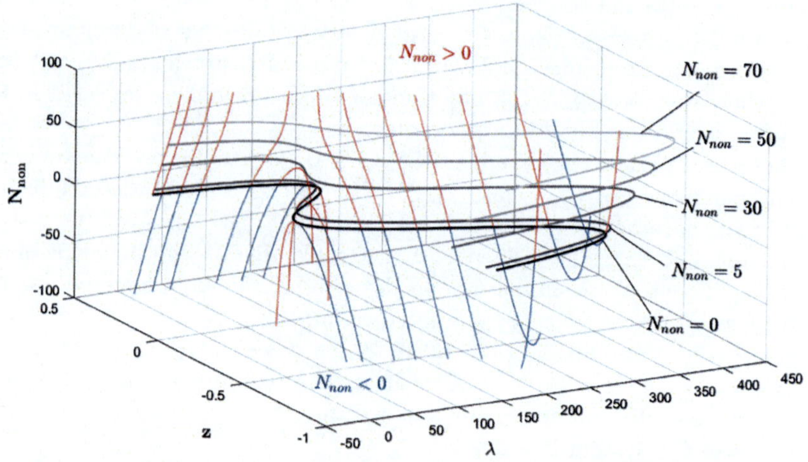

(a) first eigenform curvature, deflection in the center of the beam

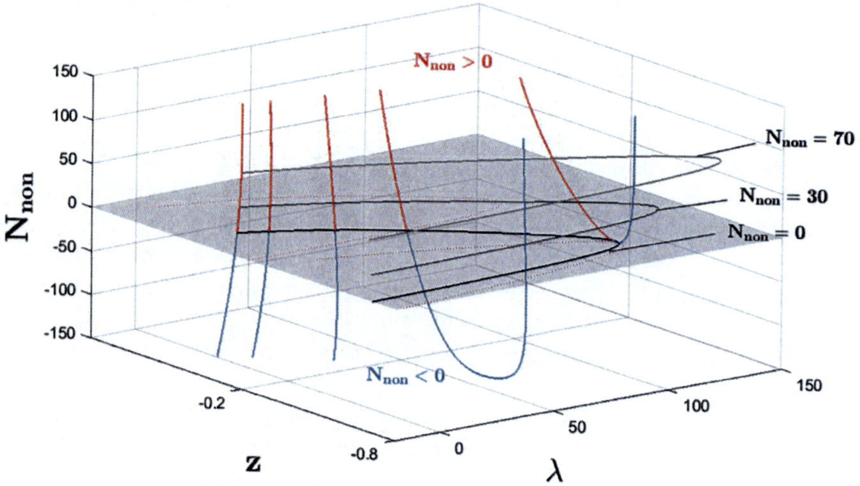

(b) second eigenform curvature, deflection in $\frac{1}{4}$ of the beam length

Fig. 6. Equilibrium diagram in three-dimensional parameter space z, N_{non}, λ

to a very complex static behavior of such system. The graph corresponding to the static diagram at $N = -200$ demonstrates the presence of static equilibrium positions near zero, that is, near the straight microbeam, however, these equilibrium positions border on instability zones and have a small basin of attraction.

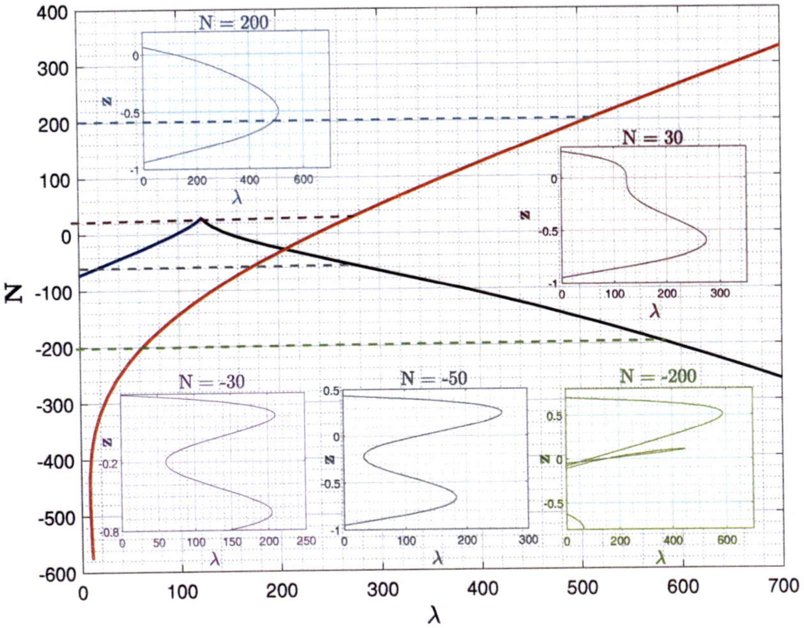

Fig. 7. Dependence of bifurcation points of the fold type in the space of parameters $N - \lambda$, $(\alpha_1 = 96)$, first eigenform curvature

A microbeam with second eigenform curvature does not have the property of bistability and the dependence graph of one bifurcation point will divide the plane into two parts—areas of stable and unstable equilibrium positions.

6 Analysis of Small Oscillations Around Equilibrium Position

In this section, we will present the results of studying the frequencies and amplitudes of small oscillations near the equilibrium positions found earlier.

6.1 Finite-Element Modeling

To verify the results obtained on the analytical calculation model, a comparison was made with the results of finite element (FE) modeling of the sensing element. For comparison, a beam was chosen that was first eigenform initial curvature. FE modeling was carried out in the COMSOL Multiphysics software package, the FE model is shown in Fig. 8.

A related problem was solved by the finite element method, which includes mechanical problems for beam elements (module "Solid Mechanics"), electrostatic (module "Electrostatics") and temperature problems (module "Heat

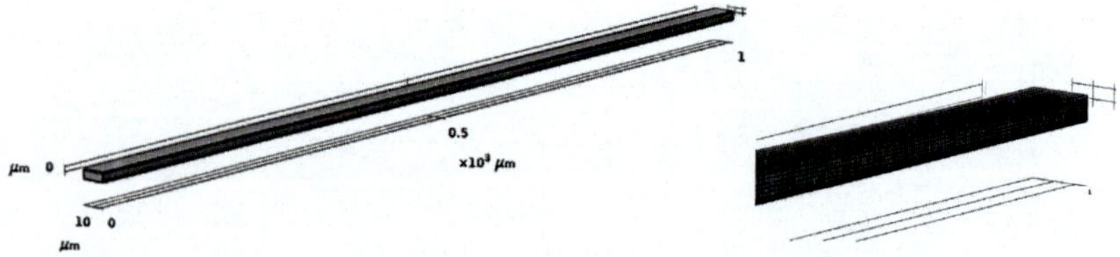

Fig. 8. Geometric model of a beam with initial curvature and an illustration of the splitting of the model into elements

Transfer in Solids") for all elements (beam and air). The boundary condition of the mechanical problem is the absence of displacements and rotations of the nodes at the ends of the beam, the electrostatic problem is the potential difference V_{DC} on the electrodes, the temperature was set on all nodes of the beam for the occurrence of a axial force according to the Eq. (5). The value of the potential difference is chosen in accordance with the selected point on the static equilibrium diagram (z_s and $\lambda = \alpha_2 V_{DC}^2$ must be substituted into the Eq. (18)).

Figure 9 shows the dependence of the first three natural frequencies of the microbeam on the axial force parameter N.

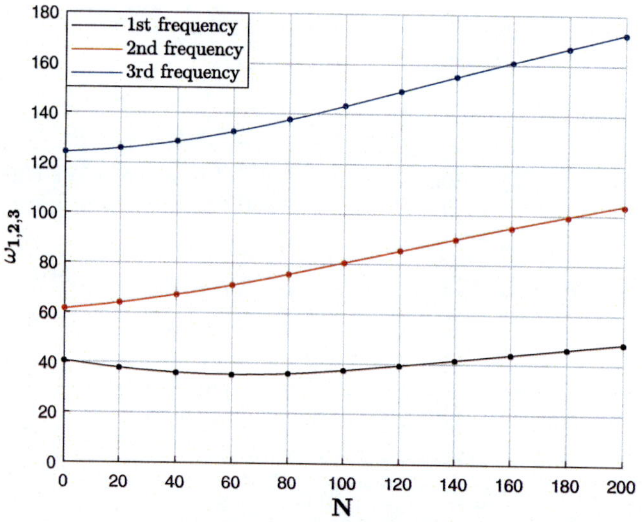

Fig. 9. Dependence of the first three natural frequencies of the microbeam on the axial force parameter N. Solid line—analytical solution, dots—FE solution

In the positive range of axial force values for a microbeam with an initial curvature, the solutions obtained analytically and using the FEM completely coincide in terms of the first three eigenforms of bending vibrations. In the negative region of values, in this case, the effect of crossing the frequency branches corresponding to the first and second eigenmodes takes place, which is confirmed by Fig. 10.

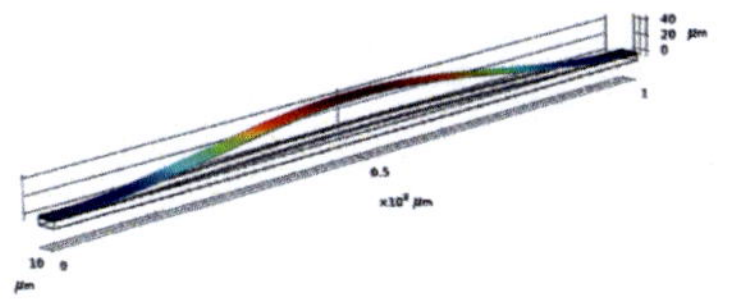

(a) First eigenform, $N = -70$

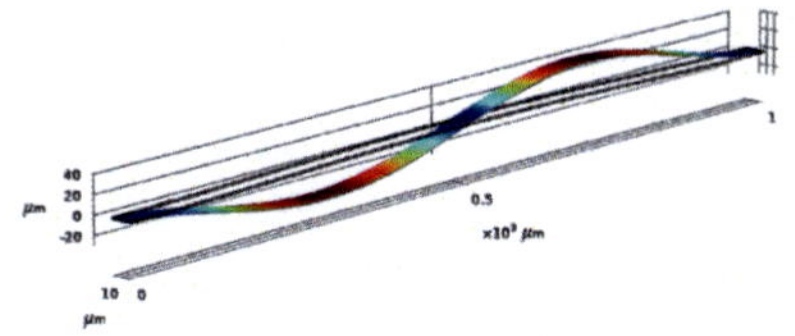

(b) First eigenform, $N = -160$

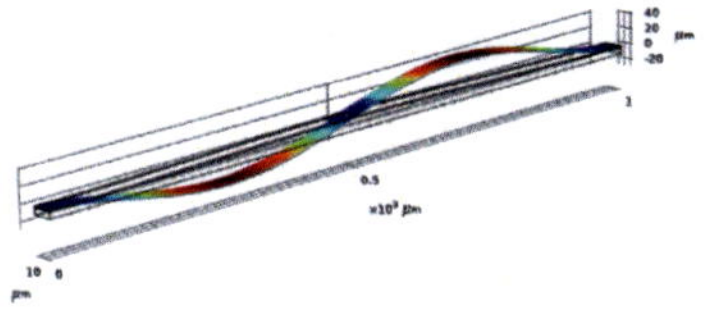

(c) Second eigenform, $N = -70$

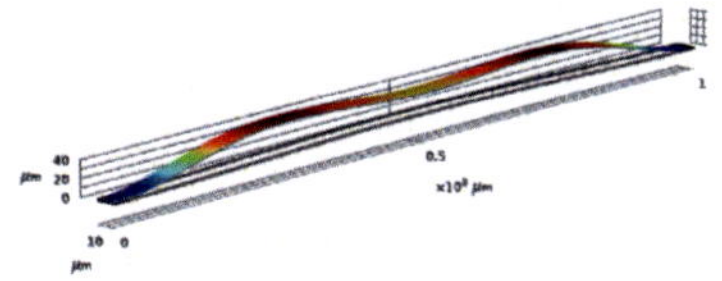

(d) Second eigenform, $N = -160$

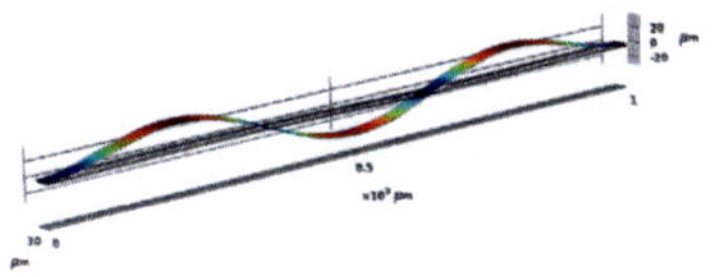

(e) Third eigenform, $N = -70$

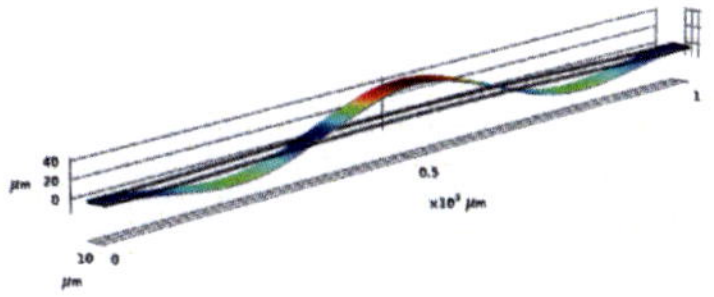

(f) Third eigenform, $N = -160$

Fig. 10. Demonstration of eigenforms reordering (crossing of frequency brahches), $N = -70$, $N = -160$

The crossing of frequency branches is typical for symmetrical systems, which is the beam under consideration with first eigenform curvature. Crossovers typically occur when variation of a control parameter entails that one frequency changes with a much greater rate than some other frequency [40]. In this case, under axial force, the symmetrical shapes frequencies of the beam increase faster due to the stiffening stretching mechanism. Whereas asymmetric shapes frequencies of the beam are pure bending modes and are not subject to stretching. Thus, the effect of crossing the branches of the symmetric and asymmetric oscillation modes is observed and the coincidence of the frequency dependences on the axial force parameter is obtained by different methods.

6.2 Dependence of Natural Frequencies on the Parameters of Static Voltage and Axial Force

Further, the results of studying the dependence of natural frequencies and forms of bending vibrations of the beam with varying parameters of electrostatic voltage λ and axial force N are presented. It should be noted that when solving the eigenvalue problem for each value of the specified parameters, the static deflection function z_s found from the Eq. (14) was used at the current values of the parameters. Figure 11 shows the dependences of the first three frequencies of the beam with an increase V_{DC} parameter.

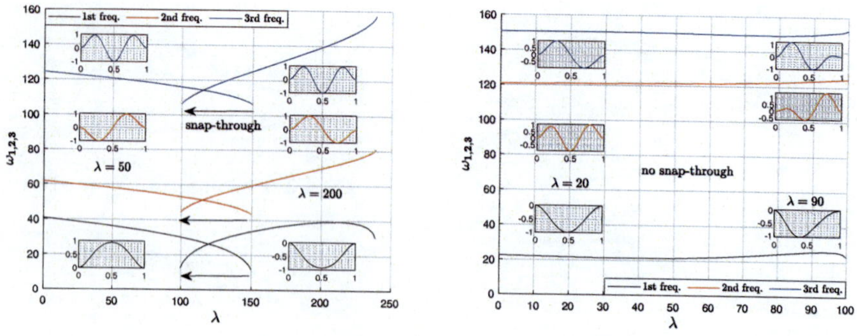

(a) first eigenform curvature

(b) second eigenform curvature

Fig. 11. Dependence of the first three natural frequencies of the microbeam on the static voltage parameter $\lambda, N = 0$

Figure 11a demonstrates the effect of a microbeam snapping from one stable position to another when a certain static voltage is reached, also illustrated in the static equilibrium diagram for a given sensing element configuration (Fig. 4a, b). This effect is not observed in Fig. 11b, since the beam with second eigenform initial curvature, does not have the effect of bistability (Fig. 4b, d).

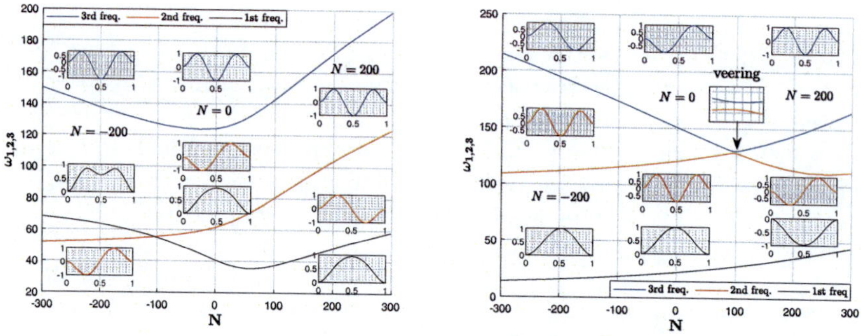

(a) first eigenform curvature

(b) second eigenform curvature

Fig. 12. Dependence of the first three eigenfrequencies of the microbeam on the axial force parameter $N, \lambda = 0$

Figure 12 shows the dependences of the first three frequencies of the bending vibrations of the beam when varying the axial force parameter for first and second eigenforms initial curvature. According to the Eq. (3), the axial force is the sum of the force of a mechanical nature caused by the prestressed state of the sensing element, the heating due to the Joule effect when current flows through the beam, and the force caused by the acceleration of the moving mass, which is a useful output signal of the device.

At the first eigenform initial curvature (Fig. 12a), in the region of negative N parameter values, which corresponds to the compression of the sensing element, the effect of crossing the frequency branches corresponding to the first and second eigenmodes of the beam is observed. Thus, with a compressive force of 100 or more dimensionless units, the frequency corresponding to the second eigenmode will be less than the frequency corresponding to the first eigenmode of bending vibrations. However, from the point of view of application for measuring transverse accelerations, this effect is of no interest, since when the frequency branches cross, there is no effect of oscillation localization and energy exchange between the forms. In the case of the second eigenform initial curvature (Fig. 12b), already in the positive region with respect to the N parameter, that is, in the tension region, there is an effect of proximity of the frequency branches corresponding to the second (first asymmetric) and third (second symmetric) eigenforms. The shape of the deflection function for the values of the parameter N before and after the proximity region shows that the frequency branches also change places after passing through this zone.
To detect the effect of localization of oscillations in this zone, the dependence of the ratio of the amplitudes of oscillations A_2/A_3 on the indicated frequency branches (second and third) on the parameter N, shown in Fig. 13, was plotted.

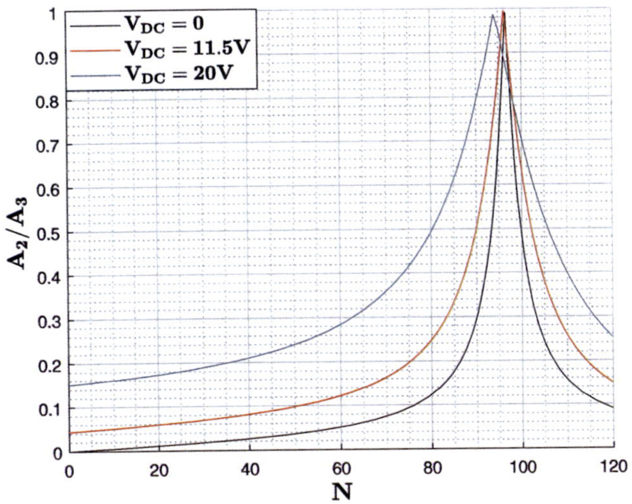

Fig. 13. Amplitude ratio of oscillations at the second and third natural frequencies, second eigenform initial curvature

An analysis of this graph shows that the system exhibits the effect of vibration localization and energy exchange between the second and third forms of flexural vibrations. However, the effect is observed in the stretching area of the sensitive element, and the adjustment to the operating mode occurs by applying a potential difference to the ends of the beam, which leads to compression. Thus, the required value of the parameter $N_{non,0}$, corresponding to the initial prestress state, is approximately 110 dimensionless units, or, in dimensional form:

$$N_0 = N_{non,0} \cdot \frac{Ebd^3}{12L^2} \approx 2.5 \cdot 10^{-4} [N]. \tag{25}$$

Next, the dependences of the frequency branches of the second and third eigenmodes are presented in dimensional form and the ratio of oscillation amplitudes according to these modes relative to the voltage value V_{th} taking into account the prestressed state N_0. The voltage V_{th} is a control voltage applied to the ends of the beam to heat it and, consequently, to change the natural frequencies of the beam and the operating point of the accelerometer.

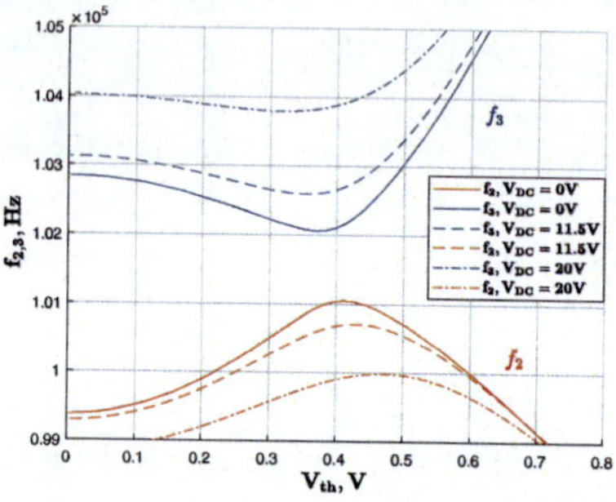

(a) dependence of frequency branches on V_{th}

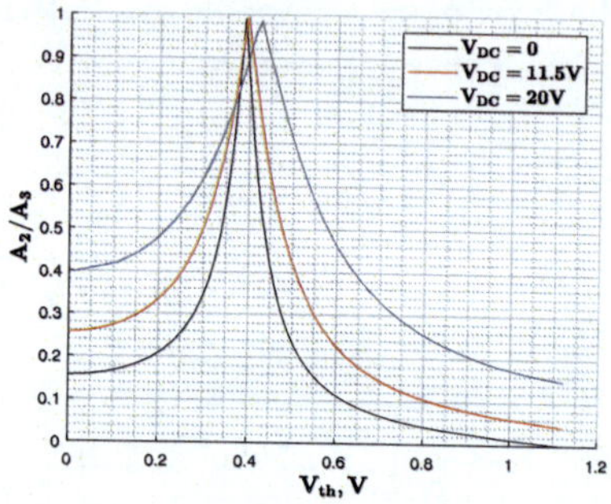

(b) dependence of amplitudes ratio on V_{th}

Fig. 14. Dependence of the second and third natural frequencies and their amplitude ratio on the potential difference at the ends of the beam

From Fig. 14 we can conclude that to shift the operating point of the device in the desired range, a relatively small value of the control voltage is required—less than 1 V.

The final stage of this work is the analysis of the sensitivity of the accelerometer, depending on the choice of the operating point and the ratio of the masses of the sensitive element and the inertial body. The sensitivity here is defined as the change in the amplitude ratio of the second and third bending modes of the beam by 1g (in ppm). The results of the study are presented in Table 4, as well as in Fig. 15 dependences of the ratio of amplitudes on the magnitude of the external measured acceleration are presented.

Table 4. Comparison of sensor output characteristics at different mass ratios and choices of operating point

Mass ratio $\frac{M}{m}$	Operating point, V	Sensitivity (A), ppm
100	$V_{th} = 0.4$	8107
	$V_{th} = 0.43$	4290
	$V_{th} = 0.46$	2227
300	$V_{th} = 0.4$	24, 321
	$V_{th} = 0.43$	12, 869
	$V_{th} = 0.46$	6682
500	$V_{th} = 0.4$	40, 535
	$V_{th} = 0.43$	21, 449
	$V_{th} = 0.46$	11, 137
1000	$V_{th} = 0.4$	81, 071
	$V_{th} = 0.43$	42, 898
	$V_{th} = 0.46$	22, 274

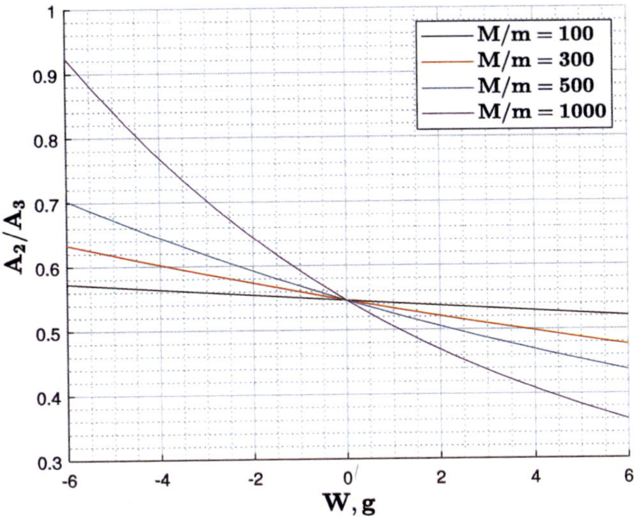

Fig. 15. Dependence of the ratio of oscillation amplitudes on the magnitude of external acceleration

Based on the results presented in the table 4, we can conclude that with the ratio of the mass of the moving part of the device to the mass of the sensitive element (beam) in the range $M/m = [100, 1000]$ the sensitivity of the amplitude output of the mode-localized accelerometer described in this paper reaches values of the order of (80,000 ppm).

7 Conclusions

In the present work, a parametric study of nonlinear problems of statics and dynamics of small oscillations is performed for an initially curved microbeam as a promising sensitive element of high-precision resonant sensors of various physical quantities. The shape factor of the initial microbeam curvature was studied. It is shown that with an asymmetric form of the initial curvature in the region of positive axial forces, there are zones of proximity of the frequency branches corresponding to the second symmetric and the first asymmetric vibration modes. If the required value of the axial tensile force in the microbeam is provided structurally, this effect can be used, in particular, to measure the axial component of the transfer acceleration according to the principle of amplitude mode localization. The possibility of heating the sensing element with the help of an electric current flowing through the microbeam, provided in the sensor layout, makes it possible to control the operating point of the oscillation mode and, thus, to vary the range of measured accelerations and the degree of sensor sensitivity within a wide range. The configuration of the oscillation excitation and output signal pickup electrodes proposed in the article makes it possible, with the help of a feedback loop, to stabilize the oscillation amplitude at the required level in the working (third) symmetrical form and, at the same time, to measure the oscillation amplitude associated with a change in the value of the measured component of the portable acceleration in the asymmetric form. Thus, a mathematical model of an original mode-localized accelerometer/gravimeter containing a single sensitive microbeam element and involving the effect of energy exchange between its various modes of vibration is proposed and investigated in the work.

Acknowledgments. The research is funded by Russian Science Foundation grant 21-71-10009, https://rscf.ru/en/project/21-71-10009/.

References

1. Amal, H., et al.: Linear and nonlinear dynamics of micro and nano-resonators: review of recent advances. Int. J. Non-Linear Mech. **119**, 103328 (2019). https://doi.org/10.1016/j.ijnonlinmec.2019
2. Belyaev, Y.V., et al.: Design of a Micromechanical Accelerometer, pp. 1–7 (2018). https://doi.org/10.23919/ICINS.2018.8405921
3. Chen, W., et al.: Micromachined accelerometers with sub-μ sg/Hz noise floor: a review. Sensors **20**, 4054 (2020). https://doi.org/10.3390/s20144054
4. Morozov, N.F., et al.: Nonlinear dynamics of mode-localized MEMS ac-celerometer with two electrostatically coupled microbeam sensing elements. Int. J. Non-Linear Mech. **138**, p. 103852 (2022). ISSN 0020–7462. https://doi.org/10.1016/j.ijnonlinmec.2021.103852. https://www.sciencedirect.com/science/article/pii/S0020746221001864
5. Morozov, N.F., et al.: A novel model of a mode-localized MEMS accelerometer. In: Doklady Physics. vol. 65. 10, pp. 371–375. Springer (2020)
6. Morozov, N.F., et al.: Effect of nonlinearity on mode localization phenomena in dynamics of MEMS resonant sensor with two electrostatically coupled microbeams. Vestnik St. Petersburg Univ. Math. **54**(2), 135–144 (2021)

7. Jing, Y., Zhong, J., Chang, H.: A closed-loop modelocalized accelerometer. J. Microelectromech. Syst. **27**(2), 210–217 (2018). https://doi.org/10.1109/JMEMS.2017.2787544
8. Erez, B., Stella, L., Slava, K.: Bistable Force/Acceleration Sensor Based on Pull-in Voltage Monitoring (2016), pp. 109–112. https://doi.org/10.1109/ISISS.2016.7435557
9. Erez, B., Stella, L., Krylov, S.: Design and implementation of a bistable force/acceleration sensing device considering fabrication tolerances. J. Microelectromech. Syst., 1–12 (2018). https://doi.org/10.1109/JMEMS.2018.2852621
10. Naftaly, K., Ilic, B., Krylov. S.: Displacement sensing based on resonant frequency monitoring of electrostatically actuated curved micro beams. J. Micromech. Microeng. **26**, 115006 (2016). https://doi.org/10.1088/0960-1317/26/11/115006
11. Yoav, K., et al.: Flow sensor based on the snap-through detection of a curved micromechanical beam. J. Microelectromech. Syst., pp. 1–3 (2018). https://doi.org/10.1109/JMEMS.2018.2868776
12. Kessler, Y., Liberzon, A., Krylov, S.: Flow velocity gradient sensing using a single curved bistable microbeam. J. Microelectromech. Syst., pp. 1–6 (2020). 2020.3012690. https://doi.org/10.1109/JMEMS.2018.2868776
13. Amal, H., et al.: A resonant gas sensor based on multimode excitation of a buckled microbeam. IEEE Sens. J., 1 (2019). https://doi.org/10.1109/JSEN.2019.2950495
14. Najar, F., Ghommem, M., Abdel-Rahman, E.: Arch microbeam bifurcation gas sensors. Nonlinear Dyn. **104**(2), 923–940 (2021). https://doi.org/10.1007/s11071-021-06319-0
15. Rui, Z., et al.: Semi-rigid ring-shaped electrode dielectric electroactive polymer membrane as buckling actuator. J. Micromech. Microeng. **29** (2019). https://doi.org/10.1088/1361-6439/ab078d
16. Sherif, T., Younis, M.I.: Toward cascadable MEMS logic device based on mode localization. Sens. Actuators A Phys. **315**, 112367 (2020). https://doi.org/10.1016/j.sna.2020.112367
17. Rega, G., Lacarbonara, W., Nayfeh, A.: Reduction Methods for Nonlinear Vibrations of Spatially Continuous Systems with Initial Curvature, pp. 235–246 (2000). ISBN 978-94-010-5809-4. https://doi.org/10.1007/978-94-011-4150-5_24
18. Cao, D.Q., Liu, D., Wang, C.: Nonlinear dynamic modelling for MEMS components via the Cosserat rod element approach. J. Micromech. Microeng. **15**, 1334. 1317/15/6/027 (2005). https://doi.org/10.1088/0960-1317/15/6/027
19. Walter, L., Yabuno, H.: Refined models of elastic beams undergoing large in-plane motions: theory and experiment. Int. J. Solids Struct. **43** (2005). https://doi.org/10.1016/j.ijsolstr.2005.07.018
20. Narakorn, S., Giuseppe, R., Somchai, C.: Two-to-one resonant multi-modal dynamics of horizontal/inclined cables. Part I: theoretical formulation and model validation. Nonlinear Dyn. **48**, 231–252 (2007). https://doi.org/10.1007/s11071-006-9086-0
21. Narakorn, S., Rega, G.: Two-to-one resonant multi-modal dynamics of horizontal/inclined cables. Part II: internal resonance activation, reduced-order models and nonlinear normal modes. Nonlinear Dyn. **48**, pp. 253–274 (2007). https://doi.org/10.1007/s11071-006-9087-z
22. Cao, D.Q., Tucker, R.: Nonlinear dynamics of elastic rods using the Cosserat theory: modelling and simulation. Int. J. Solids Struct. **45**, 460–477 (2008). https://doi.org/10.1016/j.ijsolstr.2007.08.016
23. Luo, A.C.J.: On a nonlinear theory of thin rods. Commun. Nonlinear Sci. Numer. Simul. **15** (2010). https://doi.org/10.1016/j.cnsns.2010.01.043

24. Vlajic, N., et al.: Geometrically exact planar beams with initial pre-stress and large curvature: Static configurations, natural frequencies, and mode shapes. Int. J. Solids Struct. **51**, 3361–3371 (2014). https://doi.org/10.1016/j.ijsolstr.2014.05.026
25. Lembo, M.: Exact solutions for post-buckling deformations of nanorods. Acta Mechanica **228**(6), 2283–2298 (2017). https://doi.org/10.1007/s00707-017-1834-3
26. Yang, Z., Yi, Z., Stanciulescu, I.: Nonlinear buckling and postbuckling of shallow arches with vertical elastic supports. J. Appl. Mech. **86**, 1 (2019). https://doi.org/10.1115/1.4042572
27. Rodrigo, N.: On the natural frequencies of simply supported beams curved in mode shapes. J. Sound Vibr. **485**, 115597 (2020). https://doi.org/10.1016/j.jsv.2020.115597
28. Pandurangi, S.S., Elliott, R.S., Healey, T.J., Triantafyllidis, N.: Stable Spatially Localized Configurations in a Simple Structure—A Global Symmetry-Breaking Approach. Journal of Elasticity **142**(1), 163–199 (2020). https://doi.org/10.1007/s10659-020-09794-5
29. Samir, E., Walter, L.: Buckling and postbuckling of extensible, shear-deformable beams: Some exact solutions and new insights. Int. J. Non-Linear Mech. **129**, 103667 (2021). https://doi.org/10.1016/j.ijnonlinmec.2021.103667
30. Medina, L., Gilat, R., Krylov, S.: Symmetry breaking in an initially curved micro beam loaded by a distributed electrostatic force. Int. J. Solids Struct. **49**, 1864–1876 (2012). https://doi.org/10.1016/j.ijsolstr.2012.03.040
31. Lior, M., Rivka, G., Krylov, S.: Symmetry breaking in an initially curved pre-stressed micro beam loaded by a distributed electrostatic force. Int. J. Solids Struct. **51**, 2047 (2014). https://doi.org/10.1016/j.ijsolstr.2014.02.010
32. Milad, S., Beni, Y.T., Ataei, H.: Size-dependent snap-through and pull-in instabilities of initially curved pre-stressed electrostatic nano-bridges. J. Phys. D Appl. Phys. **49** (2016). https://doi.org/10.1088/0022-3727/49/29/295303
33. Medina, L., Gilat, R., Krylov, S.: Latching in bistable electrostatically actuated curved micro beams. International Journal of Engineering Science **110**, 15–34 (2017). https://doi.org/10.1016/j.ijengsci.2016.10.001
34. Medina, L., Gilat, R., Krylov, S.: Bow Actuator: Low Voltage Switching in Electrostatically Actuated Bistable Beams (2018). https://doi.org/10.1115/DETC2018-85534
35. Sharma, A.K., Godara, R.K., Joglekar, M.M.: Static and DC dynamic pull-in analysis of curled microcantilevers with a compliant support. Microsystem Technologies **25**(3), 965–975 (2018). https://doi.org/10.1007/s00542-018-4046-4
36. Medina, L., Gilat, R., Krylov, S.: Dynamic release condition in latched curved micro beams. Commun. Nonlinear Sci. Numer. Simul. **73** (2019). https://doi.org/10.1016/j.cnsns.2019.01.022
37. Medina, L., Ashwin, S.: Bistability and simultaneous mode actuation in electrostatically actuated initially curved coupled micro beams. Int. J. Non-Linear Mech. **126**, 103549 (2020). https://doi.org/10.1016/j.ijnonlinmec.2020.103549
38. Medina, L., Ashwin, S.: Tristable properties and limit point behaviour in electrostatically actuated initially curved coupled micro beams. Int. J. Mech. Sci. **204**, 106543 (2021). https://doi.org/10.1016/j.ijmecsci.2021.106543
39. Medina, L., et al.: Single electrode bidirectional switching of latchable prestressed bistable micromechanical beams. IEEE Sens. J., 1 (2021). https://doi.org/10.1109/JSEN.2021.3103265

40. Lacarbonara, W., Arafat, H., Nayfeh, A.: Non-linear interactions in imperfect beams at veering. Int. J. Non-Linear Mech. **40**, 987–1003 (2005). https://doi.org/10.1016/j.ijnonlinmec.2004.10.006
41. Sami, A., Younis, M.: Statics and dynamics of MEMS arches under axial forces. J. Vibr. Acoust. **135**, 021007 (2013). https://doi.org/10.1115/1.4023055
42. Amal, H., Nouha, A., Younis, M.: The static and dynamic behavior of MEMS arch resonators near veering and the impact of initial shapes. Int. J. Non-Linear Mech. **95** (2017). https://doi.org/10.1016/j.ijnonlinmec.2017.07.002
43. Alqasimi, J., Ouakad, H.: Vibrational response of initially deformed bi-stable microbeams under the combined effect of mechanical shock loads and electrostatic forces. J. Vibr. Acoust. **140** (2017). https://doi.org/10.1115/1.4038107
44. Ghayesh, M.H., Farokhi, H.: Bistable nonlinear response of MEMS resonators. Nonlinear Dynamics **90**(3), 1627–1645 (2017). https://doi.org/10.1007/s11071-017-3753-1
45. Feras, A., Amal, H., Mohammad, Y.: Theoretical and experimental investigation of two-to-one internal resonance in MEMS arch resonators. J. Comput. Nonlinear Dyn. **14** (2018). https://doi.org/10.1115/1.4041771
46. Tausiff, M., Ouakad, H.M., Alqahtani, H., Alofi, A.: Local nonlinear dynamics of MEMS arches actuated by fringing-field electrostatic actuation. Nonlinear Dynamics **95**(4), 2907–2921 (2019). https://doi.org/10.1007/s11071-018-4731-y
47. Hassen, O., Fehmi, N.: Nonlinear dynamics of MEMS arches assuming out-of-plane actuation arrangement. J. Vibr. Acoust. **141**, 1 (2019). https://doi.org/10.1115/1.4043064
48. Hajjaj, A.Z., Alfosail, F.K., Jaber, N., Ilyas, S., Younis, M.I.: Theoretical and experimental investigations of the crossover phenomenon in micromachined arch resonator: part I—linear problem. Nonlinear Dynamics **99**(1), 393–405 (2019). https://doi.org/10.1007/s11071-019-05251-8
49. Hajjaj, A.Z., Alfosail, F.K., Jaber, N., Ilyas, S., Younis, M.I.: Theoretical and experimental investigations of the crossover phenomenon in micromachined arch resonator: part II—simultaneous 1:1 and 2:1 internal resonances. Nonlinear Dynamics **99**(1), 407–432 (2019). https://doi.org/10.1007/s11071-019-05242-9
50. Ayman, A., et al.: Large oscillation of electrostatically actuated curved beams. J. Micromech. Microeng. **30**, 095005 (2020). https://doi.org/10.1088/1361-6439/ab94d1
51. Nouha, A., et al.: Investigations Into the Linear Coupling Between Symmetric and Anti-Symmetric Modes of V-Shaped MEMS Resonators Under Electrostatic Perturbation (2021). https://doi.org/10.1115/IMECE2021-73535
52. Nouha, A., et al. Crossover/veering in V-shaped MEMS resonators. J. Microelectromech. Syst. PP, 1–13 (2021). https://doi.org/10.1109/JMEMS.2021.3126551
53. Rosenberg, S., Shoshani, O.: Zero-dispersion point in curved micro-mechanical beams. Nonlinear Dynamics **107**(1), 1–14 (2021). https://doi.org/10.1007/s11071-021-07015-9
54. Lior, M., et al.: Open Loop. Self-Excitation in a Bistable Microme-chanical Beam Actuated By a DC Electrostatic Load (2017). https://doi.org/10.1109/MEMSYS.2017.7863525
55. Rajaei, A., Vahidi-Moghaddam, A., Ayati, M., Baghani, M.: Integral sliding mode control for nonlinear damped model of arch microbeams. Microsystem Technologies **25**(1), 57–68 (2018). https://doi.org/10.1007/s00542-018-3931-1
56. Yoav, K., Alexander, L., Krylov, S.: On sampling rate limits in bistable microbeam sensors. J. Microelectromech. Syst., pp. 1–10 (2021). https://doi.org/10.1109/JMEMS.2021.3117838

57. Lior, M., Rivka, G., Krylov, S.: Bistable behavior of electro-statically actuated initially curved micro plate. Sens. Actuators A Phys. **248** (2016). https://doi.org/10.1016/j.sna.2016.07.027
58. Shahid, S., et al.: Initially curved microplates under electrostatic actuation: theory and experiment. J. Micromech. Microeng. **26** (2016). https://doi.org/10.1088/0960-1317/26/9/095004
59. Lior, M., Rivka, G., Slava, K.: On The Usage of Berger's Model for Electrostatically Actuated Circular Curved Micro Plates (2017). https://doi.org/10.1115/DETC2017-67523
60. Medina, L., Gilat, R., Krylov, S.: Bistability criterion for electrostatically actuated initially curved micro plates. Int. J. Eng. Sci. **130**, 75–92 (2018). https://doi.org/10.1016/j.ijengsci.2018.05.006
61. Saghir, S., Younis, M.I.: An investigation of the mechanical behavior of initially curved microplates under electrostatic actuation. Acta Mechanica **229**(7), 2909–2922 (2018). https://doi.org/10.1007/s00707-018-2141-3
62. Ghayesh, M., Farokhi, H.: Nonlinear behaviour of electrically actuated microplate-based MEMS resonators. Mech. Syst. Signal Process. **109**, 220–234 (2018). https://doi.org/10.1016/j.ymssp.2017.11.043
63. Manuel, D., et al.: Switching performance of bistable membranes activated with integrated piezoelectric thin film transducers. J. Micromech. Microeng. **29** (2019). https://doi.org/10.1088/1361-6439/ab3185
64. Asaf, A., et al.: Bistable micro caps fabricated by sheet metal forming. J. Micromech. Microeng. **30** (2020). https://doi.org/10.1088/1361-6439/ab7f52
65. Annick, D., Willy, G., Kuznetsov, Y.A.: MATCONT: a MATLAB package for numerical bifurcation analysis of ODEs. In: ACM Trans. Math. Softw. (TOMS) **29**, 141–164 (2003). https://doi.org/10.1145/980175.980184

Different Beam Configurations for Compliant Mechanism-Based MEMS Accelerometer

Nikul Jani[1(✉)], Mithlesh Kumar[1], P. Krishna Menon[2], and Ashok Kumar Pandey[1]

[1] Department of Mechanical and Aerospace Engineering, Indian Institute of Technology, Hyderabad, Telangana 502284, India
nikul.jani28@gmail.com

[2] Research Centre Imarat, Hyderabad 500069, India

Abstract. In this article, a compliance-based displacement amplifier has been analyzed. The utility of a couple of variations within a pre-existing compliant mechanism has been systematically investigated. The presented modifications within the pre-existing design can help to get a large output displacement. The suggested modification within the design doesn't improve the amplification ratio, but it can cause a large increase in the output displacement of the amplifier for the given input force. Such characteristics of an amplifier are very helpful in a device like an accelerometer in which proof mass provides a certain inertial force. Here, the modified displacement amplifier has been considered to be included in a capacitive sensing-based MEMS accelerometer. The influence of beam configurations on the performance of MEMS accelerometer is analyzed, and the dimensions are chosen to get the optimum performance. It is found that the proposed variations in the design are helpful to get larger sensitivity. They are extremely useful to be included within the design when the allowed minimum size of the beams is larger due to fabrication limitations. Natural frequencies of the MEMS accelerometer have been calculated numerically and analytically where a good agreement has been found. The presented study assures improvement in the device performance through suggested modifications within the compliant mechanism.

Keywords: Compliant mechanism · Displacement amplifier · Folded beams · Inclined beams · MEMS accelerometer

The work has been sponsored by Defence Research and Development Organization (DRDO), New Delhi, India.

A. K. Pandey et al. (Eds.): MAMM 2022, 126, pp. 119–135, 2023.
https://doi.org/10.1007/978-3-031-20353-4_8

Abbreviation

f_{SML}	SML model based approximation of natural frequency for the in-plane-axial motion
f_{FEM}	FEM based approximation of natural frequency for the in-plane-axial motion
f_{DaCM}	Fundamental natural frequency of solely DaCM for in-plane-axial motion
F_{in}	Input force
K_{in}	Input stiffness of DaCM
K_{out}	Output stiffness of DaCM
K_{sp}	Stiffness of supports connected to proof mass
K_{ss}	Stiffness of supports connected to sense-combs
M_p	Mass of the proof mass
M_s	Element mass on the sense-side
M_{DaCM}	Equivalent mass of DaCM for in-plane fundamental natural frequency and motion within axial direction
K_{DaCM}	Equivalent stiffness of DaCM
p	Parameter for the relative dimensions of folded beams and inclined beams
L_o	Height of folded beam and inclined beam sections
L_a	Length of the thick vertical section connected to the output side of DaCM
y_1	Input displacement
y_2	Output displacement
DaCM	Displacement-amplifying compliant mechanism
FoM	Figure of merit
SDOF	Single-degree-of-freedom
SML	Spring-mass-lever

1 Introduction

A compliant mechanism uses a single continuum body to transfer force or displacement applied at the input port to the output port. It can amplify the applied force or displacement, or it can be used for solely transferring the motion. The compliant structures have been demonstrated for applications like crimping mechanism, pliers, grippers, and bistable switches [1–3]. In recent times, compliance-based mechanisms have got wide acceptance in the field of MEMS devices due to their salient features of efficient transfer of work from one end to another [4,5]. The compliant structures can provide effective kinematic pairs without actually producing rotary or translatory joints, thus, allowing it to realize very complex mechanisms at the microscale. Consequently, compliant mechanisms have been used for micro-newton force sensors, tiny mirror-based optical devices, MEMS accelerometers, etc. The class of compliant mechanisms which amplify a given displacement has been extensively used for micro-motion devices [10]. These can also be employed to improve the sensitivity of MEMS actuators and sensors [7,8]. It is called displacement-amplifying compliant mechanism (DaCM). Typically, micro-actuators have a very small displacement range, for example, piezoelectric, electromagnetic or electrostatic actuators. Here, the concept of a compliant mechanism can become extremely useful to get amplified motion with high resolution. Micro-gripper-based manipulation is a particularly good example of that. DaCM can be classified based on the design strategy to obtain amplification. They are mainly of two types, lever-based mechanism, and bridge-based mechanism [13]. It can also be combination of the two types, for example, compound lever type mechanism [6], lever-bridge type mechanism [9], or double bridge-type mechanism [12].

The domain of MEMS accelerometer has been found with potential applications for DaCMs. From the literature study, it is found that the compliant mechanism-based displacement amplification can provide larger capacitance change in MEMS accelerometer in which sensing is carried out electrostatically. It's a potential approach for sensitivity enhancement. Providing supports for the proof mass with exceedingly small stiffness can also improve the sensitivity to some extent, but the natural frequencies of the device will reduce. Here also, the DaCM is a useful alternative to get large sensitivity without compromising with natural frequency [11].

In this article, potential modifications within the pre-existing DaCM have been proposed which can increase the sensitivity of the MEMS accelerometer. They are found highly advantageous especially when the allowed minimum size of beams is larger. Configurations of folded beams and inclined beams have been suggested to be considered in the selected DaCM model which can improve the performance of the MEMS accelerometer. For the different dimensions of beam configurations, rigorous FEA analysis has been conducted to optimize the performance of MEMS accelerometer. Detailed results presenting various aspects of DaCM based accelerometer are discussed. The structure of the paper is as follows. In the next section, the inclusion of both the beam configuration in DaCM is discussed. In Sect. 3, a single-degree-of-freedom (SDOF) model with lumped mass and stiffness is presented for the in-plane-axial motion of the MEMS accelerometer. Natural frequency is calculated for the SDOF model, which is the fundamental natural frequency for the MEMS accelerometer. Results of FEA analysis are presented in Sect. 4, where characteristics of solely DaCM and modified DaCM based MEMS accelerometer are presented. The results of the numerical analysis show that for desired dimensions of MEMS accelerometer, sensitivity can be improved while using the proposed beam configurations.

2 DaCM in MEMS Accelerometer

As discussed in the previous section, including DaCM within MEMS accelerometer, sensitivity can be increased without reducing the natural frequency. A

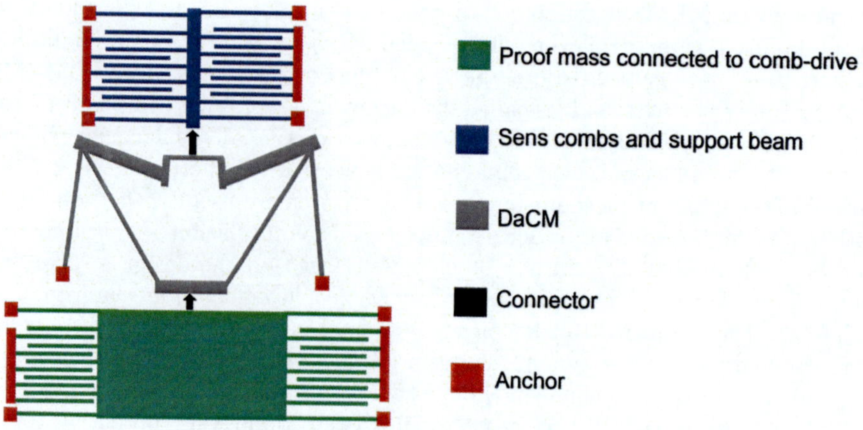

Fig. 1. Schematic of MEMS accelerometer with DaCM

schematic of electrostatic sensing-based MEMS accelerometer with DaCM is shown in figure Fig. 1. Different components of the device are shown using different colors. Due to the input acceleration, there will certain displacement of the proof mass, and it will be amplified using DaCM, and the increased displacement will be determined using electrostatic comb-drives on the sense side. The sense-combs are shown to have the differential arrangement. Suspension is provided to the proof mass and also to the sense-combs. To improve the sensitivity, it is desirable to have larger displacement of the sense combs which will induce larger change in capacitance. Here, efforts have been made to reduce the stiffness of DaCM up to certain extent so that the fundamental natural frequency (in-plane-axial motion) stays within the desirable range. Certain reduction in stiffness will cause larger displacement at the sense combs. A couple of configurations for the output side of DaCM have been considered here instead of using a straight horizontal beam. The configurations to be included at the output end of DaCM are shown below. Figure Fig. 2b presents folded beam structure, and Fig. 2c presents inclined beam structure. Including these configurations within DaCM can cause larger displacement of the combs in comparison to the straight beam-based arrangement which is shown in Fig. 2a. While carrying out the numerical study, a parameter $p = \frac{L_o}{L_a}$ has been used which provides the dimension of folded beams and inclined beams relative to L_a.

3 Analytical and Numerical Study

The contribution of different elements of DaCM based MEMS accelerometer is analyzed using lumped parameter modelling [11]. A simplified, lumped spring-mass-lever (SML) model is shown in Fig. 3. While calculating M_s and M_p, mass of combs is also considered. The support connected to the proof mass will drive the proof mass for input acceleration. Here, two terms, K_{DaCM} and M_{DaCM}

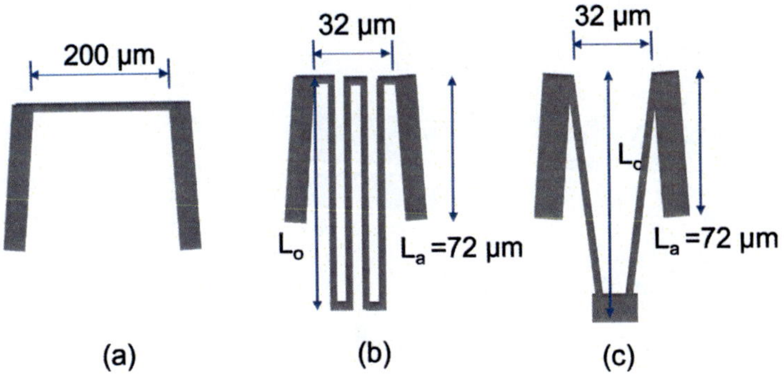

Fig. 2. Configurations for the output end of DaCM, (a) straight, (b) folded, and (c) inclined

which are related to DaCM have been introduced. K_{DaCM} is the effective stiffness of DaCM for axial displacement of DaCM. It can be calculated as,

$$K_{DaCM} = \frac{F_{in}}{y_2 - y_1} \tag{1}$$

Here, F_{in} is the input force given to the input node of DaCM. K_{DaCM} has been calculated using FEM simulation for which force of 100 μ N has been applied at the input node, and input and output displacements (y_1 and y_2) have been determined. Value of K_{DaCM} can provide very useful information to study the effect of different length of inclined and folded beam configurations on the compliant behavior of DaCM. M_{DaCM} is the equivalent mass of DaCM for in-plane-axial motion. It can be calculated using the following relationship,

$$M_{DaCM} = \frac{K_{DaCM}}{4\pi^2 f_{DaCM}^2}, \tag{2}$$

Here, f_{DaCM} is the fundamental natural frequency of solely DaCM for in-plane-axial motion. Further, SML model has been utilized to approximate the fundamental natural frequency for in-plane-axial motion, f_{SML}. Using SML model, one can get the kinetic energy and potential energy for the system as following.

$$V = \frac{1}{2}\left(M_p + \frac{M_{DaCM}}{2}\right)\frac{\mathrm{d}y_1}{\mathrm{d}t}^2 + \frac{1}{2}\left(M_s + \frac{M_{DaCM}}{2}\right)\frac{\mathrm{d}y_1}{\mathrm{d}t}^2 \tag{3}$$

$$T = \frac{1}{2}(K_{sp} + K_{in})y_1^2 + \frac{1}{2}(K_{ss} + K_{out})y_2^2 \tag{4}$$

Here, four pairs of folded beam springs have been used for the supports for both the excitation and sense sides. The stiffness of the supports for proof mass and sense mass can be calculated using the following formula [14].

$$K_{ss} = \frac{4}{3}\frac{12EI}{L^3}$$

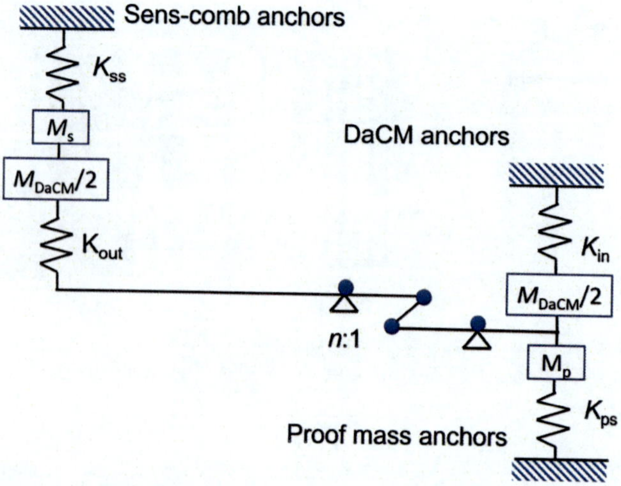

Fig. 3. Spring-mass-lever model for DaCM based MEMS accelerometer, [11]

One can easily get the equation of motion using the Euler-Lagrange equation. Equation of motion for the free vibration of fundamental mode (in-plane-axial motion) is given below.

$$\frac{\mathrm{d}^2 y_1}{\mathrm{d}t^2} M_{eq} + K_{eq} y_1 = 0 \tag{5}$$

Equivalent mass and stiffness of DaCM based MEMS accelerometer can be given as

$$M_{eq} = \left(M_p + \frac{M_{DaCM}}{2}\right) + n^2\left(M_s + \frac{M_{DaCM}}{2}\right), \tag{6a}$$

$$K_{eq} = (K_{sp} + K_{in}) + n^2 \frac{K_{ss} K_{out}}{K_{ss} + K_{out}}. \tag{6b}$$

Fundamental natural frequency for in-plane-axial motion can be approximately calculated as,

$$f_{SML} = \frac{1}{2\pi}\sqrt{\frac{K_{eq}}{M_{eq}}} \tag{7}$$

4 Results and Discussion

The selected design of DaCM with the proposed configurations has been analyzed using COMSOL. First, solely DaCM model has been analyzed to characterize the displacement amplifier, for example, amplification ratio, equivalent stiffness of DaCM, etc. Further, the performance of MEMS accelerometer with the inclusion of DaCM has been evaluated, and improvement in the sensitivity has been reported. Here, while carrying out the discussed studies different dimensions of folded beam and inclined beam configurations have been chosen to determine their influence on the performance of MEMS accelerometer.

The parameters of spring-mass-lever model, $K_{in}, K_{out}, K_{DaCM}, M_{DaCM}$, etc. have also been determined and in-plane natural frequency of DaCM based MEMS accelerometer (f_{SML}) has been calculated analytically. First of all, the results of the numerical analysis carried out for solely DaCM have been presented which is followed by the results for MEMS accelerometer combined with DaCM.

4.1 Influence of p on the Characteristics of DaCM

As we discussed in Sect. 2, the proposed configurations causes reduction in the stiffness of displacement amplifier. Output displacement of DaCM, y_2 has been calculated for different values of p and input load, F_{in}. For both the cases, the displacement characteristics for the given input load are shown in Fig. 4a, b. One can see that the slope of graph reduces with increase in p value.

Further, the effective stiffness of DaCM, K_{DaCM} is analyzed here, which can be used to study the compliant behavior of DaCM more specifically. It has been defined in Sect. 3. Here, instead of K_{DaCM}, compliance of the displacement amplifier mechanism has been presented on the y-axis. It can be calculated as $C = K_{DaCM}^{-1}$. For both the configurations, C has been determined for a range of p. The amplification ratio of DaCM, n is also calculated here. Variation in C and n is shown in Fig. 5. One can see that C increases rapidly as p increases. It is found that folded beam configuration provides larger compliance than inclined beams. For $p = 0.5$ and $p > 4$, amplification ratio for both the cases is found very close to each other, however, for $1 < p < 3$, folded beam configuration provides a smaller amplification ratio.

4.2 Variation in K_{eq} and M_{eq} Due to Folded Beam Configuration

Spring-mass-lever model has been discussed before in Sect. 3. Using the simplified model, formulae for equivalent stiffness, K_{eq} and equivalent mass, M_{eq} for MEMS accelerometer have been derived (Eqs. 6a and 6b). K_{eq} and M_{eq} have been calculated for different values of p. It has been found that K_{eq} and M_{eq}, both decrease as p increases, however, the decrement in K_{eq} is much sharper than reduction in M_{eq}. One can see that for $p > 2$, the change in K_{eq} and M_{eq} are very small. One can note that the stiffness of the supports, K_{sp} and K_{ss} are very small compared to the stiffness of DaCM. So, the DaCM strongly influences the overall stiffness of MEMS accelerometer. For $p < 2$, the beam configuration at the output side stiffens the mechanism. As p increases, the contribution of beam configurations for stiffness of DaCM or MEMS accelerometer decreases, and for larger values of $p (> 3.5)$, the contribution becomes negligible.

The fundamental natural frequency of MEMS accelerometer using SML model, f_{SML} combined with DaCM is also calculated here, and they are compared with the natural frequency values calculated using COMSOL. The comparison is shown in Fig. 6b. For $1 < p < 2$, a very good agreement is found in

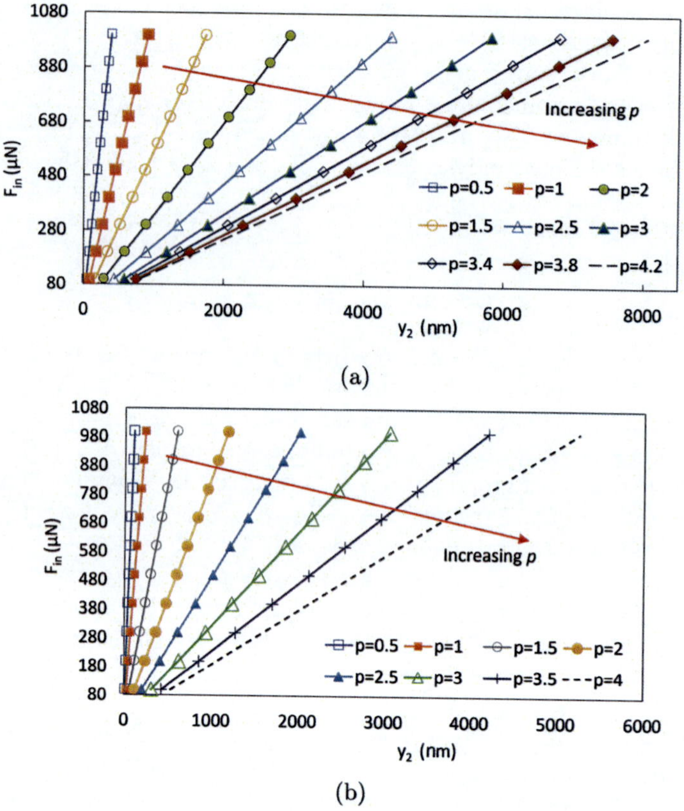

Fig. 4. Output displacement for different values of p for a large range of applied load at the input node of DaCM for (a) folded beam and (b) inclined beam configurations

between the f_{SML} and f_{FEA}. However, for larger p, f_{SML} is found 14–16 % smaller than f_{FEM}. SML model-based approximation of natural frequency has been carried out for folded beam-based configuration only, however for inclined beam configuration also it can provide approximation with a satisfactory agreement with FEM results.

FEM based approximation of fundamental natural frequency (for in-plane-axial motion) for both the configurations is shown in Fig. 7a. Here, discrete points with markers show the results of numerical analysis, and these points are connected with continuous lines. It shows that inclined beam configuration can provide the fundamental natural frequency larger for the same value of p, and it is advantageous for the MEMS accelerometer. Mode shapes for the first mode of MEMS accelerometer for all the configurations are shown in Fig. 7b.

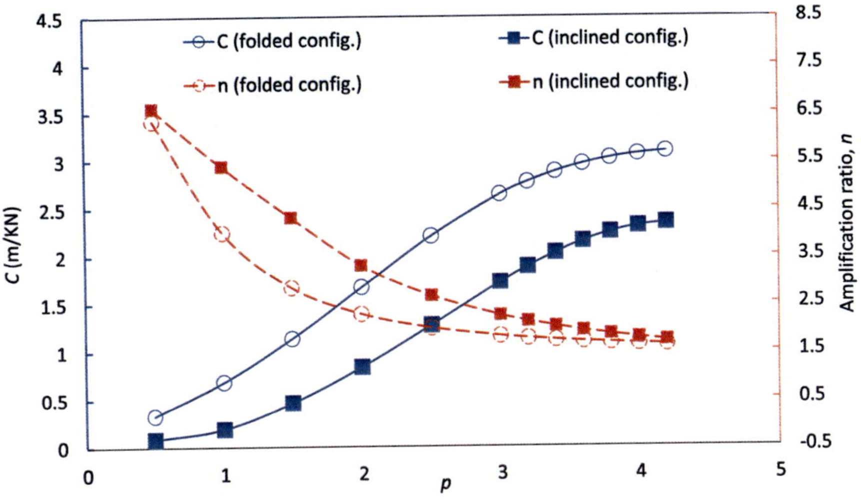

Fig. 5. Variation in compliance of DaCM (K_{DaCM}^{-1}) and amplification ratio for different values of p

4.3 Characteristics of MEMS Accelerometer Combined with DaCM

Numerical analysis of MEMS accelerometer including the displacement amplifier with proposed variations in the design can provide very useful insights about performance of the device. To analyze the influence of the height of both the configurations (L_o), for a fixed value of input acceleration ($30g$), input and output displacements (y_1 and y_2) are determined. The results of numerical analysis are shown in Fig. 8a. It is found that for similar height (L_o) folded beam configuration can provide relatively larger output displacement. For the results presented in Fig. 8a, static structural analysis is carried out. So electrostatic effect is not considered.

For calculating the change in the capacitance for the given input acceleration, electromechanics analysis has been carried out, and results are shown in Fig. 8b. For a certain value of p, folded beam configuration provides larger output displacement, however, values of change in capacitance are found very similar for both the configurations. That is due to the electrostatic stiffness of the sense-combs.

As shown in Fig. 7a, inclined beam configuration can provide larger value of fundamental natural frequency. For the considered design of MEMS accelerometer, fundamental natural frequency should be higher than or close to 6 KHz. Considering this criterion, lengths of beam configuration have been chosen taking $p = 2$ and $p = 3$ for folded and inclined beam configurations, respectively. Electromechanics analysis for the MEMS accelerometer with the selected dimensions has been conducted for different values of input acceleration. Change in capacitance is shown in Fig. 9. Here, discrete data points show the results of numerical analysis, and the points are connected with continuous lines.

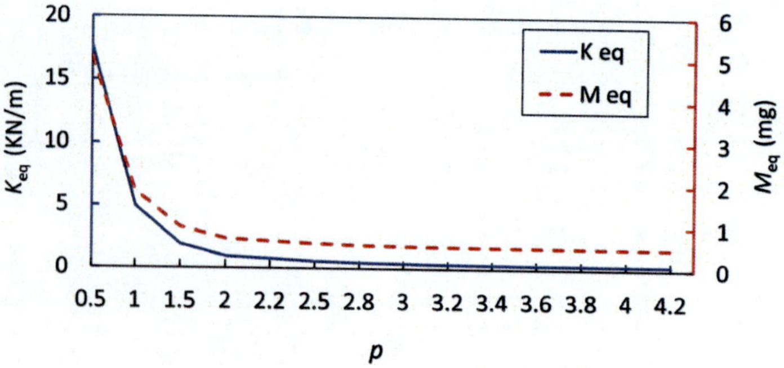

(a)

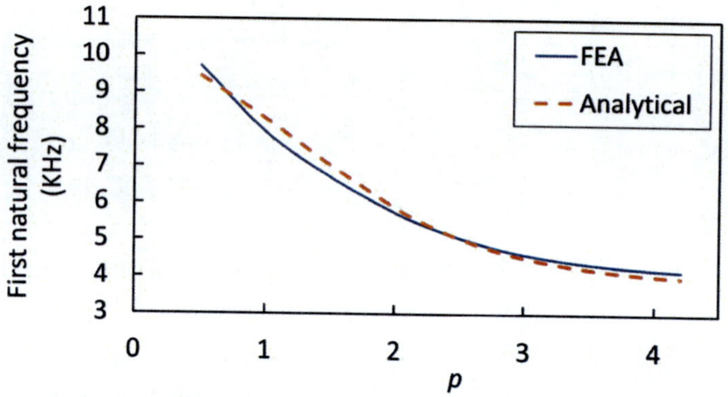

(b)

Fig. 6. (a) Equivalent mass and stiffness of MEMS accelerometer according to SML model, and (b) comparison between f_{SML} and f_{FEM} for different values of p

Further, the effect of slanting on the performance of MEMS accelerometer has been analyzed using Coventorware. Results of the analysis are presented in Sect. 4.3 which shows that the device characteristics can be strongly influenced due to slanting effect. With increase in sidewall angle, fundamental natural frequency of MEMS accelerometer (f_{FEM}) increases, and output displacement (y_2) decreases.

Sidewall angle	Folded		Inclined	
	f_{FEM} (kHz)	y_2(nm)	f_{FEM} (kHz)	y_2(nm)
0	5.8	330	5.986	310
0.5	6.76	250	6.885	230
1	7.26	190	7.978	170

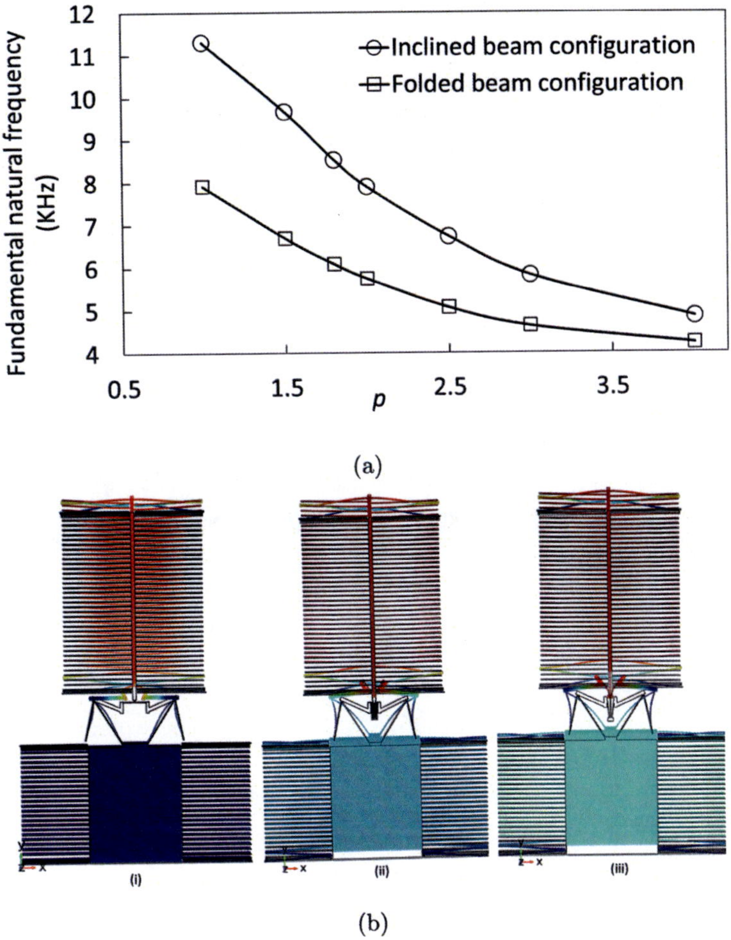

Fig. 7. (a) f_{FEM} for inclined beam and folded beam configuration for different values of p and (b) mode shapes for fundamental mode of MEMS accelerometer for (i) straight, (ii) folded and (iii) inclined beam configurations

Profiles generated using DRIE includes a small inclination [14]. That is called slanting effect which stiffens the structure. In such case, the designer needs to choose a larger value of p to get larger output displacement even the fundamental natural frequency calculated using FEM simulations is smaller, as due to the expected slanting effect, f_{FEM} will increase, and y_2 will decrease. For slanting or side-wall angle of 1°, outcome of modal analysis for fundamental natural frequency and output displacement of MEMS accelerometer for $30g$ input acceleration is shown in Figs. 10 and 11, respectively.

The MEMS accelerometer has several elements which undergo large deformation, for example supports for the output side of MEMS accelerometer, thin

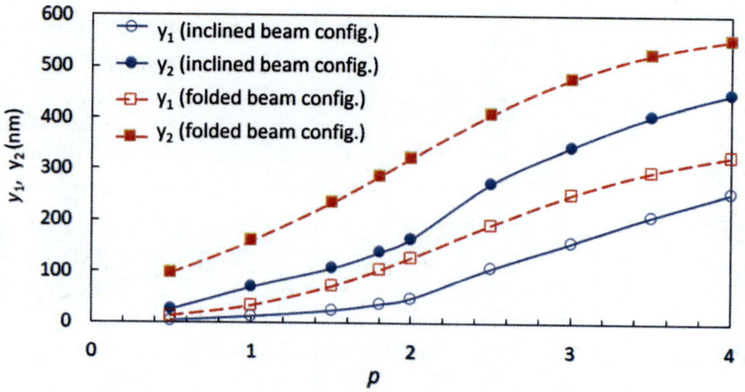

(a) Displacements at the input and output port of DaCM for the given inertia force respective to $30g$ input acceleration

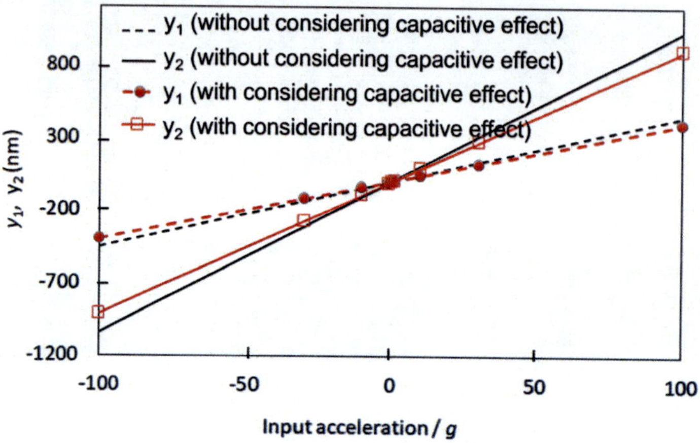

(b) Reduction in y_1 and y_2 due to electrostatic force

Fig. 8. Analysis of DaCM combined with proof mass and sense combs

beams of DaCM, etc. For safety of the device, it is essential that the maximum stress within the device should be less than the yield stress value for a large value of input acceleration. Here, FEM simulations have been carried for the input acceleration taken as $500g$ for both the configurations. Maximum values of Von Mises stress found for folded beam ($p = 2$) and inclined beam ($p = 3$) configuration are 110 MPa and 121 MPa, respectively. Results of the stress analysis are shown in Fig. 12 where the locations of maximum stress are also shown. For comparison of the different configurations, FE simulations have been carried out while choosing minimum thickness of the beams as 7µm. To calculate the output displacement and change in the capacitance, input acceleration of $30g$ has been given to the MEMS accelerometer, and the results are shown in Sect. 4.3. Here,

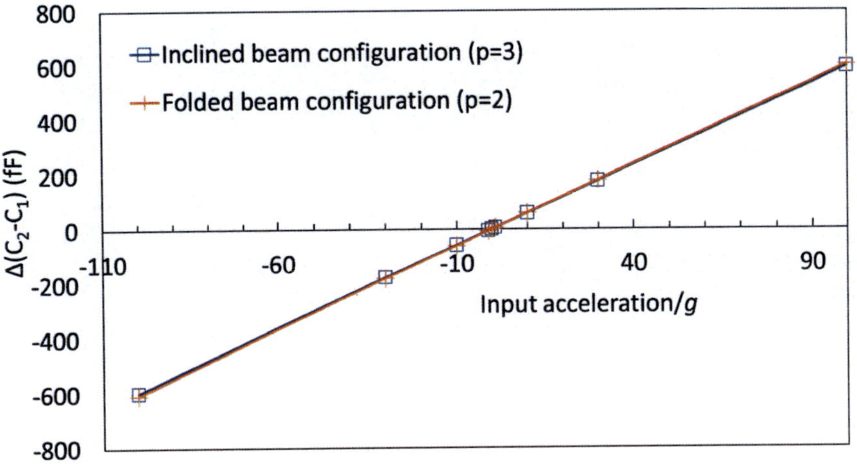

Fig. 9. Change in capacitance for folded ($p = 2$) and inclined ($p = 3$) beam configurations

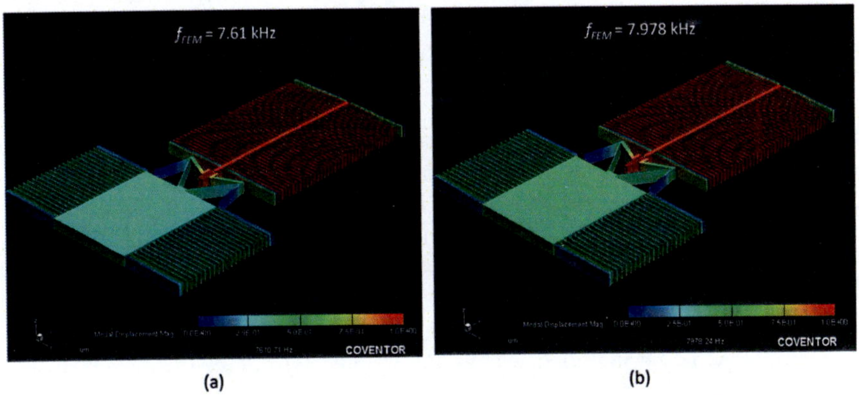

Fig. 10. Analysis of fundamental natural frequency of MEMS accelerometer for (a) folded beam ($p = 2$) and (b) inclined beam ($p = 2.8$) configurations for 1° of sidewall angle

the comparison is also carried out with the outcomes of numerical analysis given in the pre-existing literature [15]. They had chosen minimum thickness of the beams as 4 μ m. Value of fundamental natural frequency was found 6.7 kHz, and larger output displacement was also achieved. However, if the minimum thickness of beams is needed to be chosen as 7 μm, then the output displacement decreases by a large amount. In such case, the proposed configurations provides satisfactory results for larger output displacement, y_2. Further, the change in capacitance can also be greatly increased. Figure of merit has been calculated as $4\pi^2 \frac{y_2}{30g} f_{SML}^2$. It has been defined in literature for overall comparison of DaCM based MEMS accelerometer [15]. Here, y_2 is the output displacement for $30g$

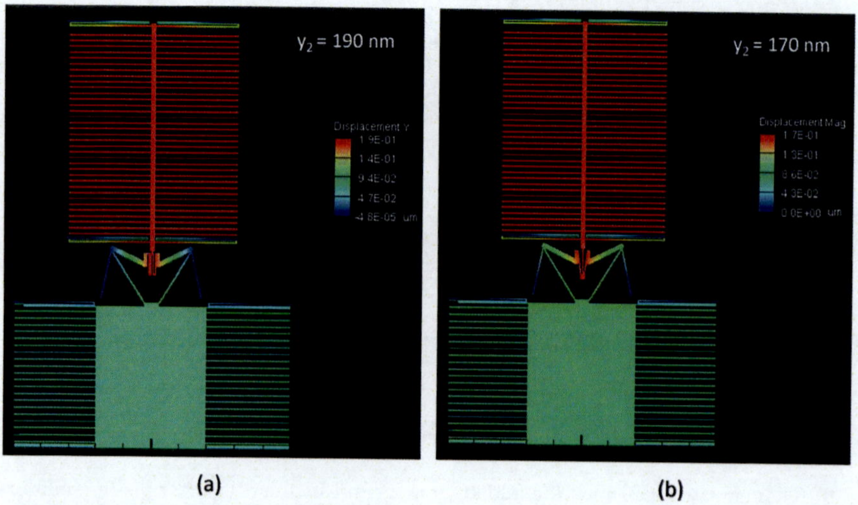

Fig. 11. Analysis of output displacement of MEMS accelerometer $30g$ input acceleration for (a) folded beam ($p = 2$) and (b) inclined beam ($p = 2.8$) configurations for $1°$ of sidewall angle

input acceleration. It is found that choosing minimum thickness of microbeam larger (due to fabrication limitations) performance of DaCM reduces. Here, the proposed beam configurations can be a useful methodology to improve the characteristics of DaCM while maintaining the requirement of fabrication. Values of fundamental natural frequency for in-plane-axial motion are found close to 6 kHz for folded beam (taking $p = 2$) and inclined beam (taking $p = 3$) configurations, respectively. Inclined beam configuration (taking $p = 3$) can be chosen for the design of DaCM to be included within MEMS accelerometer (Table 1).

Table 1. Comparison between the performance of MEMS accelerometer including DaCM with different configuration

Configuration	Straight, [15]	Straight (7 µm)	Folded (7 µm)	Inclined (7 µm)
y_2 (nm)	261	62	322	343
n	15.26	11.46	2.53	2.18
$f_{FEM}(kHz)$	6.7	9.88	5.75	5.85
ΔC (fF)	–	34	179.4	175
FoM	1.57	0.811	1.42	1.57

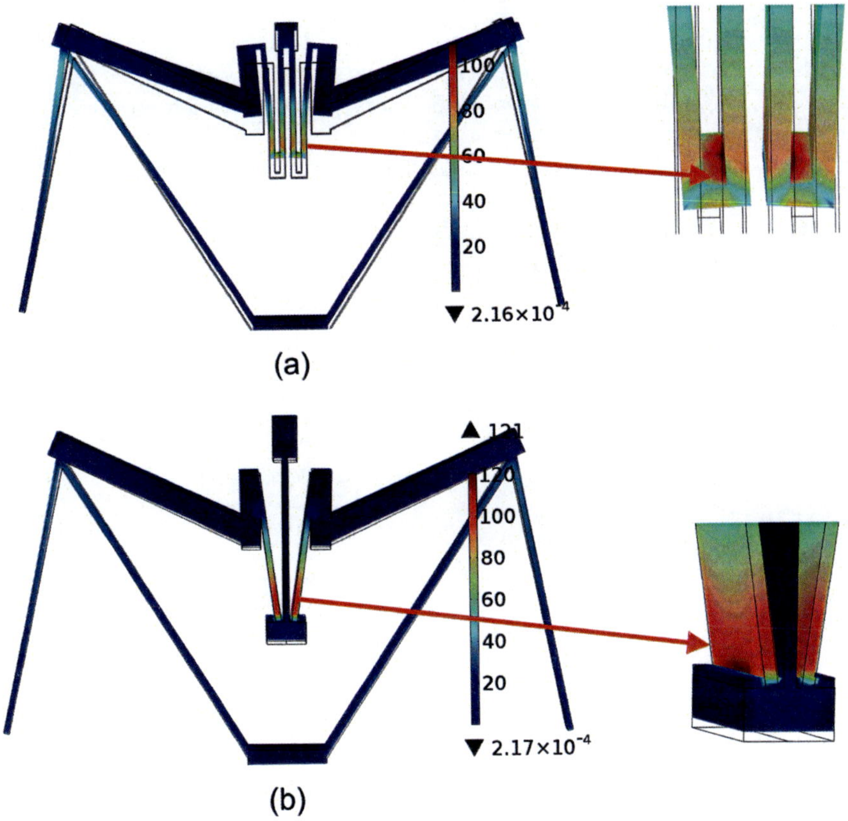

Fig. 12. Indicating the value and location of maximum stress in DaCM for 500g input acceleration for (a) folded and (b) inclined beam configurations

5 Conclusion

Displacement amplifier based on compliant mechanism has been analyzed for enhancing the performance of MEMS accelerometer. Here, configurations of folded beams and inclined beams have been proposed to be included at the output end of the compliant mechanism. The suggested modification within the design can provide a large increase in the output displacement of the amplifier for the given input acceleration. If the required minimum thickness of the beams in DaCM is larger due to fabrication limitations, then the proposed configurations are extremely helpful for better performance.

The effect of the length of both the beam configurations on the fundamental natural frequency for in-plane-axial motion has also been studied. It is found that inclined beam configuration can provide a higher value of natural frequency for in-plane-axial motion. For the different heights of beam configurations, the characteristics of the displacement amplifier and MEMS accelerometer are studied.

Here, a couple of terms have been introduced, namely effective mass and effective stiffness of DaCM. They have been used to study the characteristics of displacement amplifier for different dimensions of beam configurations. Using a spring-mass-lever model, the equivalent mass and spring of MEMS accelerometer combined with DaCM have been calculated. Values of fundamental natural frequencies of the MEMS accelerometer have also been calculated for both configurations. While considering both configurations, the inclined beam configuration is found to give better performance.

Acknowledgments. The authors would like to acknowledge DRDO, New Delhi, India for funding the research work through the grant number DRDO/./IITHRC-011. We also like to thank Prof. Ananthsuresh, IISc Bangalore, and Dr. Safvan, IIT Hyderabad for useful discussion on the Mechanical Amplifiers. The first author is very grateful to Sai Kishore Jujjuvarapu for his kind help for Coventorware analysis.

References

1. Howell, L.L.: Compliant mechanisms. In: 21st Century Kinematics. Springer, London (2013)
2. Wang, D.-A., Pham, H.-T., Hsieh, Y.-H.: Dynamical switching of an electromagnetically driven compliant bistable mechanism. Sens. Actuators A Phys. **149**(1), 143–151 (2009)
3. Chen, X., et al.: Design of a compliant mechanism based four-stage amplification piezoelectric-driven asymmetric microgripper. Micromachines **11**(1), 25 (2019)
4. Gan, J., et al.: Designs of compliant mechanism-based force sensors: a review. IEEE Sens. J. (2022)
5. Iqbal, S., Malik, A.: A review on MEMS based micro displacement amplification mechanisms. Sens. Actuators A Phys. **300**, 111666 (2019)
6. Tang, H., Li, Y.: A new flexure-based $Y\theta$ nanomanipulator with nanometer-scale resolution and millimeter-scale workspace. IEEE/ASME Trans. Mechatron. **20**(3), 1320–1330 (2014)
7. Schmitt, P., Hoffmann, M.: Engineering a compliant mechanical amplifier for MEMS sensor applications. J. Microelectromech. Syst. **29**(2), 214–227 (2020)
8. Kota, S., Joo, J., Li, Z., Rodgers, S.M., Sniegowski, J.: Design of compliant mechanisms: applications to MEMS. Analog Integr. Circ. Sig. Process. **29**(1), 7–15 (2001)
9. Chen, X., Xie, Z., Tan, H.: Design, analysis, and test of a novel cylinder-driven mode applied to microgripper. J. Mech. Des. **144**(5) (2022)
10. Ouyang, P.R., Tjiptoprodjo, R.C., Zhang, W.J., Yang, G.S.: Micro-motion devices technology: the state of arts review. Int. J. Adv. Manuf. Technol. **38**(5), 463–478 (2008)
11. Khan, S., Ananthasuresh, G.K.: Improving the sensitivity and bandwidth of in-plane capacitive microaccelerometers using compliant mechanical amplifiers. J. Microelectromech. Syst. **23**(4), 871–887 (2014)
12. Iqbal, S., Lai, Y.J., Shakoor, R.I., Raffi, M., Bazaz, S.A..: Design, analysis, and experimental investigation of micro-displacement amplification compliant mechanism for micro-transducers. Rev. Sci. Instrum. **92**(10), 105007 (2021)
13. Dong, W., Chen, F., Gao, F., Yang, M., Sun, L., Zhijiang, D., Tang, J., Zhang, D.: Development and analysis of a bridge-lever-type displacement amplifier based on hybrid flexure hinges. Precis. Eng. **54**, 171–181 (2018)

14. Biswas, A., et al.: Influence of fabrication tolerances on performance characteristics of a MEMS gyroscope. Microsyst. Technol. **27**(7), 2679–2693 (2021)
15. Khan, S.: Development of Micromachined and Meso-scale Multi-axis Accelerometers with Displacement-amplifying Compliant Mechanisms (Doctoral dissertation, Doctoral Thesis, Department of Electronic Systems Engineering, Indian Institute of Science, Bangalore, India) (2013)

Check for updates

Design and Analysis of the Two-Level Accelerometer

Hoang Bao Thien Nguyen[1], Thanh Lam Tran[2], and Ngoc Dang Khoa Tran[1](✉)

[1] Faculty of Mechanical Engineering, Industrial University of Ho Chi Minh City, Ho Chi Minh City, Vietnam
trannngocdangkhoa@iuh.edu.vn

[2] Faculty of Mechanical Engineering, Ho Chi Minh City University of Technology and Education, Ho Chi Minh City, Vietnam
lamtt@hcmute.edu.vn

Abstract. The accelerometer is the sensor that detects the change of acceleration, has many applications in transportation, industry, navigation, and medical devices. This paper introduces a new design of the micro accelerometer that detects two distinct acceleration values. The design model is a micro-electro-mechanical system (MEMS). The device is a monolithic structure comprising two compliant bistable mechanisms that connect series to generate two-level distinct positions in the motion. The accelerometer operates based on the capacitive spring-mass system. The comb drives assist in indicating the acceleration value with the sensing capacitor utilized to archive the electrical output signal. The accelerometer works only in a direction; therefore, the reset mechanism is introduced to install the initial stage of the device. The electrothermal V-beam is employed to move the mechanism from the first or second level position to the initial position. The testing mechanism is also presented, which has the function of examining the operation of the device. The nonlinear behavior of the mechanism is investigated and analyzed by finite element method. The device can detect the two accelerations, 13g and 48g. The simulation of the mechanism assists in evaluating the operation of the mechanism. The accelerometer has great potential for energy-saving applications.

Keywords: Acceleration · Comb drives · Electrothermal · Nonlinear mechanism

1 Introduction

Accelerometers are used to detect different levels of acceleration based on inertia. They are used in fall detection [1], rotary switches [2], etc. The accelerometer measures a variety of acceleration values to protect against previous life events [3]. Accelerometers are used in everyday sports activities as well as in patient monitoring [4]. Industrial systems use accelerometers for precise control of high-speed actuators and controllers [5]. Designing accelerometers with different operating values can provide an economical and highly reliable solution.

Many latching mechanisms can be implemented to design the acceleration sensors. Sun et al. [6] introduce an accelerometer constructed by an array of tether notches

A. K. Pandey et al. (Eds.): MAMM 2022, 126, pp. 136–146, 2023.
https://doi.org/10.1007/978-3-031-20353-4_9

and a cantilever beam. The sensors are capable detect the acceleration from 60g to thousand g value. An acceleration sensor with two cantilever beams hooked together in the structure introduced by Ciarlo [7] can detect high acceleration. Currano et al. [3] also utilize the latching mechanism in designing the impact sensor. The advantage of their mechanism is monitoring the acceleration for an extended period during the available power. Reddy et al. [8] demonstrated a shock sensor with high reliability and saves energy. This mechanism is used to measure many levels of acceleration by the interlock of the seismic mass and the latching mechanism. The development of the latching mechanism is the ratcheting mechanism that firmly keeps the interlock. Mehner et al. [9] introduce an accelerometer work in multi-value based on the hook of the mass and the ratcheting structure. However, the challenge of applying the latching mechanism is the damage from the impact contact and adhesion between the parts.

Compliant mechanisms are the innovative structures that fabricate a monolithic part, jointless, and reduce the resistance during the operation. Accelerometers also employed these mechanisms in design. Bistable mechanisms have two stable positions within their range of motion, which have applications in many designs such as switches, valves, and sensors [10–12]. The mechanism demands no energy input to remain in positions at these stages. Furthermore, the mechanisms also achieve high response in operation and improve the accuracy in positioning. Due to the advantages of the bistable mechanism, many researchers have attracted this structure in the design of accelerometers. Todd et al. [13] introduce the threshold accelerometer, which is comprised of a bistable mechanism. The mass of the structure can be modified to detect much acceleration value. The integration of the bistable mechanism and the radio frequency identification chip assist the readout and increase the precision of the sensor [14]. An accelerometer detects the acceleration from a compliant bistable mechanism is illustrated by Zhao et al. [15]. The application of magnetic force improves repeatability. A compliant tristable mechanism is also employed to design the threshold acceleration sensor [16]. However, a microdevice is considered to evaluate the characteristic of the structure.

This study uses a new application of the two bistable mechanisms to detect the two acceleration. The sensing of the mechanism is the electrostatic mechanism, and the reset structure is presented. The finite element method is employed to characterize the behavior of the mechanism. The response behavior is also investigated.

2 Design

Figure 1 displays the overall structure of the two-level accelerometer. The coordinate system is demonstrated in this figure. In this design, the seismic mass connects the anchor by the springs. Two comb drive is connected to pad a and pad b, which anchor in the substrate. The comb drive drives the sensing capacitor. When the seismic mass receives the acceleration, the mass translates in the y-direction and brings the comb drive to move together. When the movable comb drives the mass movement, the capacitor of the fixed comb drive change, and the voltage output occur. The detect signal is transferred to the readout system. The testing capacitor is an actuator that has the function of pulling the seismic mass in the y-direction to check the measurement of the capacitance of comb drives in the sensing mechanism. The reset mechanism is the electrothermal actuator, which employs the V-beam shape in operation.

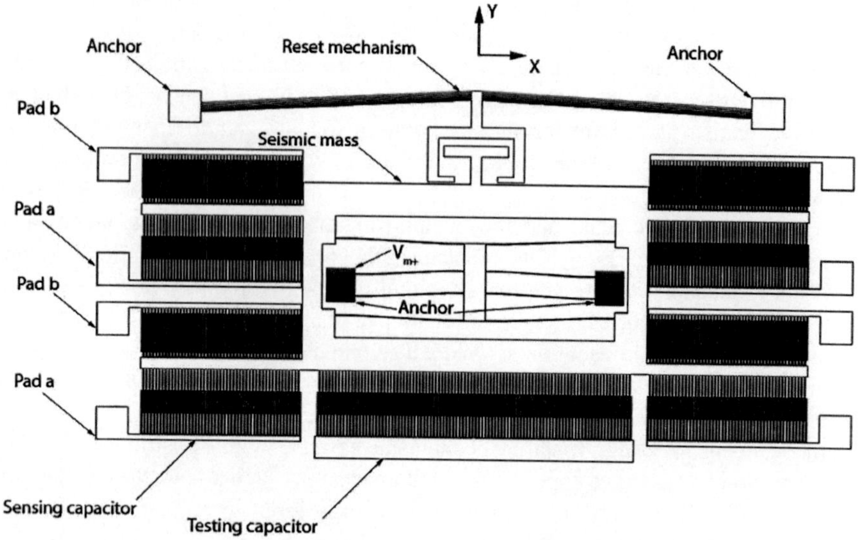

Fig. 1. Design of the two-level accelerometer.

Figure 2a reveals the details of the seismic mass without comb drive. The mass is fixed to the substrate by two anchors. Spring 2 connects the anchors and the support mass. Spring 1 links the support mass and the seismic mass. Spring 1 and spring 2 are flexible beams, which are deformed based on the compression and tension of the beam. The reset hook is also introduced in the T-shape profile. The reset mechanism allows to grip and pull the seismic mass to the initial stage. Figure 2a also displays the seismic mass in the initial stage. When the acceleration is applied to the seismic mass in the y-direction, the mass moves down and is stable in the position. Because of the nonlinear behavior of the flexible beam, spring 2 is deformed and brings the support mass move together. Figure 2b is the position of the seismic mass when it archives the first level acceleration. Continually increasing the acceleration to the second level value, spring 1 is compressed and released, making the seismic mass move down in the y-direction and stable in the position following the Fig. 2c.

The reset mechanism is demonstrated in Fig. 3. The electrothermal actuator [17] is employed to drive this mechanism. The V-shape beams are attached to the gripper arm. The gripper arm has the function of gripping and pulling the reset hook in the y-direction, as shown in the figure. Two anchors are fixed to the substrate. When the accelerometer operates, the seismic mass stays in the first or second-level positions. The reset mechanism is active to install the seismic mass to the initial position. The electric current is applied to the anchors. The current goes through the V beams following the directions shown in the figure, increasing the beam's temperate. Then, the beams bent and created the force to pull the gripper arm move in the y-direction, which is the direction demonstrated in the figure. The gripper arm keeps the seismic mass and moves to the initial stage. Belong to the position of the seismic mass, the current applied to the reset mechanism is different. The current is higher than the first level position when the seismic mass is in the second value position. After the installation is finished, the current is interrupted.

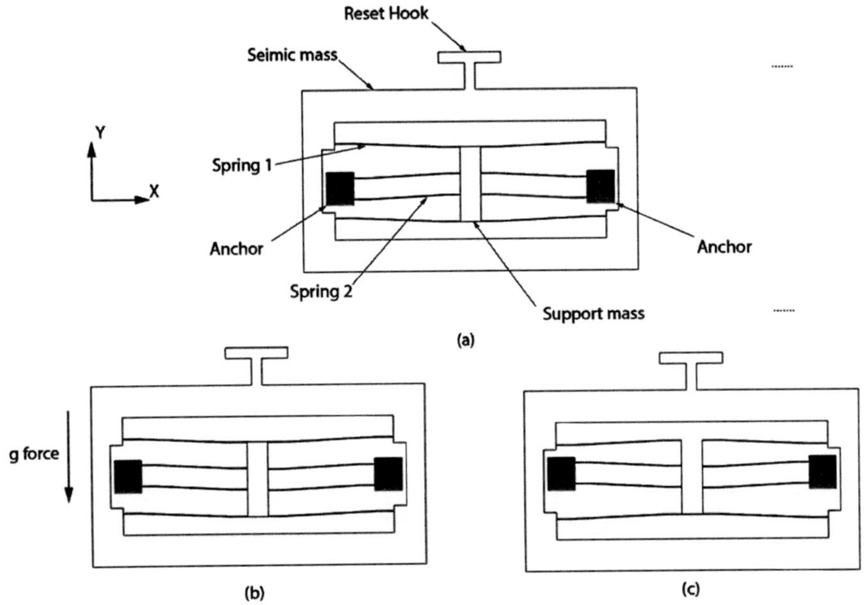

Fig. 2. Seismic mass structure. **a**) Initial position, **b**) first level position, **c**) second level position.

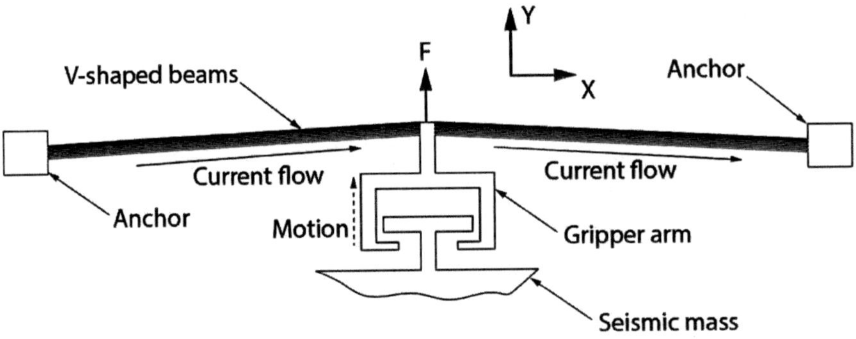

Fig. 3. Electothermal reset mechanism.

Figure 4 is the large view of the sensing mechanism. Movable comb drive attaches to the seismic mass and moves together. The upper comb drive is fixed and connected to pad a where electric wire contact. Besides, pad b links to the lower comb drive and fixes to the substrate. Figure 4a shows the movable comb drive staying in the initial position. The capacitance C_1 appears between the upper comb drive and movable comb drive. C_2 is the capacitance between movable comb drive and lower comb drive. The movable comb drive is close to the upper comb drive in the initial stage, so the Cu occurs. The C_1 has the highest value, and C_2 has the lowest value because of the overlap area of the fingers. When the seismic mass move to the first level position, as shown in Fig. 4b, the movable comb drive goes far from the upper comb drive, and C_U disappears. C_1 and C_2

value also change because the area of plate overlap decrease in the upper comb drive and increase in the lower comb drive. Figure 4c depicts the movable comb drive in the second-level position. In this case, the movable fingers close to the lower comb drive and appear the capacitance C_L. The C_1 archives the smallest value because the overlap area is miniature. Based on the difference of the C_1, C_2, C_U, and C_L, the sensing mechanism measures the device's acceleration value.

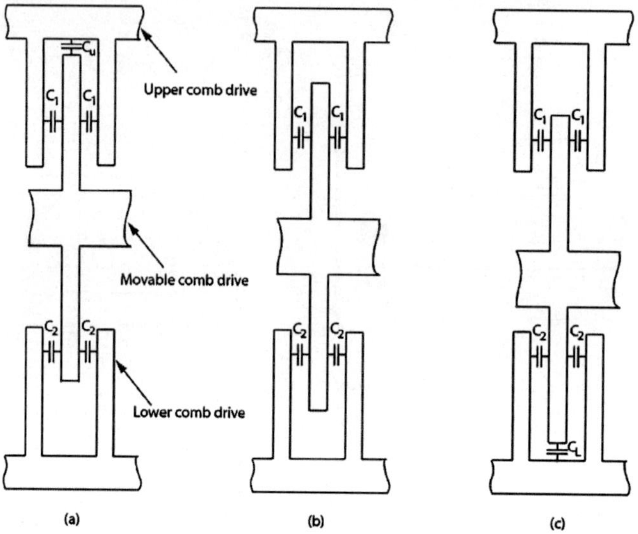

Fig. 4. Capacitance sensing mechanism. **a**) Initial stage, **b**) first-level acceleration, **c**) second-level acceleration.

The equation for capacitance is given:

$$C = \frac{\varepsilon_0 \varepsilon_r A}{d} \tag{1}$$

C is the capacitance value, ε_0 is the dielectric constant of the material. In this case, Nickel is the material of the comb drive. ε_r is the dielectric constant of free space with air is the material of this case. d is the gap between two fingers of the comb drive. The gap of C_1 and C_2 is constant during the operation. A is the overlap area of the comb drive, area values will change in every case of the seismic mass movement.

Figure 5 zooms the testing mechanism of the sensor. The mechanism is the fixed capacitor, anchors to the substrate. When the current applies to the fixed capacitor, the capacitor attracts the movable capacitor in the seismic mass. The seismic mass moves down, and the sensing has the output signal. The testing mechanism assists in checking the sensor still found in the experiment.

3 Analysis and Discussions

The finite element method is employed to predict the behavior of the device. Commercial Abaqus software is used for analysis. Due to the symmetry of the design model, a

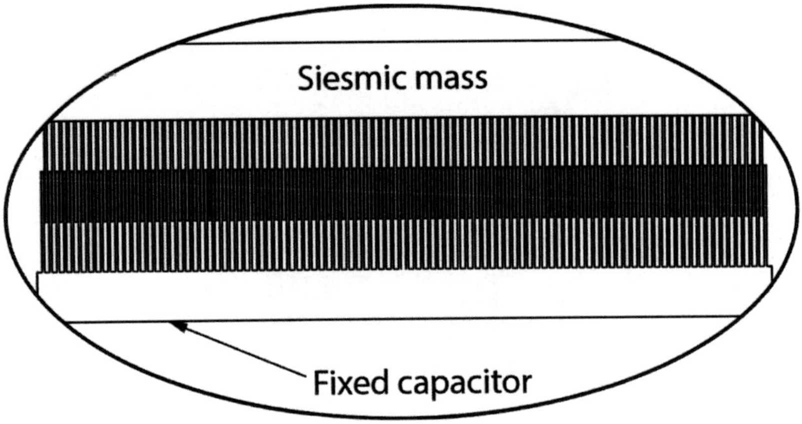

Fig. 5. Testing mechanism of the sensor

half model is used for analysis. Figure 6 shows the 3D mesh model of the device. A coordinate system is shown in this figure. The anchor is fixed, and the symmetry plane also demonstrates the half model in the analysis. Nickel is applied to the material of the accelerometer due to the high tensile strength and good at electrical properties [18]. The Young modulus is 200 GPa, and the Poisson ratio is 0.25. The density of Nickel is 8.9 Mg/m^3. In order to investigate the nonlinear behavior of the spring, static analysis is applied. The reaction force in the anchor is acquired, and the seismic is put the displacement in the y-direction. Figure 6a depicts the seismic in the original position, Fig. 6b is the second-level position of the mechanism.

Figure 7 illustrates the deflection of the springs. Spring 1 and spring 2 are compressed and tense to make the seismic mass archives two-level positions in operation. As shown in the figure, spring 1 and spring 2 are flexible beams and deform to seismic mass reaches the second level position. The stress of the beams is also investigated in this analysis. The maximum stress focus in the fixed end of the beams with the value is 28 Mpa. It is satisfied the Yield strength of the Nickel is 70 Mpa.

Figure 8 shows the force-displacement relation of the sensor. In the operation of the sensor, the force is increased and reaches the maximum value at 0.25 mN. After that, the force decrease to zero, and spring 2 is released, and the seismic mass moves quickly to the first-level position, 117 μm. The force continually increases and archives the second maximum force is 0.35 mN. While spring 1 is compressed. Then the force reduces to zero, spring 2 releases the energy, and the seismic mass jumps quickly to the second-level position at 244 μm. The support mass also forces together. The spring of the system has a nonlinear characteristic. For the opposing force displayed in the figure, the force is not accurate in the world and is unpredicted. The weight of the seismic mass is 178 mg, and the weight of support mass is 34 mg. Based on the maximum forces of the device, the two-level acceleration values are indicated. They are 13g for the first-level value and 48g for the second-level value, respectively.

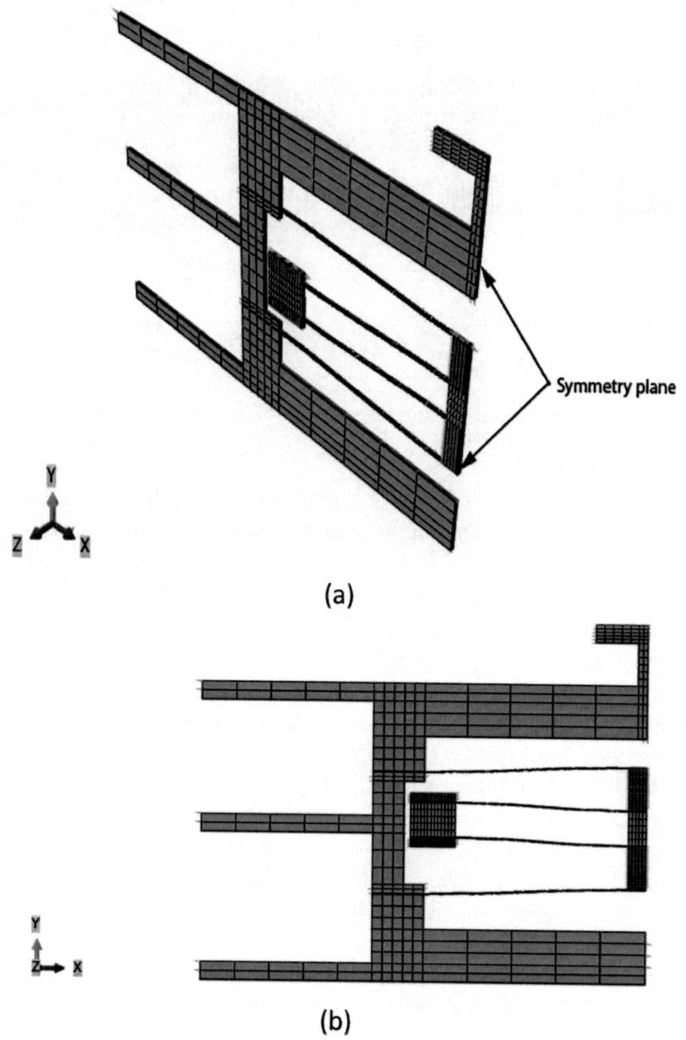

Fig. 6. Meshing model of the compliant bistable mechanism. **a**) Initial stage, **b**) deformed stage.

Figure 9 investigates the rise time of the accelerometer. The sensor cannot respond immediately and will indicate transient responses when subjected to inputs or disturbances. The rise time required for its response to rising from 0 to 100% of its threshold value to a unit-step acceleration input is collected by finite element analyses. The damping ratio used in the analyses is determined by measuring the decay rate of free oscillations of the device. Assume the Rayleigh damping has the value $\beta = 2e{-}4$ and declares in the FEA method. Figure 9a displays the input value of the acceleration signal with a value of 13g. Figure 9b depicts the second-level acceleration input to the device at 48g. The time responses of the device acceleration input signal of 13g and 48g are

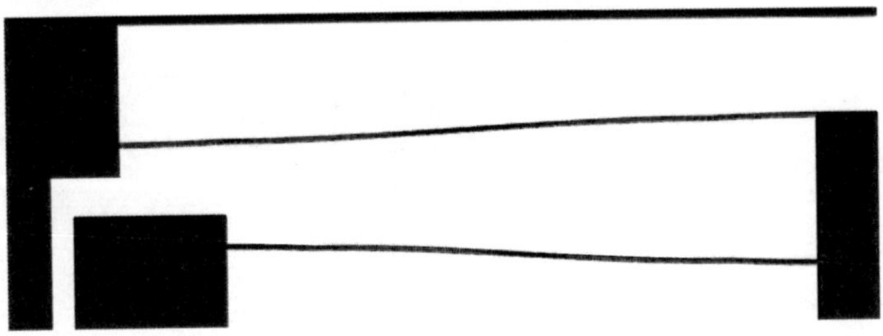

Fig. 7. Deformed of the springs.

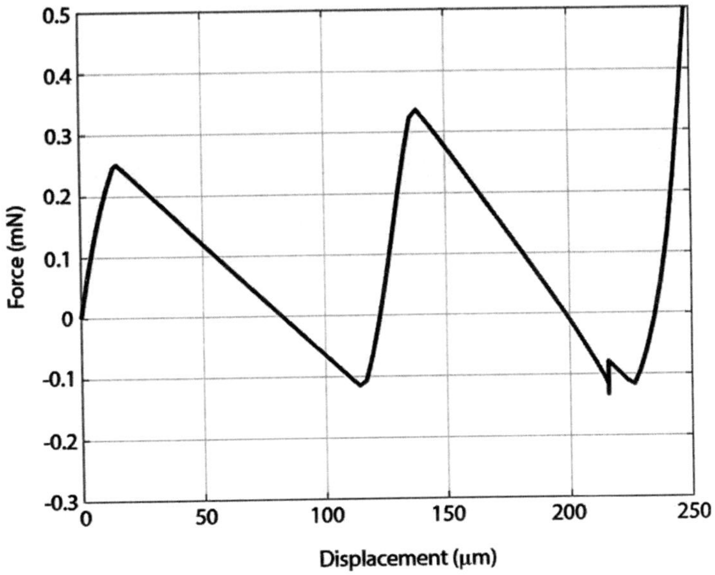

Fig. 8. Force-displacement of the model.

demonstrated in Fig. 9c, d, respectively. The rise time of the acceleration signals of 13g and 48g is 0.125 s and 0.126 s, respectively.

Figure 10 investigates the difference in the static and dynamic of the sensor when it works a sequence. The dashed line shows the displacement of the seismic mass in the static analysis, and the black line shows the value in the dynamic analysis. The damping of the system causes the difference in the displacement. The first level position in dynamic is larger than the static. The displacement is 140 µm compared with the static is 117 µm. The seismic mass required the rise time to reach the stable position. The second level value also needs time to stable, and the displacement is 247 µm for the dynamic.

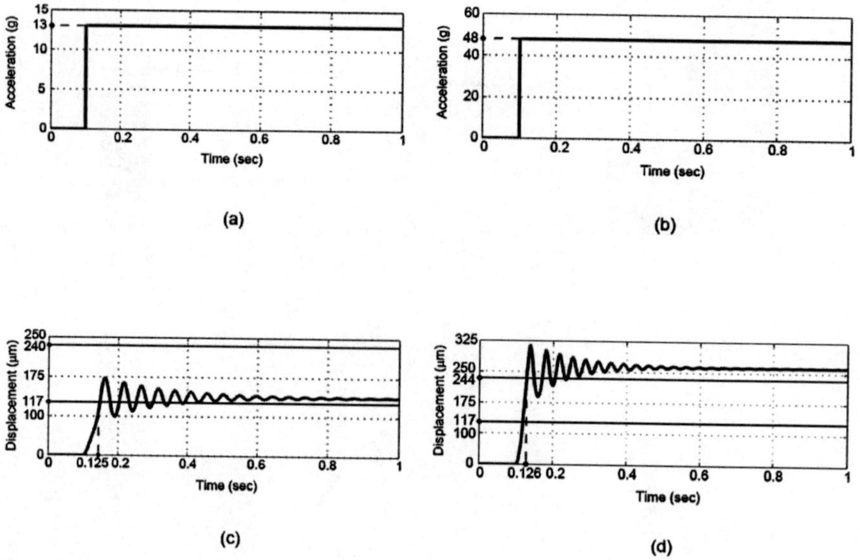

Fig. 9. **a**) Acceleration inputs at 13g, **b**) acceleration input at 48g, **c**) transient response of the acceleration 13g, **d**) transient response of the acceleration 48g.

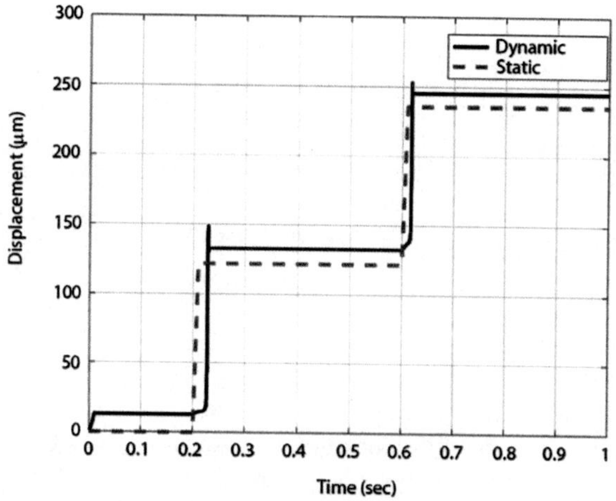

Fig. 10. The comparison of the static and dynamic analysis.

4 Conclusion

This paper introduces a new sensor design, which detects two value acceleration values. The structure includes many mechanisms. The main sensor is the two spring-mass systems with three distinct positions in operation. In order to reset the mechanism, an electrothermal actuator is a design, and the V-shape beams are applied. The sensing capacitor is also presented which the different capacitance values for each level position. The application of comb drives assists in archiving the output signal and support for the testing sensor. The finite element method is used to analyze the system's nonlinear behavior. The displacement of the seismic mass is 117 μm for the first-level position and 244 μm for the second-level position. Based on the analysis, the acceleration for each level is predicted with 13g and 48g values. The time response analysis indicates the rise time of the sensor, around 0.125 s, to reach the first and second level acceleration. The sensors save energy and increase accuracy in the operation.

Acknowledgments. The authors are thankful for support from colleagues at Faculty of Mechanical Engineering, Industrial University of Ho Chi Minh City, Vietnam and Faculty of Mechanical Engineering, Ho Chi Minh City University of Technology and Education, Vietnam.

References

1. Bourke, A.K., O'Brien, J.V., Lyons, G.M.: Evaluation of a threshold-based tri-axial accelerometer fall detection algorithm. Gait Posture **26**, 194–199 (2007)
2. Kruglic, E., Pister, K.: Bistable MEMS relays and contact characterization. In: Proceedings of the Solid State Sensors and Actuators Workshop, pp. 333–337 (1998)
3. Currano, L.J., et al.: Latching ultra-low power MEMS shock sensors for acceleration monitoring. Sens. Actuator A Phys. **147**, 490–497 (2008)
4. Veltink, P.H., Bussmann, H.B.J., Vries, W., Martens, W.L.J., Lummel, R.C.V.: Detection of static and dynamic activities using uniaxial accelerometers. IEEE Trans. Rehabil. Eng. **4**(4), 375–385 (1996)
5. Stoesslein, M., Axinte, D.A., Guillerna, A.B.: Pulsed laser ablation as a tool for in-situ balancing of rotating parts. Mechatronics **38**, 54–67 (2016)
6. Sun, X.Q., Zhou, S., Carr, W.N.: A surface micromachined latching accelerometer. In: Proceedings of 1997 International Conference on Solid-States Sensors and Actuators, pp. 1189–1192 (1997)
7. Ciarlo, D.R.: A latching accelerometer fabricated by the anisotropic etching of (110) oriented silicon wafers. J. Micromech. Microeng. **2**, 10–13 (1992)
8. Reddy, R.R., Komeda, K., Okamoto, Y., Lebrausseur, E., Higo, A., Mita, Y.: A zero-power sensing MEMS shock sensor with a latch-reset mechanism for multi-threshold events monitoring. Sens. Actuator A Phys. **295**, 1–10 (2019)
9. Mehner, H., Weise, C., Schwebke, S., Hampl, S., Hoffmann, M.: A passive microsystem for detecting multiple acceleration events beyond a threshold. Microelectron. Eng. **145**, 104–111 (2015)
10. Tanner, J.D., Jensen, B.D.: Power-free bistable threshold accelerometer made from a carbon nanotube framework. Mech. Sci. **4**(2), 397–405 (2013)
11. Zhao, J., Yang, Y., Fan, K., Hu, P., Wang, H.A.: Bistable threshold accelerometer with fully compliant clamped-clamped mechanism. IEEE Sens. J. **10**(5), 1019–1024 (2010)
12. Vilorio, C., Stark, B., Hawkins, A.R., Frogget, K., Jensen, B.: Stress relaxation insensitive designs for metal compliant mechanism threshold accelerometers. Sens. Bio-Sens. Res. **6**, 33–38 (2015)
13. Todd, B., Phillips, M., Schultz, S.M., Hawkins, A.R. Jensen, B.D.: RFID threshold accelerometer. IEEE Autotestcon, 55–59 (2008)
14. Todd, B., Phillips, M., Schultz, S.M., Hawkins, A.R., Jensen, B.D.: Low-cost RFID threshold shock sensors. IEEE Sens. J. **9**(4), 464–469 (2009)

15. Zhao, J., Gao, R., Yang, Y., Huang, Y., Hu, P.A.: Bidirectional acceleration switch incorporating magnetic fields based tristable mechanism. IEEE ASME Trans. Mechatron. **18**, 113–120 (2013)
16. Tran, H.V., Ngo, T.H., Tran, N.D.K., Dang, T.N., Dao, T.P., Wang, D.A.: A threshold accelerometer based on a tristable mechanism. Mechatronics **53**, 39–55 (2018)
17. Lee, C.K., Wu, C.Y.: Study of electrothermal V-beam actuators and latched mechanism for optical switch. J. Micromech. Microeng. **15**(1) (2004)
18. Pratap, R., Arunkumar, A.: Material selection for MEMS devices. Indian J. Pure Appl. Phys. **45**, 358–367 (2007)

Microchannel Induced Tailoring of Bandwidth of Push-Pull Capacitive MEMS Accelerometer

Nidhi Gupta[1,2](✉), Siva Rama Krishna Vanjari[2], and Shankar Dutta[1]

[1] Solid State Physics Laboratory, DRDO, Lucknow Road, Timarpur, Delhi 110054, India
nidhi.hcst.g@gmail.com

[2] Department Electrical Engineering, Indian Institute of Technology Hyderabad, Hyderabad 502285, India

Abstract. High quality capacitive accelerometers with appreciable operation bandwidth are being employed in different high-end defence applications. This paper discusses a novel concept of introducing microchannels in the push-pull capacitive accelerometer structure to control the damping arising due to squeeze film effect. The width and depth of the microchannels are varied to study the damping in the accelerometer structure. The damping factor of accelerometer structure is reduced from 50.2 to 2.51 and its corresponding 3 dB bandwidth is improved from 62.24 to 1.26 kHz due to the introduction of the microchannels.

Keywords: Accelerometer · Push-pull capacitor · Microchannel · Squeeze film damping · Bandwidth

1 Introduction

High performance capacitive MEMS accelerometers are being widely used for inertial navigation applications and other defence applications [1–6]. In capacitive accelerometer, change in capacitance can be sensed by change in gap or change in area approach. Among them, change in gap approach is commonly used due to the ease of fabrication. Higher sensitivity in the capacitive accelerometer can be realized either by increasing the overlap area of the capacitor plates or by reducing the air gap between them [5, 6].

However, increasing the area or reducing the gap can influence the squeeze film damping arising due to the trapped air (or gas) between the capacitor plates through the relations [7, 8]:

$$c = 0.42\mu L^2/d^3 \tag{1}$$

where, the value of air viscosity (μ) at atmospheric pressure is 1.87×10^{-5} Pa-sec (at room temperature); c = damping coefficient; L = the proof-mass dimensions; and d = the gap between capacitive electrodes. From Eq. (1), it is clear that the value of c is bound to be high for large area to gap ratio.

Corresponding damping factor can be defined as (ς) the ratio of damping coefficient (c) to critical damping coefficient ($c_c = 2M\omega_n$). Here, M = mass of the vibrating structure (proof-mass), ω_n = the resonant frequency.

A. K. Pandey et al. (Eds.): MAMM 2022, 126, pp. 147–151, 2023.
https://doi.org/10.1007/978-3-031-20353-4_10

It is well-known that the value of damping factor in the accelerometer structure plays a very crucial role in determining the operational bandwidth of the device. To achieve an substantial bandwidth (> 200 Hz) in capacitive accelerometer, the damping factor should not be very high (preferably < 20) [5, 6].

To counter the challenges of high damping factor in MEMS devices, researchers have either introduced large number of tiny holes (~6–7 μm diameter size) in the proof mass [5–10], or employed reduced air pressure (vacuum) during wafer level packaging of push-pull type capacitive accelerometer structure [11]. Fabrication of the tiny holes in the thick proof-mass requires deep reactive ion etching (DRIE) system; whereas, the dynamic characteristic of the capacitive accelerometer may alter over time (reliability issue) due to the leakage induced change in vacuum level. As an alternative to the hole-structures or vacuum packaging, microchannels may also be introduced in the accelerometer structure to control the squeeze film damping between the push-pull type capacitor electrodes.

This paper presents a novel concept of introducing microchannels in the push-pull capacitive accelerometer structure to control the damping arising due to squeeze film effect. The width and depth of the microchannels are varied to study the damping effect. Dynamic behavior of the push-pull capacitive accelerometer structure is studied and operation bandwidth (3dB) of the accelerometer structure is estimated.

2 Device Structure

Design and simulation of all silicon process technology-based MEMS capacitive accelerometer is presented in this paper for 30g acceleration range. The Z-axis sensitive MEMS capacitive accelerometer structure comprises of a silicon proof mass (1500 μm × 1500 μm × 400 μm) over hanged by four narrow beams of dimension: 1000 μm × 350 μm × 25 μm as shown by Fig. 1a. The structure is sandwiched between two fixed electrodes having a gap of 2 μm. Five microchannels (1500 μm × 30 μm × 30 μm) are also introduced to minimize the damping effect of structure as shown in Fig. 1b. The structure is sandwiched between two fixed electrodes having a gap of 2 μm.

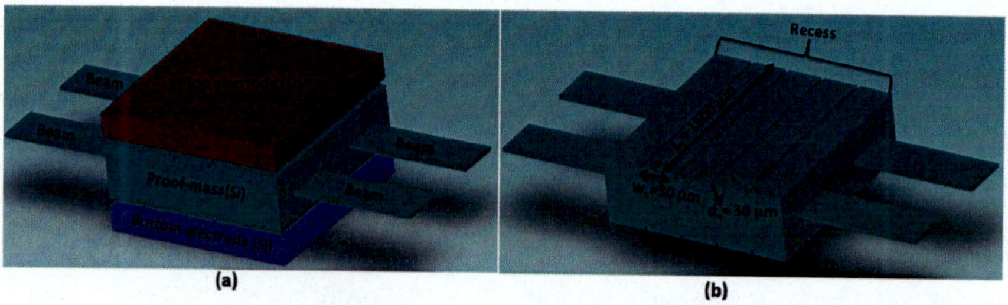

Fig. 1. (**a**), (**b**) Schematic of capacitive MEMS accelerometer

This structure consists of following features: (1) improved linearity and high temperature stability due to push-pull approach; (2) bridge type structure having less cross axis sensitivity and high stability; (3) all silicon process technology to avoid thermal mismatch and residual stress issues.

3 Results and Discussions

The performance of accelerometer structure is evaluated by FEM analysis based software (Coventorware 10.1). The modal patterns of the accelerometer structure are simulated to visualize the favourable deflection modes. The first three modal patterns of the structure are shown in Fig. 2. The first three modal frequencies of the accelerometer structure are found to be 6.6 kHz (1st mode), 7.6 kHz (2nd mode) and 18.5 kHz (3rd mode) respectively. Deflection and Von-mises is simulated with input acceleration 30g as shown by Fig. 3. Figure 4 shows the variation of change in capacitance with applied acceleration from (0 to 30g). The change in capacitance is found to be 1.83551 pF at 30g. Thus, corresponding scale factor sensitivity is found to be 61.1 fF/g.

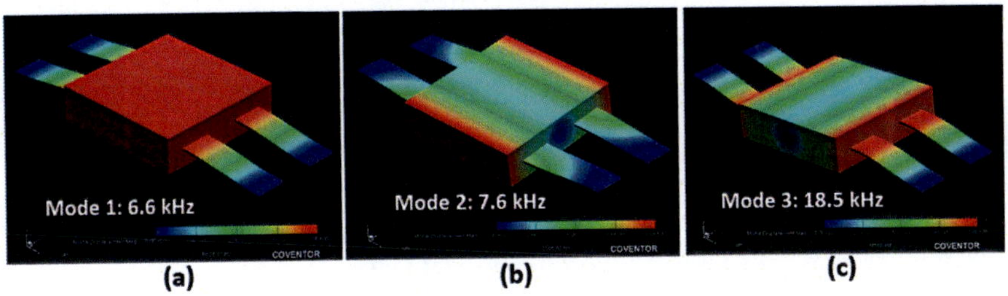

Fig. 2. (**a**)–(**c**) Modal patterns of accelerometer

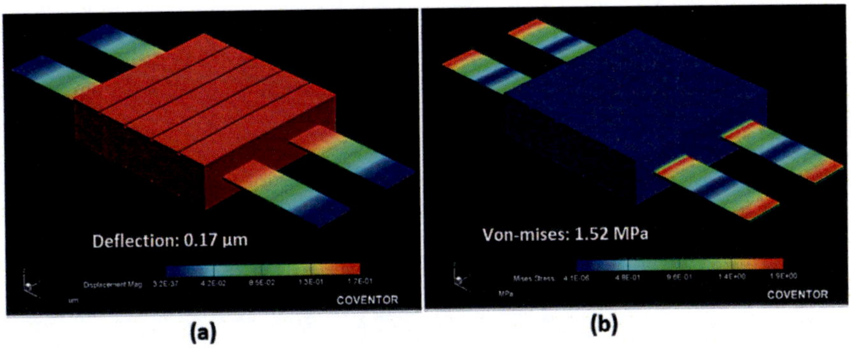

Fig. 3. (**a**), (**b**) Deflection and Von-mises at 30g

Frequency response of the accelerometer greatly depends on the extent of damping arises due to the squeeze film effect of the trapped air between the movable plate and fixed plate. The squeeze film damping coefficient and its corresponding damping factor has been estimated. At atmospheric pressure, the accelerometer structure exhibits a damping factor ($\zeta = c/2m\ \omega_n$) of ~ 50.2 having air gap of 2 μm. To reduce the damping effect, channel type recess structure is introduced in thick proof-mass instead of introducing through hole. Introduction of microchannel in thick proof mass structure has reduced the damping factor to ~ 3.05. Hence the bandwidth is found to be drastically increasing from 62.24 Hz to 1 kHz as shown in Fig. 5b. However, the 3 dB bandwidth of structure

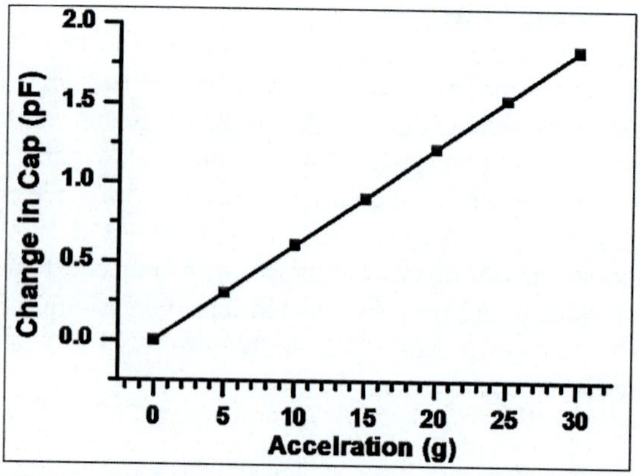

Fig. 4. Variation of change in capacitance with acceleration

can further be improved by increasing the width and depth of microchannel. Figure 5b, c shows the variation of damping factor and 3 dB bandwidth of structure with microchannel width and depth. The corresponding damping factor reduces from 3.05 to 2.51and 3 dB bandwidth is found to be increasing from 1 to 1.26 kHz.

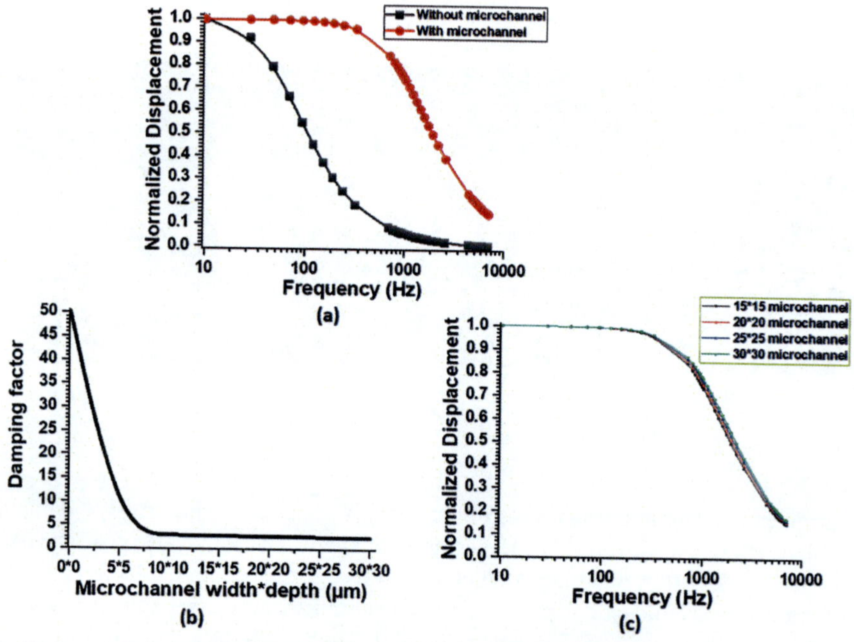

Fig. 5. **(a)** 3 dB bandwidth of structure with and without microchannel; **(b)**, **(c)** variation of damping factor and 3dB bandwidth with microchannel width and depth

4 Conclusion

In this work, microchannels are introduced to tune the operation bandwidth in push-pull type capacitive accelerometer structure of futuristic high-performance defence applications. The width and the depth of the microchannels are varied to study its effect on the dynamic behavior of the accelerometer structure. The Z-axis sensitive (30g) MEMS capacitive accelerometer structure exhibited a scale factor sensitivity of 60.1 fF/g. The damping factor arising due to the squeeze film damping between the capacitor plates are simulated based on the FEM simulation. The operation bandwidth of the accelerometer (with and without microchannels) is also simulated. The estimated damping factor is diminished from 50.2 to 2.51 and the corresponding 3 dB bandwidth of structure is increased from to 60.24 Hz to 1.26 kHz.

Acknowledgment. The authors acknowledge Director SSPL for his continuous support and permission to publish this work.

References

1. Aydemir, A., Terzioglu, Y., Akin, T.: A new design and a fabrication approach to realize a high performance three axes capacitive MEMS accelerometer. Sens Act A **244**, 324–330 (2016)
2. Mohammed, Z., Dushaq, G., Chatterjee, A., Rasras, M.: An optimization technique for performance improvement of gap-changeable MEMS accelerometers. Mechatronics **54**, 203–216 (2018)
3. Xu, W., Tang, B., Xie, G., Yang, J.: An all-silicon double differential MEMS accelerometer with improved thermal stability. IEEE Sens., 1–4 (2018)
4. Kalaiselvi, S., Sujatha, L., Sundar, R.: Analysis of damping optimization through perforations in proof-mass of SOI capacitive accelerometer. Analog Integr. Circ. Sig. Process **102**(3), 605–615 (2019). https://doi.org/10.1007/s10470-019-01560-5
5. Zhou, X., Che, L., Wu, J., Li, X., Wang, Y.: A novel sandwich capacitive accelerometer with a symmetrical structure fabricated from a D-SOI wafer. J. Micromech. Microeng. **22**, 085031 (2012)
6. Dutta, S., et al.: Lapping assisted dissolved wafer process of silicon for MEMS structures. J. Mater. Sci. Mater. Electron. **25**(4), 1984–1990 (2014). https://doi.org/10.1007/s10854-014-1833-2
7. Xu, W., et al.: Design and fabrication of a slanted-beam MEMS accelerometer. Micromachines **8**(3), 77 (2017)
8. Hu, Q., Li, N., Xing, C., Mei, W., Sun, P.: Design, fabrication, and calibration of a full silicon WLP MEMS sandwich accelerometer. In: International Conference ICEPT, Shanghai, China, pp. 919–933 (2018)
9. Bao, M.: Analysis and Design Principles of MEMS Devices. Elsevier Publication (2005)
10. Mo, Y., Du, L., Qu, B., Peng, B., Yang, J.: Squeeze film air damping ratio analysis of a silicon capacitive micromechanical accelerometer. Microsyst. Technol. **24**(2), 1089–1095 (2017). https://doi.org/10.1007/s00542-017-3464-z
11. Dutta, S., Saxena, P., Panchal, A., Pal, R., Jain, K.K., Bhattacharya, D.K.: Effect of vacuum packaging on bandwidth of push-pull type capacitive accelerometer structure. Microsyst. Technol. **24**, 4855–4862 (2018)

A Comparison of Ring and Disk Resonator Gyroscopes Based on Their Degenerate Eigenmode Shapes

Kandula Eswara Sai Kumar[1], Karthik Raveendranath[2], Sudhanshu Sekhar[1], and G. K. Ananthasuresh[1](✉)

[1] Indian Institute of Science, Bengaluru, India
suresh@iisc.ac.in
[2] ISRO Inertial Systems Unit, Thiruvananthapuram, India

Abstract. Micromachined gyroscopes are beginning to compete with their macro-scale counterparts in terms of performance. This has been made possible because of a remarkable design change that has come to be known as a *Disk Resonator Gyroscope* (DRG). The design of a DRG can be thought of as an extension of the Ring Resonator Gyroscope (RRG) or vibratory ring gyroscope (VRG). After experimenting with various ring configurations and their suspensions attached to the ring, research has now moved to multiple configurations of DRGs. Central to all these designs are the degenerate eigenmode shapes: a pair of mode shapes for the same frequency. As one mode shape is used for resonant electrostatic actuation and another for sensing the energy transferred to it due to Coriolis acceleration, mode shapes decide the change in capacitance per unit angular rate. The objective of maximizing the capacitance change is to be balanced with the performance constraints on keeping the eigenfrequency low (but not too low) and quality factor high. In this paper, we compare different designs of ring and disk resonators from the literature in terms of eigenfrequency and mode shapes. We conclude that it is beneficial to have the mode shape of a gyroscope design close to the elliptical mode shape of a free ring.

Keywords: Micromachined gyroscope · Mode shape · Capacitive sensing

1 Introduction

Micromachined gyroscopes are preferred in aerospace, automotive, and consumer products due to their low weight, small size, and low cost [1]. Beginning with the tuning fork and dual-mass designs, several configurations were explored. All of them work on the principle of Coriolis acceleration and depend on two degenerate mode shapes belonging to the same eigenfrequency. One mode is used to set the device into resonant vibration. This is often done using the electrostatic force. The other mode is used to sense the motion that results due to the transfer of energy from the first mode when the base, on which the gyroscope is mounted, begins to rotate. Apart from these, ring vibratory gyroscopes have also received special attention from researchers because of their symmetry and because they have improved resistance to external vibrations and shocks [2].

A. K. Pandey et al. (Eds.): MAMM 2022, 126, pp. 152–161, 2023.
https://doi.org/10.1007/978-3-031-20353-4_11

Just like the Foucault's pendulum (Fig. 1a), and a hemispherical bowl (a wineglass, Fig. 1b) [3], a ring or a disk also have two degenerate mode shapes (Fig. 1c) [4]. A free ring needs to be suspended by attaching it to a central post where it is anchored to the substrate (Fig. 1d). The same is true with a disk (Fig. 1e). Different configurations of ring resonators (Fig. 1f) and disk resonators (Fig. 1g) are shown. In ring resonator gyroscopes, the central post needs to be connected with the ring. In disk resonators, holes are made in the disk to make it flexible by introducing concentric rings attached to one another in various ways using short connecting beams. This structure helps increase the number of sense electrodes and thereby the change in capacitance (ΔC). The quality factor (Q-factor) of a DRG is further improved by adding several lumped masses to it (stiffness-mass decoupling technique), but it reduces the available capacitance area [5, 6]. A DRG also has structural imperfections that create high relative frequency errors after fabrication. This led to the development of a new gyroscope that is based on a honeycomb structure. The honeycomb-like disk resonator gyroscope (HDRG) has high immunity to fabrication errors [7], but the Q-factor is much lower than the DRG [8]. Furthermore, researchers developed a cobweb-like disk resonator gyroscope (CDRG) with linear beams instead of circular beams to reduce fabrication errors [9]. A high Q-factor for CDRG is obtained with the stiffness-mass decoupling technique. Apart from the above, researchers developed a gear-like disk resonator gyroscope (GDRG) and stated that it has high immunity to fabrication errors through computational results [10], but the fabrication trials were not yet done on GDRG. There are also other studies reporting new geometries for gyroscope applications to improve the Q-factor, mechanical sensitivity, and high immunity to the fabrication errors individually. There is scope for a new design that can improve all three of the above-mentioned performance measures. A comparison study among the existing designs provides some insights to lead further development. We limit the comparison criteria to frequency and mode shapes.

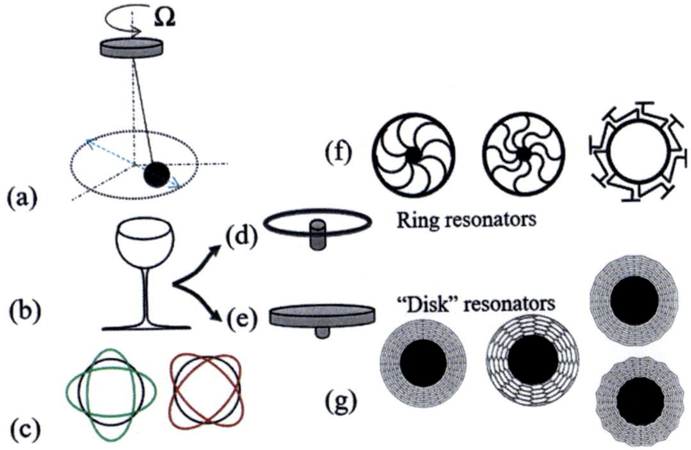

Fig. 1. The working principle and design configurations of resonator gyroscopes (**a**) Focoult pendulum; (**b**) a wineglass with hemispherical shell; (**c**) degenerate mode shapes of a ring; (**d**) a ring gyroscope; (**e**) a disk gyroscope; (**f**) ring resonators; and (**g**) disk resonators.

Unlike previous works, we analyzed the gyroscope designs by calculating the eigenfrequency that contains stiffness information; and the change in sense capacitance based on the eigenmodes. Relative sense capacitance ($\Delta C / C_0$) is defined as the ratio between the change in sense capacitance (ΔC) and the base capacitance (C_0) before the excitation. These two are tunable by geometric design. Furthermore, we feel that the total displacement of a point on a ring is a good measure of the change in capacitance. As shown in Fig. 2, the total displacement, $u_t = \sqrt{u_r^2 + u_\theta^2}$, can be plotted for all points, where u_r is the radial displacement and u_θ is the tangential displacement. The total displacement is plotted for the two degenerate mode shapes of a free ring in Fig. 2. They are perfect ellipses for a free ring. For all other designs, ring or disk resonators, they will be distorted ellipses. The distortion plays a role in change in capacitance. Hence, the total displacement plotted against θ is a good measure for comparison. It helps to see the change in gap between corresponding points on the ring and the mode shape. We assumed a certain modal amplitude, which will be proportional to the angular rate in reality and use that to compute the change in capacitance. As shown in Fig. 2, we consider the displaced point on the mode shape and use a parallel-plate approximation for computing the capacitance in the changed configuration.

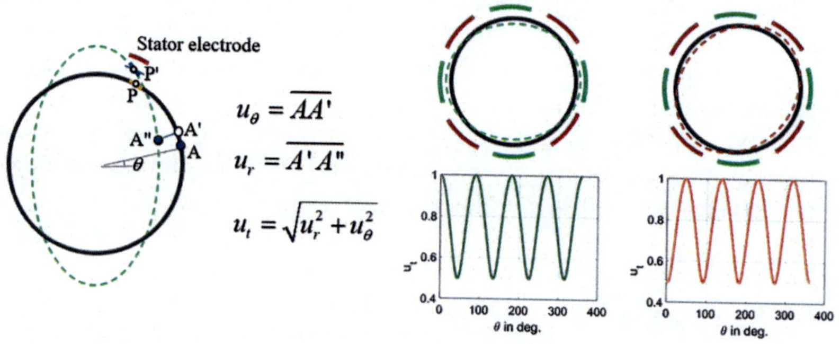

Fig. 2. Total displacement of a point on the mode shape of a free ring for the two degenerate mode shapes. This is compared with considered designs to quantify the deviation from a perfect ellipse for all other designs. The method of computing the changed capacitance is also illustrated pictorially.

2 Methodology

Five different designs shown in Fig. 3 as RRG, DRG, HDRG, CDRG, and GDRG are selected to compare their performance. Further details of these are given in Table 1. All five gyroscopes are surrounded by 16 external stator electrodes, out of which four are used for driving, four are used for sensing, and the remaining eight are for tuning. Additionally, DRG, HDRG, CDRG, and GDRG have inner electrodes in between the concentric rings. For fair comparison, the five gyroscopes have the same structural parameters as follows: outer diameter = 8 mm, in-plane ring width = 10 μm, spoke width = 10 μm, number of concentric rings = 9, spoke length = 200 μm, out-of-plane thickness = 150 μm, and initial gap between the static electrode and the moving gyroscope structure = 10 μm.

The eigenfrequency and relative sense capacitance are computed by COMSOL Multiphysics 6.0 and MATLAB. The following procedure is used for computing the relative sense capacitance in a MATLAB script, using parallel-plate capacitance assumption:

(a) The eigenmode is normalized to a maximum displacement of 1 μm to obtain the normalized eigenmode.
(b) The sense capacitance is calculated from the normalized eigenmode with the reference capacitive gap of 10 microns (see also Fig. 2)

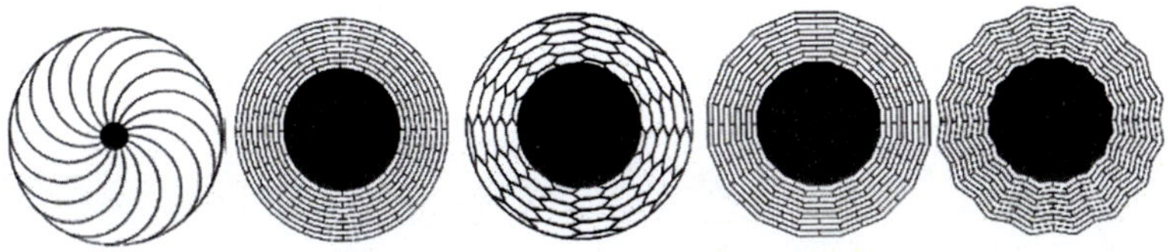

Fig. 3. Five selected gyroscopes for the comparative study of eigenfrequency and relative sense capacitance; from left to right: ring resonator gyroscope (RRG) with semi-circular spokes, disk resonator gyroscope (DRG), honey-comb disk resonator gyroscope (HDRG), cob-web disk resonator gyroscope (CDRG), and gear-like disk resonator gyroscope (GDRG).

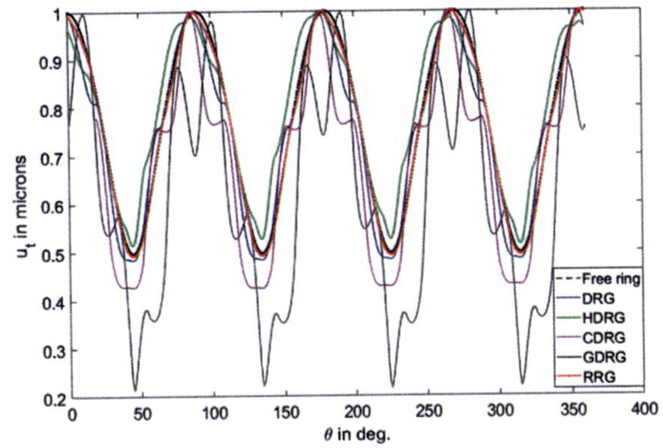

Fig. 4. u_t versus θ plot for the outer ring of the selected five gyroscopes.

Table 1 shows the stator electrodes for all five designs in the second column. In the third and fourth columns, the two degenerate mode shapes are shown. As stated earlier, except in RRG, the others have stator electrodes inside the concentric rings. For the sake of clarity of illustration, the static electrodes are shown only for one sector. We draw attention to the distortion of the mode shapes to notice how much they deviate from an ellipse. We presume that the closer the mode shape is to the ellipse the better the performance. It need not be true. Ideally, the capacitance change needs to be computed. Doing it using parallel-plate assumption is not accurate, but it does give an indication.

Table 1. Five different gyroscopes along with static electrodes are presented in the first two columns respectively. The static electrodes which are in green colour are used for actuating, red colour is used for sensing and blue colour are for tuning. Third and fourth columns contain the two degenerated modes.

Type	*Schematic with stator electrodes*	*Mode 1*	*Mode 2*
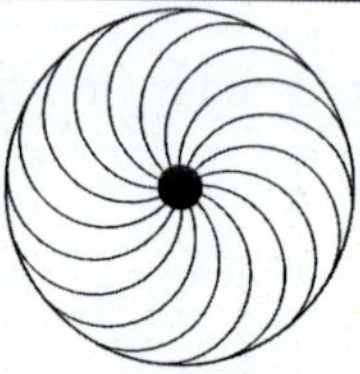	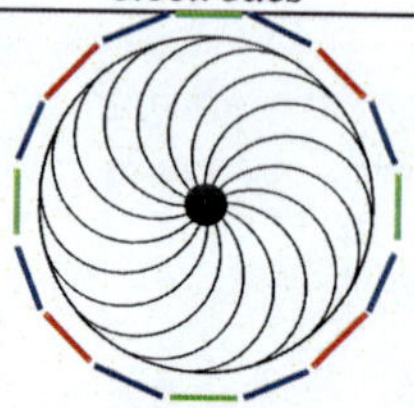	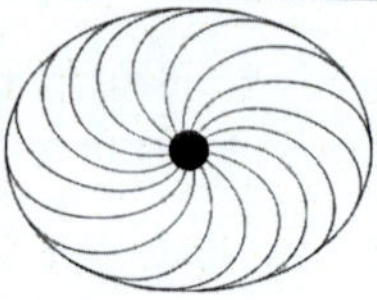	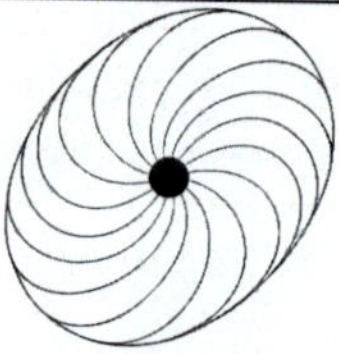
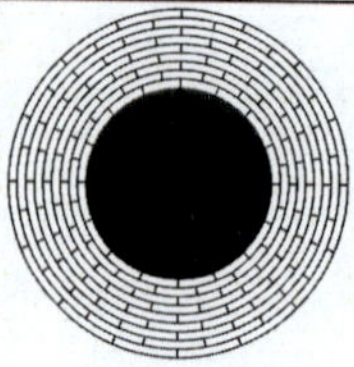	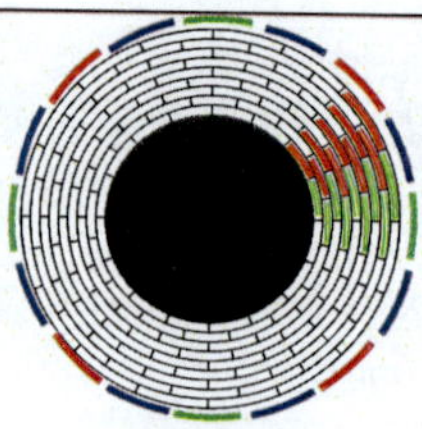	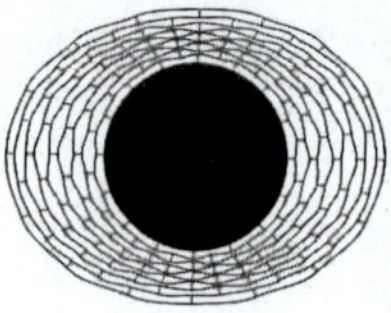	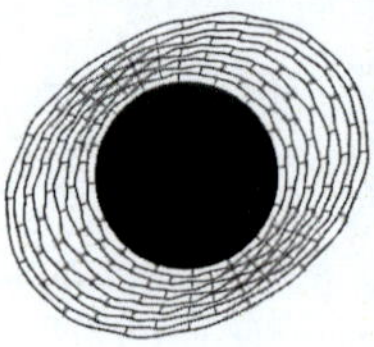
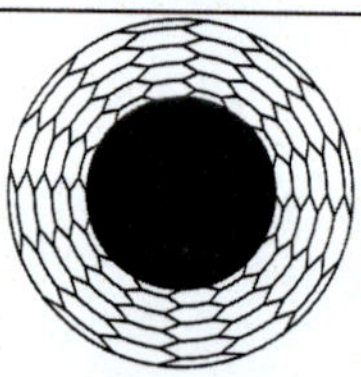	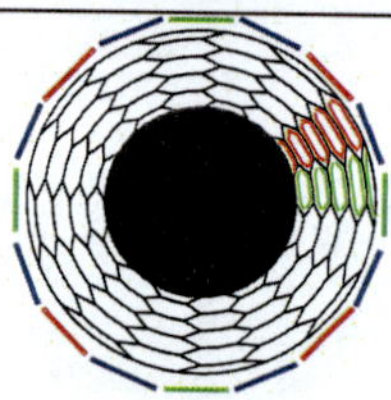	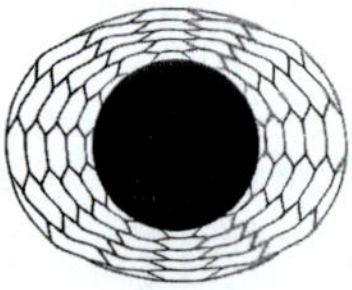	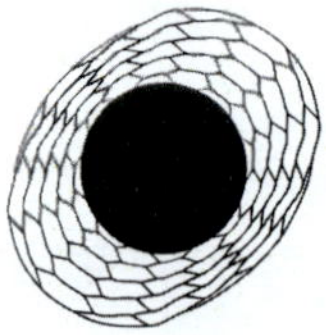
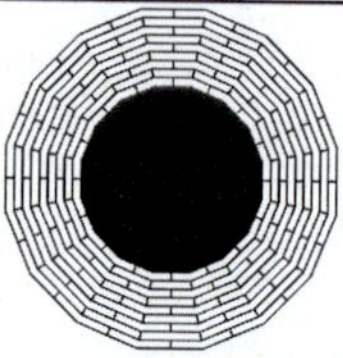	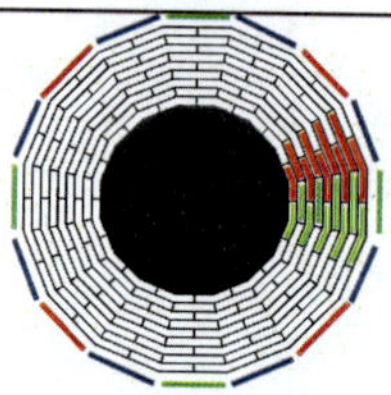	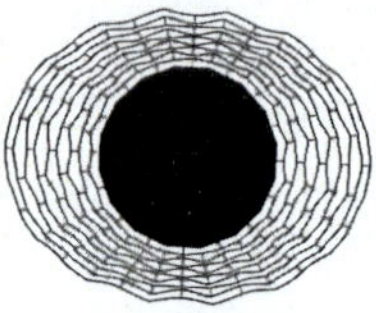	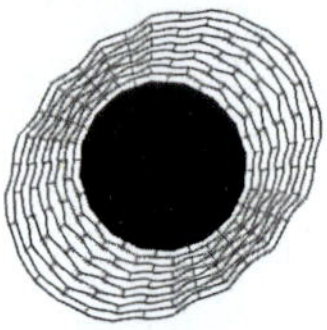
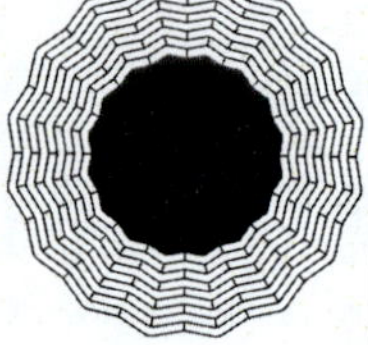	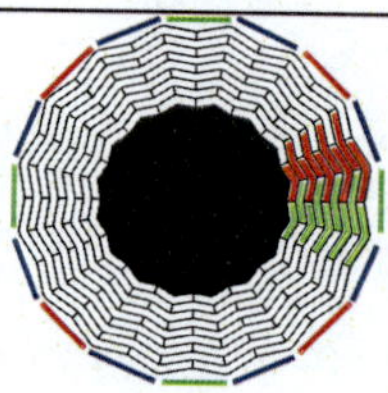	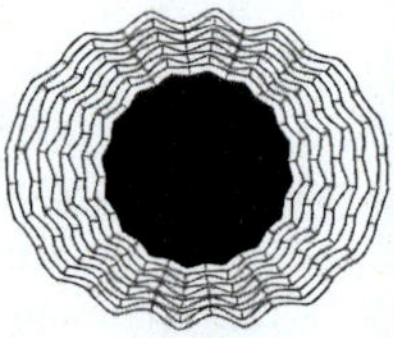	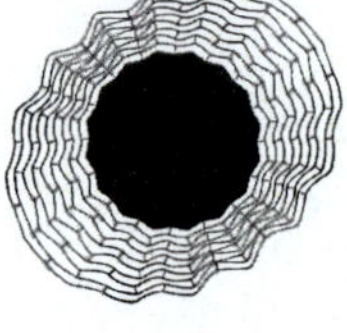

Computing it using a finite element analysis software is accurate as it accounts for fringing fields and the actual 3D geometry. In lieu of this, to understand.

The change in capacitance computed in this manner, we wanted to see how the mode shapes compare to the ellipse. This is shown in Fig. 4. It can be seen that mode shapes of RRG and HDRG are closer to the ellipse than that of the others. A similar trend was seen when this was plotted for inner rings too. In this case, RRG needs to be excluded because it lacks inner electrodes.

3 Results and Discussions

Table 2 contains a comparison of different designs. Also included is the free ring, which forms the basis of comparison in terms of mode shape. It would have a perfect ellipse as its mode shape. A free ring has the lowest eigenfrequency. One implication of this is that it has the least stiffness. But it is impractical to use a free ring.

In terms of eigenfrequency, the best design is RRG to limit it under 2k Hz for the practical consideration of tracking the displacement of the electronically. But RRG has lower change in capacitance and lower relative sense capacitance than HDRG. DRG, which is the trend-setting design that triggered the competitive performance of micro-machined gyroscopes, is surprisingly not better than HRDG or RRG, but it is close. The recently reported CDRG and GDRG fare poorly because their unique features are not in terms of relative sense capacitance. Since RRG and free ring are included here, the capacitance is calculated based only on the outer stator electrodes. Table 3 illustrates how the numbers in Table 2 can be interpreted by considering the mode shapes plotted in terms of total displacement versus θ.

Table 2. Comparison of eigenfrequency, base capacitance, relative sense capacitance for all five selected gyroscopes with proper elliptical eigenmode when only external electrodes are present.

Gyroscope	Eigen frequency ω_0 (Hz)	Base capacitance C_0 (pF)	Sense capacitance ΔC (fF)	Relative sense capacitance $\Delta C/C_0$
RRG	**1744**	3.21	44.2	0.0138
DRG	12,250	3.27	44.4	0.0136
HDRG	11,210	3.21	**49.9**	**0.0156**
CDRG	12,573	3.32	39.7	0.0112
GDRG	15,080	**3.46**	33.8	0.0097
Free ring	641	3.30	46.4	0.0141

In Table 3, one degenerate mode shape, the plot of the total displacement, and two performance measures are indicated. The first is ratio of the eigenfrequency of a gyroscope design to that of the free ring. It needs to be low, preferably below 2000 Hz. Since the free ring has a frequency of 641, the frequency ratio should be below 3.1. Only RRG satisfies this requirement. It is easy to see why this is so. All designs in the DRG-category have much more stiffness than RRG. When it comes to relative capacitance, RRG fares well but when only the outer electrodes are used to compute the capacitance. DRG class of gyroscopes have inner electrodes too. So, RRG does not fare well when we consider inner electrodes that fill the "disk" in DRGs.

Table 3. Comparison of eigenmode, total displacement versus angle plot, ratio of resonant frequency of gyroscope with respect to that of the free ring, and relative sense capacitance of all five gyroscopes when only outer electrodes are considered.

Type	Eigenmode shape	Total displacement versus angle plot	$\frac{\omega_0}{\omega_{free}}$	$\frac{\Delta C}{C_0} \times 10^3$
RRG	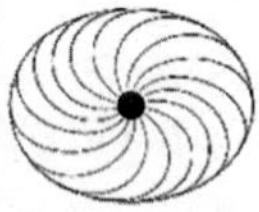	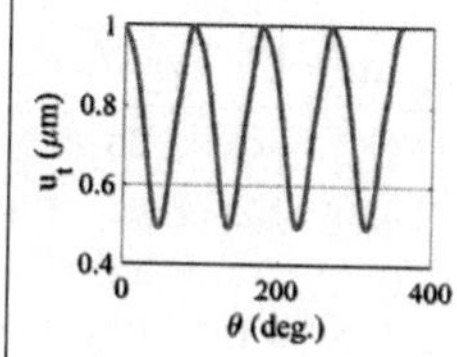	2.7	13.8
HDRG	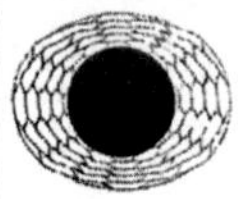	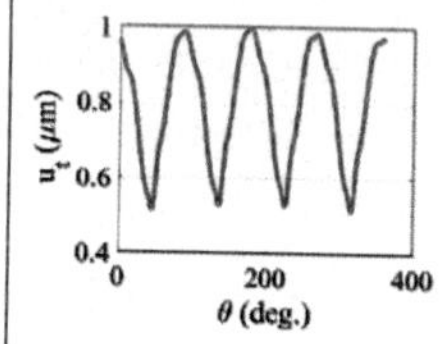	17.5	15.6
DRG	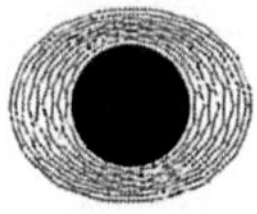	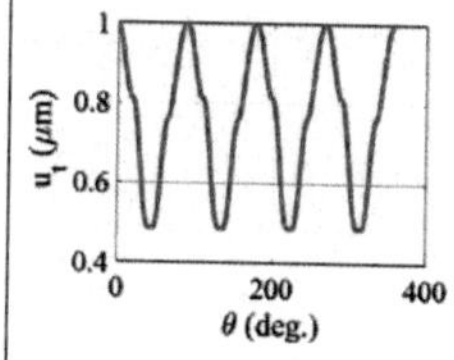	19.1	13.6
CDRG		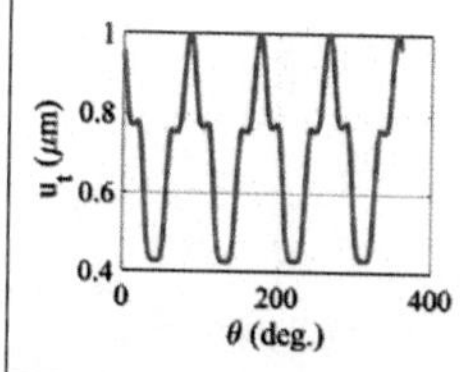	19.6	11.2
GDRG	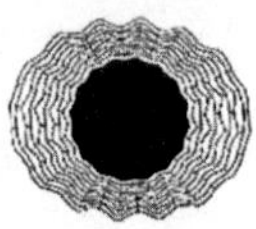	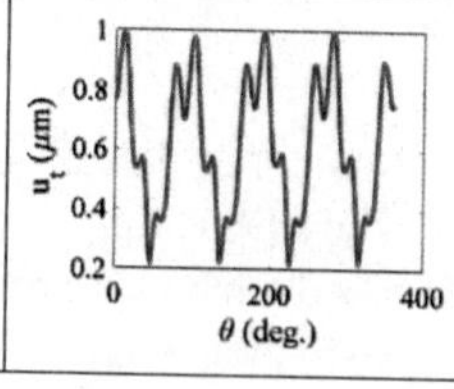	23.5	9.7

In Table 4, we compare the DRG class of gyroscopes by considering all stator electrodes, outer and inner. HDRG is a clear winner here in terms of the frequency ratio as well as the relative sense capacitance. The former has to do with the stiffness and the latter has to do with how close the mode shape is to the ellipse. It can be seen in Table 3 that, the mode shape of HDRG's outer ring has the least deviation from the ellipse. This feature holds for the inner rings also (a figure of this not included here). Thus, future designs should pay attention the mode shape as well as the frequency.

Table 4. Comparison of eigenmode, ratio of resonant frequency of gyroscope with respect to free ring, relative sense capacitance all four disk resonators when both inner and outer electrodes are considered.

Type	Eigenmode shape	$\frac{\omega_0}{\omega_{free}}$	$\frac{\Delta C}{C_0} \times 10^3$
HDRG	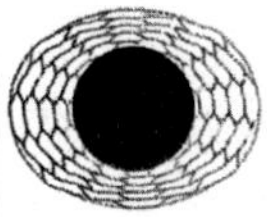	**17.5**	**28.3**
DRG	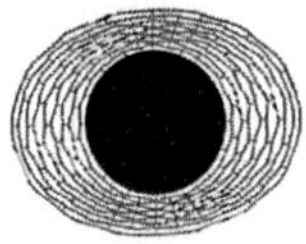	19.1	24.7
CDRG	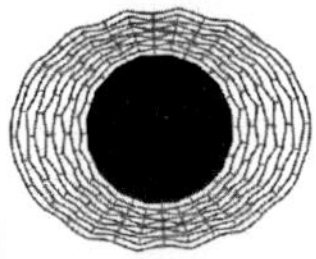	19.6	20.3
GDRG	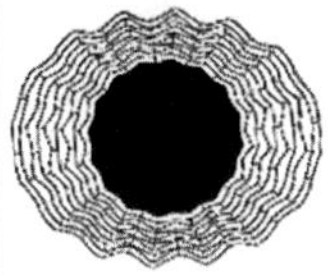	23.5	18.4

Any of the microfabricated DRG designs have degenerate eigenmodes only in principle. There is often a frequency mismatch in the two mode shapes due to uncertainties in microfabrication. This mismatch can be modelled under a spring-mass-damper framework as damping mismatch and stiffness mismatch assuming material uniformity. Both the mismatches are modelled as absolute difference in values and axis difference from the preferred axis of excitation and sensing. Static stiffness matching can be done by electrostatic tuning, but the mismatches vary with respect to temperature and aging. This renders tuning difficult. Damping mismatches are not tuned generally but compensated after extensive calibration and testing. Since the performance will be deteriorated due to the presence of these asymmetries, it is always preferred to tackle the problem at the source itself by envisaging novel geometries which reduces dependencies on fabrication precision. The current fabrication technologies (photolithography and etching) are developed based on linear structures, which may create issues while fabricating other geometries. To increase the "fabrication immunity", one can modify the arc structures which are presented in the geometry to linear structures. The focus of the possible design is to minimize the fabrication errors, thus helping us to reduce the effort for frequency mismatch.

4 Conclusions

In this work, we compared the performance of the five different resonator gyroscopes based on the eigenfrequency and eigenmodes. This study was undertaken because the reports of various designs directly went to consider quality factor and overall sensitivity rather than studying the intrinsic geometric feature. It is true that the quality factor also is influenced by the geometric design, but it is a derived quantity. Here, we considered only the mode shape to highlight the fact that a mode shape that stays close to the ellipse (which is the mode shape of a free ring) helps in increasing the relative sense capacitance. The honeycomb disk resonator gyroscope (HDRG) emerged as the best among the DRG class of designs. However, DRG class of designs have much larger eigenfrequency values as compared to the free ring and ring resonator gyroscope (RRG). This comparative analysis thus points to the need for a different design that keeps the eigenfrequency low and yet has a mode shape that remains close to an ellipse.

Future work points to fabrication and mechanical characterization of these designs to benchmark the simulation results. Comparison of other performance measures such as scale factor, frequency mismatch, quality factor, and angle random walk may provide insights for a better of design gyroscope. Our ongoing work is focused on a possible candidate design, which needs to be evaluated for these performance measures along with detailed finite element analysis.

References

1. Zhanshe, G., Fucheng, C., Boyu, L., Le, C., Chao, L., Ke, S.: Research development of silicon MEMS gyroscopes: a review. Microsyst. Technol. **21**(10), 2053–2066 (2015). https://doi.org/10.1007/s00542-015-2645-x
2. Jia, J., et al.: Overview and analysis of MEMS Coriolis vibratory ring gyroscope. Measurement **182**, 109704 (2021)
3. Cho, J.Y., Singh, S., Woo, J.-K., He, G., Najafi, K.: 0.00016 deg/$\sqrt{}$hr Angle random walk (ARW) and 0.0014 deg/hr bias instability (BI) from a 5.2M-Q and 1-cm precision shell integrating (PSI) gyroscope. In: Proceedings of IEEE Inertial 2020, Hiroshima (2020)
4. Ayazi, F., Najafi, K.: Design and fabrication of a high-performance polysilicon vibrating ring gyroscope. In: Proceedings of the IEEE Micro Electro Mechanical Systems (MEMS), pp. 621–626. Germany (1998)
5. Challoner, A., Ge, H., Liu, J.: Boeing disc resonator gyroscope. In: IEEE/ION Position, Location and Navigation Symposium (PLANS) 2014, pp. 504–514. Monterey (2014)
6. Xiao, D., Zhou, X., Li, Q., Hou, Z., Xi. X., Wu, Y., Wu, X.: Design of a disk resonator gyroscope with high mechanical sensitivity by optimizing the ring thickness distribution. J. Microelectromech. Syst. **25**(4), 606–616 (2016)

7. Li, Q., Xiao, D., Zhou, X., Xu, Y., Zhuo, M., Hou, Z., He, K., Zhang, Y., Wu, X.: 0.04 degree-per-hour MEMS disk resonator gyroscope with high-quality factor (510 k) and long decaying time constant (74.9 s). Microsyst. Nanoeng. 4(32), 2018
8. Li, Q., Xiao, D., Zhou, X., Hou, Z., Xu, Y., Hu. Q., Wu, X.: Research on the high fabrication error immunity of the honeycomb-like disk resonator gyroscope. In: IEEE Micro Electro Mechanical Systems (MEMS), pp. 1016–1019. Belfast (2018)
9. Xu, Y., Li, Q., Wang, P., Zhang, Y., Zhou, X., Yu, L., Wu, X., Xiao, D.: 0.015 Degree-per-hour honeycomb disk resonator gyroscope. IEEE Sens. J. **21**(6), 7326–7338 (2021)
10. Fan, B., et al.: A novel high-symmetry cobweb-like disk resonator gyroscope. IEEE Sens. J. **19**(22), 10289–10297 (2019)
11. Feng, J., Zhang, W., Gu, L., Liu, Z.: Design of a novel gear-like disk resonator gyroscope with high mechanical sensitivity. Microsyst. Technol. **27**(7), 2715–2722 (2020). https://doi.org/10.1007/s00542-020-05047-6

Parametric Tuning of Natural Frequencies of Tuning Fork Gyroscope

Rakesha Chandra Dash[1](✉), Rakesh Tirupathi[1], P. Krishna Menon[2], and Ashok Kumar Pandey[1]

[1] Indian Institute of Technology Hyderabad, Kandi, Sangareddy 502285, Telangana, India
[2] Research Centre Imarat, Hyderabad 500069, Telangana, India
{rakesh.chandradash,ashok}@mae.iith.ac.in

Abstract. The sequence of mode shapes play a vital role in designing a dual mass tuning fork gyroscope (TFG). To avoid loss of energy, a desired separation of frequencies between operating modes (out-of-phase drive and sense) and parasitic modes is required. Hence, regulation of mode shapes is an essential criterion in TFG design. In the present work, the influence of several crucial parameters such as coupling mechanisms and dimensions of folded beams on the in-plane frequencies are studied numerically by using finite element based COMSOL software.

Keywords: Tuning fork gyroscope · MEMS

1 Introduction

Microelectromechanical systems (MEMS) have been of significant interest for the design of inertial sensors because of their low cost, small size, and ruggedness and easy integration with the CMOS based signal conditioning readout interface [1–3]. MEMS inertial sensors like accelerometer and gyroscope find applications in consumer electronics, automobiles, defence, navigation, medical and other fields [4–7]. MEMS gyroscopes can be classified as tuning fork gyroscopes (TFG) [8–11], comb type [12,13], disk type [14,15], and wine glass type gyroscopes [16]. Applications like inertial navigation systems for ships, aircraft, submarines, and other spacecraft require very high sensitivity and resolution [17–19]. However, the resolution of inertial MEMS devices is mainly limited by Brownian noise in micromechanical structures and interface electronics circuit noise [20–22]. Low mechanical sensitivity of MEMS devices may cause the electronics noise to dominate and limit the resolution [23]. This challenge can be addressed by exploiting different techniques like mechanical amplification using

Supported by Defence Research and Development Organisation, New Delhi, India.

A. K. Pandey et al. (Eds.): MAMM 2022, 126, pp. 162–171, 2023.
https://doi.org/10.1007/978-3-031-20353-4_12

compliant mechanisms, geometric anti-spring design, mechanical coupling, and mode-localized [26] designs toward the sensitivity improvement of MEMS gyroscope. However, while using any of these techniques, frequencies of gyroscopes also get affected. The variation of frequencies due to coupling mechanisms and their dimensions, we take tuning fork gyroscope (TFG) from [24] for further analysis.

The TFG is mainly categorized as single mass gyroscope, dual-mass gyroscope and quad mass gyroscope. Single mass gyroscope consists of an overlap between drive and sense vibrations, which requires a complex interface to distinguish the required Coriolis force. In the case of a quad mass gyroscope, the mismatch between the four masses and stiffness introduces unwanted errors to the measurement. In the case of a dual-mass gyroscope, the two proof masses vibrate either in-phase or out-of-phase to capture the Coriolis force. Among these two approaches, out-of-phase oscillations (operating modes) eliminate the common vibrations.

Therefore, it is a necessary criterion to design dual-mass TFG such that the out-of-phase drive and sense modes are away from other modes to reduce interference. The present study utilizes various important parameters to regulate the in-plane mode shapes to get the desired output.

2 Dual Mass Tuning Fork Gyroscope

Dual mass tuning fork gyroscopes consist of two masses oscillating in out-of-phase mode. The basic design of the gyroscope is shown in the Fig. 1. The performance of the TFG can be changed based on the coupling mechanisms that will be described in further sections. In this section, various critical design parameters and their effects on modulating natural frequencies are studied. The fundamental frequencies and mode shapes are found out by using finite element based COMSOL software.

Table 1. Element convergence study for TFG with spring type coupling

Element numbers	1,013,967	541,457	395,478	305,898	240,391
Element size (Lowest, highest)	(1,30)	(3,50)	(10,50)	(15,70)	(20,90)
1st mode (Hz)	8960	8963	8986	8997	9011
2nd mode (Hz)	9401	9406	9428	9440	9452
3rd mode (Hz)	10,587	10,591	10,612	10,624	10,637
4th mode (Hz)	14,001	14,020	14,038	14,058	14,067

2.1 Dual Mass TFG with Spring Coupling

The design of TFG with spring coupling is shown in Fig. 1. A convergence study based on the natural frequencies is first done to decide the minimum element size without compromising the accuracy of the solution. COMSOL software is used for this purpose and results are shown for different element sizes in Table 1. An element size of (1,30) is selected for the further modal analysis in Fig. 2. The Y-axis and X - axis are taken as drive and sense directions, respectively. In the first mode, an in-phase motion of the mass is observed in the sense direction. An out-of-phase motion in the sense direction is seen in the second mode. The third mode consists of an out-of-phase drive motion and the fourth mode is in-phase drive motion.

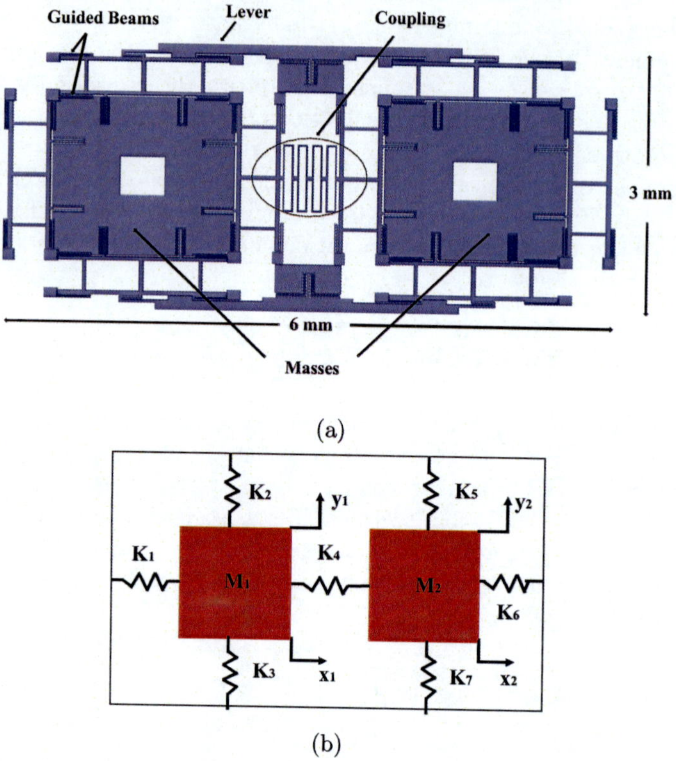

Fig. 1. (a) Dual mass TFG with spring coupling and (b) its lumped mass model

A free body diagram of the whole structure is drawn in Fig. 1b to derive the analytical expressions. Each mass possesses of 2 degree of freedom (dof) and, hence, the system will have 4 dofs. Therefore, there will be four governing differential equations given in Eqs. 1–4. The force F_1, F_2, F_3, F_4 are applied forces for each degree of freedoms. The stiffness K_1, K_2, K_3, K_4, K_5, , K_6 and K_7 are found analytically. There are three type of spring used in this gyroscope shown in Fig. 3. The stiffness of the guided beam is given by $K_G = 12EI_1/L_1^3$, where EI is Flexural rigidity of the beam. For the shorter beam like one shown with the length L_2 in Fig. 3b, the stiffness is obtained from $K_B = 3EI_1/L_1^3$.

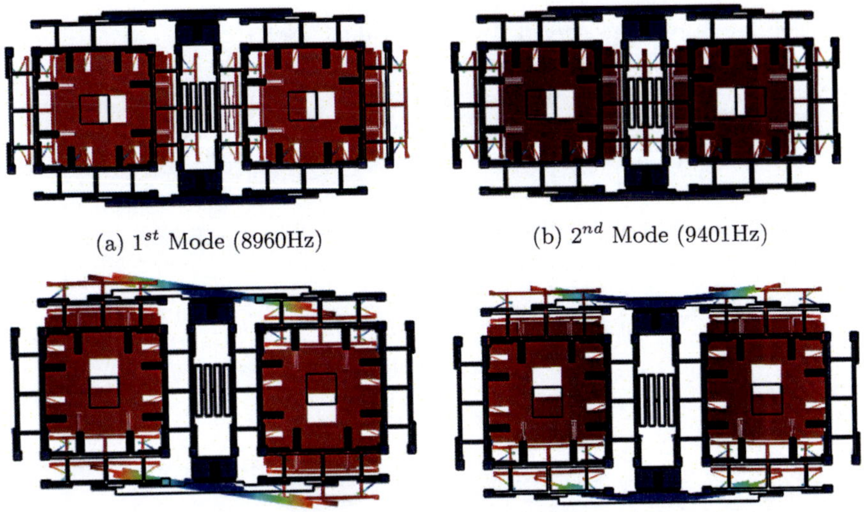

(a) 1st Mode (8960Hz)

(b) 2nd Mode (9401Hz)

(c) 3rd Mode (10587Hz)

(d) 4th Mode (14001 Hz)

Fig. 2. Mode shapes of TFG with spring coupling

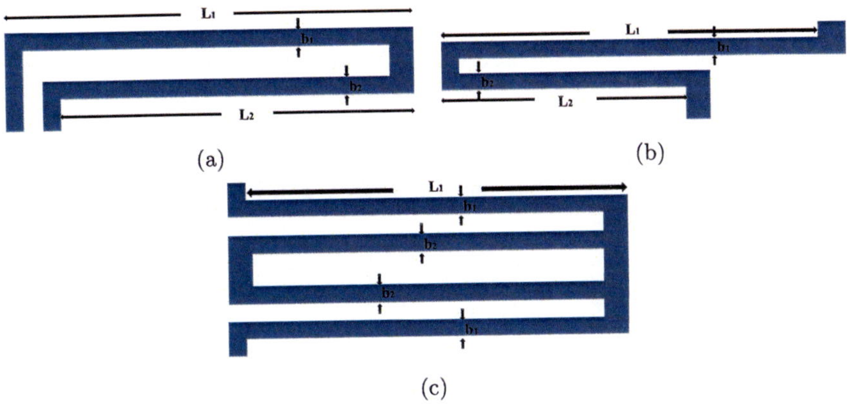

Fig. 3. Various type of springs used in TFG

After the calculation of the beam stiffness, series and parallel combinations are used to find the total stiffness of Fig. 3a–c structures.

$$M_1\ddot{x}_1 + K_1x_1 + K_4(x_2 - x_1) = F_1, \tag{1}$$

$$M_2\ddot{x}_2 + K_6x_2 + K_4(x_1 - x_2) = F_2, \tag{2}$$

$$M_1\ddot{y}_1 + (K_2 + K_3)y_1 = F_3, \tag{3}$$

$$M_2\ddot{y}_2 + (K_5 + K_7)y_2 = F_4. \tag{4}$$

The above equations are solved for eigen values and the natural frequencies are compared with COMSOL solutions in Table 2. The difference between the analytical and COMSOL is due to the consideration of simplified model and guided beam assumptions in the stiffness calculation. It is vital to regulate these mode shapes according to our need. Various regulating parameters and their effects on the sense and drive modes are described. Regulating in-phase sense mode needs a different approach so it will be described at the last.

Table 2. Comparison of natural frequencies from analytical model and COMSOL

Analytical (Hz)	COMSOL (Hz)
9275.6	8960
9681.1	9401
10,062	10,587
10,126	14,001

(a) Modulating Out-of-Phase Sense Mode The dimensions of coupling spring play a significant role in deciding the 2nd mode frequency. The length of spring effects the 2nd mode drastically and can be seen from Fig. 4a. The increase in coupling spring length reduces the stiffness in out-of-phase sense mode, thus, it reduces the corresponding frequency, while, the other modes are unaffected.

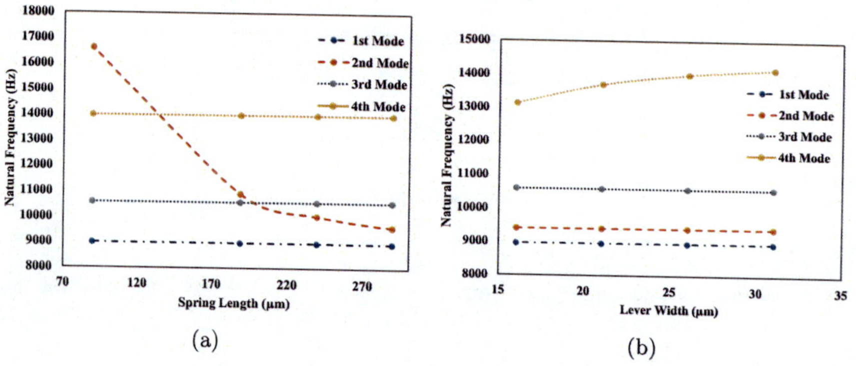

Fig. 4. Variation of natural frequencies for dual mass TFG with spring coupling with (a) change in length of spring (a) change in width of lever

(b) Modulating Out-of-Phase Drive Mode The out-of-phase drive mode can be regulated by changing dimensions of the folded beams in drive direction or by changing various coupling mechanisms. The other coupling mechanisms used in our study are shown in Fig. 5. It can be observed by comparing the mode shapes in Figs. 1, 6, and 7 that out-of-phase drive mode has shifted its position from 2nd mode to 3rd mode for elliptical and diamond coupling, respectively. Thus, coupling mechanisms can influence out-of-phase drive mode and out-of-phase sense mode drastically.

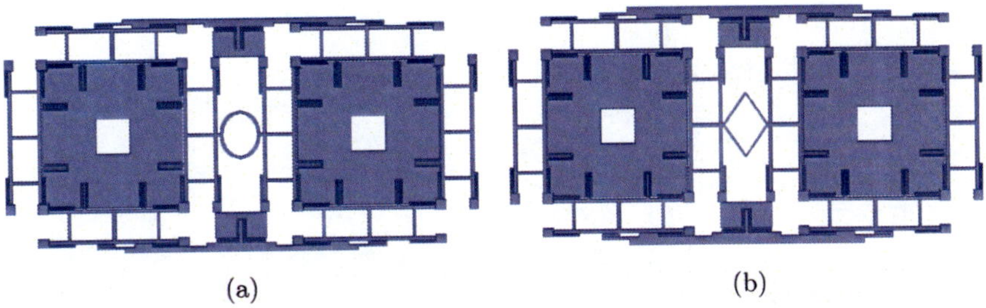

Fig. 5. Dual mass TFG with (a) elliptical coupling (a) diamond coupling

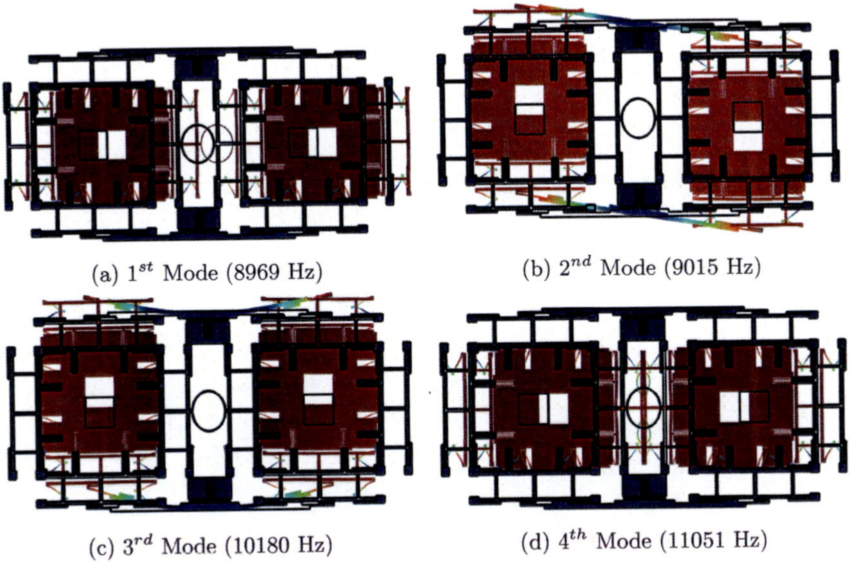

Fig. 6. Modeshapes of TFG with elliptical coupling

(c) Modulating In-Phase Drive Mode The dimensions of lever can be changed to regulate the in-phase drive mode in TFG as shown in Fig. 4b. It can be seen that with increase in the width of the lever, the stiffness of in-phase drive mode increases, and, hence the natural frequency also increases.

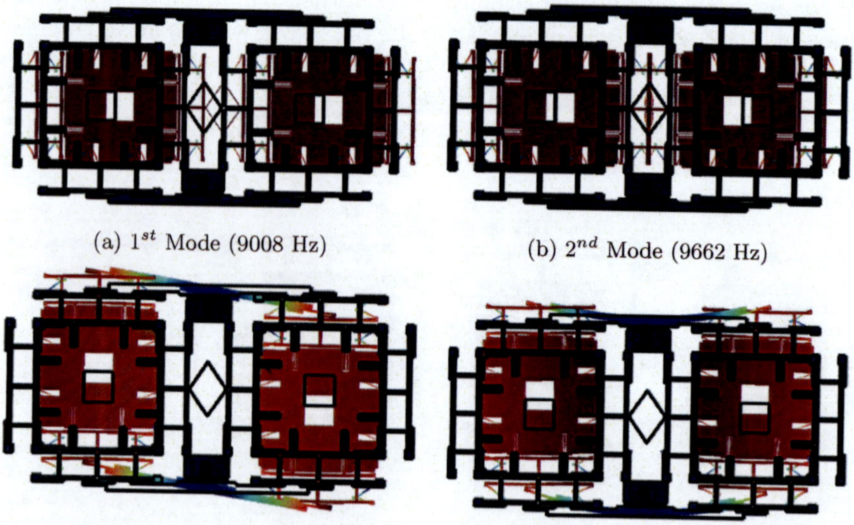

(a) 1st Mode (9008 Hz) (b) 2nd Mode (9662 Hz)

(c) 3rd Mode (10587 Hz) (d) 4th Mode (14001 Hz)

Fig. 7. Mode shapes of TFG with diamond coupling

(d) Modulating In-Phase Sense Mode It can be seen form above analysis that in-phase sense mode remains the same irrespective of change in parameters. Therefore, a special folded beam [25] is used as shown in Fig. 8. These beams will only affect the in-phase sense mode during the motion. There is very small effect on the frequency of out-of-phase sense mode. Consequently, it can be seen from the Fig. 9 that the frequency of first mode is increased 21,168 Hz from 9008 Hz (see Fig. 7).

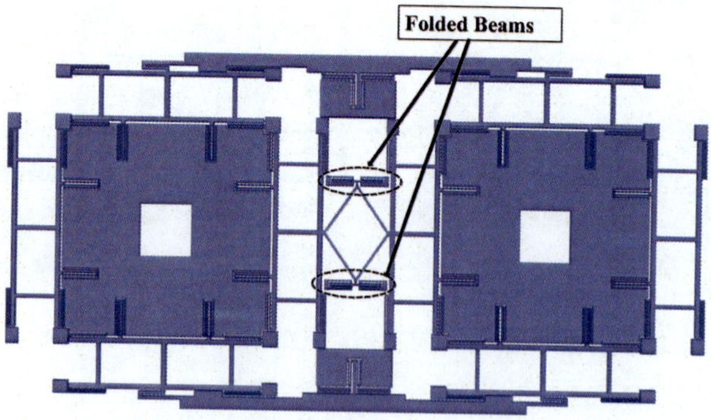

Fig. 8. Attachment of additional folded beam to regulate in-phase sense mode

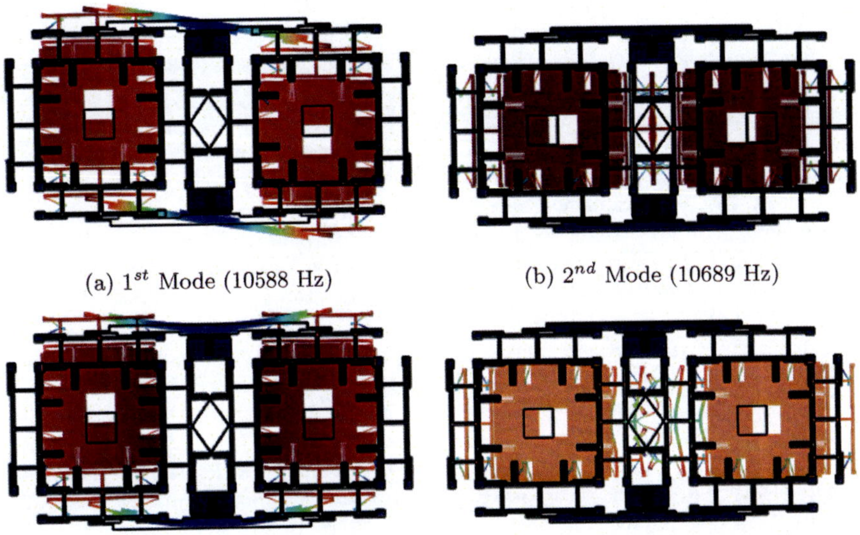

(a) 1st Mode (10588 Hz)

(b) 2nd Mode (10689 Hz)

(c) 3rd Mode (14001 Hz)

(d) 4th Mode (21168 Hz)

Fig. 9. Mode shapes of TFG with diamond coupling with folded beam attachments

3 Conclusions

In this study, the modal analysis of dual mass TFG is done by the help of COMSOL software and validated by using analytical modeling. The effects of various coupling mechanisms and folded beams dimensions on mode shapes are explored. The main aim is to regulate the mode shapes of the TFG. It is found that in-phase drive, out-of-phase drive and out-of-phase sense mode shapes can easily be interchanged by changing the coupling mechanisms and dimensions of the guided beams. The modulation of the in-phase sense mode shape require special guided beam arrangements near the coupling mechanism. These mechanisms only effects the in-phase sense mode.

Acknowledgments. The authors would like to acknowledge DRDO, New Delhi, India for funding the research work through the grant number DRDO/./IITHRC-011.

References

1. Yazdi, N., Ayazi, F., Najafi, K.: Micromachined inertial sensors. Proc. IEEE **86**(8), 1640–1659 (1998)
2. Xia, D., Yu, C., Sensors, L.K.: undefined,: the development of micromachined gyroscope structure and circuitry technology. Mdpi. Com. **14**, 1394–1473 (2014)
3. Söderkvist, J.: Micromachined gyroscopes. Sens. Actuators A Phys. **43**, 65–71 (1994)

4. Passaro, V.M.N., Cuccovillo, A., Vaiani, L., De Carlo, M., Campanella, C.E.: Gyroscope technology and applications: a review in the industrial perspective. Sensors **17**, 2284 (2017)
5. Ma, W., Lin, Y., Liu, S., Zheng, X., Jin, Z.: A novel oscillation control for MEMS vibratory gyroscopes using a modified electromechanical amplitude modulation technique. J. Micromech. Microeng. Iopscience. Iop. Org. 27(2) (2016)
6. Pang, G., Liu, H.: Evaluation of a low-cost MEMS accelerometer for distance measurement. J. Intell. Robot. Syst. Theory Appl. **30**, 249–265 (2001)
7. Wu, J., Zhou, Z., Fourati, H., Cheng, Y.: A super fast attitude determination algorithm for consumer-level accelerometer and magnetometer. IEEE Trans. Consum. Electron. **64**(3), 375–381 (2018)
8. Nguyen, M.N., Ha, N.S., Nguyen, L.Q., Chu, H.M., Vu, H.N.: Z-axis micromachined tuning fork gyroscope with low air damping. Micromachines **8**, 42 (2017)
9. Yang, C., Li, H.: Digital control system for the MEMS tuning fork gyroscope based on synchronous integral demodulator. IEEE Sens. J. **15**(10), 5755–5764 (2015)
10. Guan, Y., Gao, S., Liu, H., Jin, L., Niu, S.: Design and vibration sensitivity analysis of a MEMS tuning fork gyroscope with an anchored diamond coupling mechanism. Sensors **16**, 468 (2016)
11. Prikhodko, I., Zotov, S., Trusov, A., Shkel, A.M.: Foucault pendulum on a chip: rate integrating silicon MEMS gyroscope. Elsevier. **177**(2012), 67–78 (2012)
12. Tatar, E., Mukherjee, T., Fedder, G.K.: Stress effects and compensation of bias drift in a MEMS vibratory-rate gyroscope. J. Microelectromech. Syst. 26(3), 569–579
13. Park, B., Han, K., Lee, S., Yu, M.-J.: Analysis of compensation for a g-sensitivity scale-factor error for a MEMS vibratory gyroscope. Iopscience. Iop. Org **25**(11), 115006 (2015)
14. Sonmezoglu, S., Alper, S., Akin, T.: An automatically mode-matched MEMS gyroscope with wide and tunable bandwidth. Ieeexplore. Ieee, Org (2014)
15. Zhou, X., Xiao, D., Wu, X., Wu, Y., Hou, Z., He, K., Li, Q.: Stiffness-mass decoupled silicon disk resonator for high resolution gyroscopic application with long decay time constant (8.695 s). Appl. Phys. Lett. 109 (2016)
16. Guan, Y., Gao, S., Jin, L., Cao, L.: Design and vibration sensitivity of a MEMS tuning fork gyroscope with anchored coupling mechanism. Microsyst. Technol. 22, 247–254
17. Nusbaum, U., Rusnak, I., Klein, I.: Angular accelerometer-based inertial navigation system. Navigation. **66**, 681–693 (2019)
18. He, Q., Zeng, C., He, X., Xu, X., Lin, Z.: Measurement, undefined 2018, Calibrating accelerometers for space-stable inertial navigation systems at system level. Elsevier
19. El-Sheimy, N., Youssef, A.: Inertial sensors technologies for navigation applications: state of the art and future trends. Satell. Navig. 1 (2020)
20. Petritoli, E., Leccese, F., Leccese, M.: Inertial navigation systems for UAV: Uncertainty and error measurements. Ieeexplore. IEEE, Org (2019)
21. Handtmann, M., Aigner, R., Meckes, A., Wachutka, G.K.M.: Sensitivity enhancement of MEMS inertial sensors using negative springs and active control. Sens. Actuators A Phys. **97–98**, 153–160 (2002)
22. Masu, K., Machida, K., Yamane, D., Ito, H., Ishihara, N., Chang, T.-F.M., Sone, M., Shigeyama, R., Ogata, T., Miyake, Y.: (Invited) CMOS-MEMS based micro-gravity sensor and its application. ECS Trans. **97**, 91–108 (2020)
23. Gabrielson, T.G.: Mechanical-thermal noise in micromachined acoustic and vibration sensors. Ieeexplore. IEEE Trans. Electronic, Dev (1993)

24. Cao, L., Li, J., Liu, X., Sun, F.Y.: Research on an anchor point lever beam coupling type tuning fork micro-gyroscope. Int. J. Precis. Eng. Manuf. **21**, 1099–1111 (2020)
25. Li, Z., Gao, S., Jin, L., Liu, H., Guan, Y., Peng, S.: Design and mechanical sensitivity analysis of a MEMS tuning fork gyroscope with an anchored leverage mechanism. Sensors (Basel). **19**(16), 3455 (2019)
26. Bukhari, S.A.R., Saleem, M.M., Hamza, A., Bazaz, S.A.: A novel design of high resolution MEMS gyroscope using mode-localization in weakly coupled resonators. IEEE Access **9**, 157597–157608 (2021)

Design and Analysis of Single Drive Tri-Axis MEMS Gyroscope

Pradnya Chabbi, Anant Khedkar, and Venkatesh K. P. Rao(✉)

Department of Mechanical Engineering, Birla Institute of Technology and Science, Pilani, Rajasthan 333031, India
venkateshkp.rao@pilani.bits-pilani.ac.in

Abstract. MEMS based single and dual-axis gyroscopes have been widely explored for potential application in automotive, space, defense, and consumer electronics sectors. Tri-axis gyroscopes based on MEMS, however, have been sparsely studied. This work presents a novel design for tri-axis MEMS gyroscope and an analytical model to obtain the natural frequencies in drive and sense modes. These frequency values have been compared with the numerically obtained frequencies using Finite Element Analysis (FEA). The analytical results lie within 10% of their numerically obtained values. The frequency matching process involves many iterations of geometric dimensions if the end application requires minor design changes. The proposed analytical model will make the design customization easy as the frequencies of each mode will be expressed as a function of critical geometrical parameters saving multiple numerical runs required for design optimization.

Keywords: MEMS · Gyroscope · Natural frequency · Frequency tuning · Drive mode · Sense mode

1 Introduction

Gyroscopes are resonant sensors that can measure angular rates along one or more coordinate axis. They work on the principle of the Coriolis Effect [1]. Draper Labs first reported a microelectromechanical Systems (MEMS) based single axis gyroscope in 1991 [2]. Since then many variations of single and dual axis gyroscopes have been reported [3–11]. Currently, MEMS-based gyroscopes are being used for space [12–14], defense [15,16], automotive [17,18] and consumer electronics [19–21] applications. As gyroscopes are resonant sensors, it is important to have the drive and sense mode frequencies as close to each other as possible for high sensitivity applications. However, the bandwidth of operation is a factor of the difference between the drive and sense mode frequencies. Thus,

This work is supported by Naval Research Board, DRDO NRB-471/MAR/2020-21.

A. K. Pandey et al. (Eds.): MAMM 2022, 126, pp. 172–182, 2023.
https://doi.org/10.1007/978-3-031-20353-4_13

the modal frequencies should not be too apart to ensure better sensitivity and they should not be too close to ensure wider operational bandwidth. Thus, it is important to carefully optimize the drive and sense mode frequencies. These frequency values directly depend on the structural geometry. As a result, variation of the modal frequencies with changes in geometric parameters becomes a significant study. The optimization of frequencies is simple in single-axis gyroscopes as there are only two modes to look at, the drive mode and the sense mode. For tri-axis gyroscopes, frequency optimization needs to be done for one drive and three sense modes; hence, the process becomes more complicated. As a result, tri-axis gyroscopes are not explored to the scale of single and dual-axis gyroscopes. A novel design of MEMS gyroscope was proposed by Tsai et al. [22,23] which can be manufactured by conventional SOI MEMS process, but fails to address mode decoupling [24]. Cazzinga et al. demonstrated a novel decoupled tri-axis gyroscope with elastic coupling springs in 2013 [25]. Mode decoupling was further studied in depth by Tingkai et al. [24]. Further Shah et al. [26] demonstrated improved coupling spring configuration for tri-axis gyroscopes. Although some numerical and experimental work for tri-axis gyroscopes is available across the literature, design-specific analytical models are not readily available. In this work, we have presented a novel design for the tri-axis gyroscope. We have provided an analytical frequency model that directly monitors the effect of change in geometric dimensions on the modal frequencies. The results have also been compared with values obtained from a numerical model.

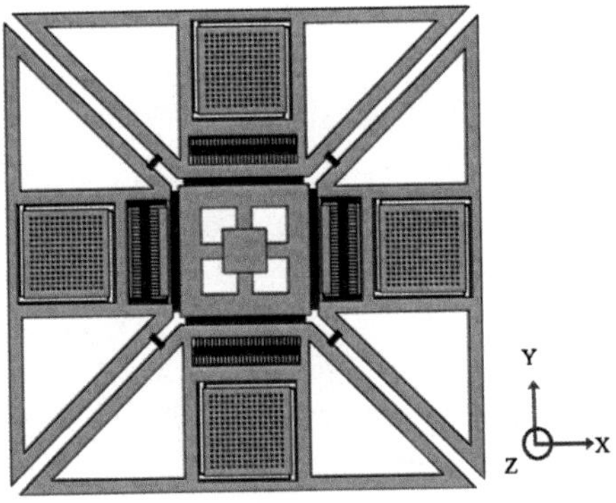

Fig. 1. Structure of proposed gyroscope

Figure 1 shows the structure of gyroscope and Fig. 2a shows the zoomed components of the structure. The structure consists of a centrally anchored component connected to four trapezoidal flanges employing central connecting beams as shown in Figs. 1 and 2a. Meandering coupling beams connect these four trapezoidal flanges. The main connecting beams allow the flanges to vibrate in an in-plane motion. The coupling beams ensure this in-plane motion to be of a push-pull configuration that cancels the linear component of acceleration. The

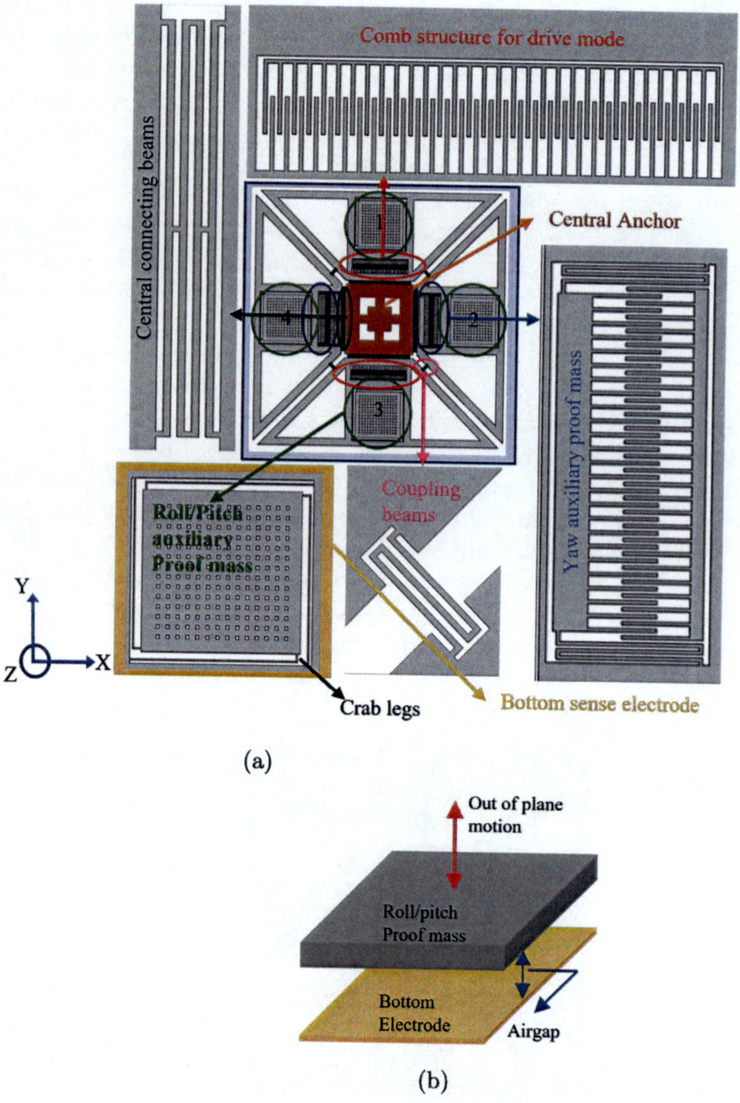

Fig. 2. (a) Structural components of the proposed gyroscope. (b) Roll/pitch proof mass with bottom electrode for capacitive sensing

structure is electrostatically driven using a set of comb finger structures etched on one of the opposite pairs of the flanges. The other flanges consist of auxiliary proof mass with a comb finger configuration responsible for yaw mode detection. It is connected to the primary flanges using meandering beams on either side, as shown in figure 1. Four auxiliary proof masses are etched into the structure to detect the roll and pitch motion, one on each flange. These proof masses are connected to the central flange using modified crab leg flexures, as shown in figure 1. The crab leg combination restricts the motion of these auxiliary proof masses in the in-plane X and Y directions and allows them to move only in the out-of-plane Z direction. Metal electrodes are patterned on the bottom substrate to detect this out-of-plane motion capacitively as shown in Fig. 2b. Proof masses 2 and 4 detect the roll mode, while proof masses 1 and 3 detect the pitch mode of the angular input. The dimensions of critical geometric components are summarized in Table 1.

Table 1. Optimized structural dimensions for updated design

Parameter	Value
Central anchor outer square	720 µm × 720 µm
Connecting beam length	660 µm
Connecting beam width	5 µm
Coupling beam length	75 µm
Coupling beam width	5 µm
Roll, pitch proof mass	480 µm × 480 µm
Yaw proof mass	535 µm × 30 µm
Comb finger width and gap	3.5 µm
Gap between structure and electrode	1 µm
Structural thickness	30 µm

2 Analytical Studies

In this section, we obtain the expression for natural frequency in drive and three sense modes.

2.1 Drive Mode

The drive mode consists of an in-plane motion of the structure. Four central connecting beams and four coupling beam configurations contribute to this in-plane drive mode motion, as shown in Fig. 3. The central connecting beam can be considered as a combination of two meandering beams (Fig. 4a) in parallel, as shown in Fig. 4c. Coupling beams are just meandering beams operating at the component of the spring force at an angle of 45 degrees. The expression for stiffness of meandering beams k_m is provided by G. Fedder [27].

$$k_m = \frac{48EI_{z,b}[(\tilde{a}+b)n-b]}{b^2(n-1)[(3\tilde{a}^2+4\tilde{a}b+b^2)n+3\tilde{a}^2-b^2]} \tag{1}$$

where, $\tilde{a} \equiv \frac{aI_{z,b}}{I_{z,a}}$, $I_{z,a} = \frac{tw_a^3}{12}$, $I_{z,b} = \frac{tw_b^3}{12}$ and $n =$ number of turns of the meander.

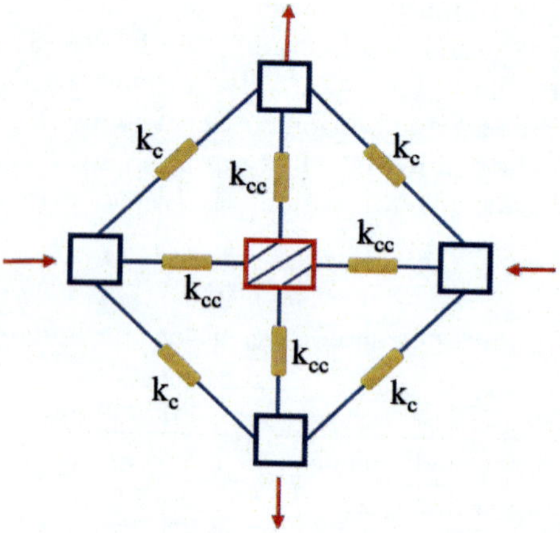

Fig. 3. Stiffness components for drive mode

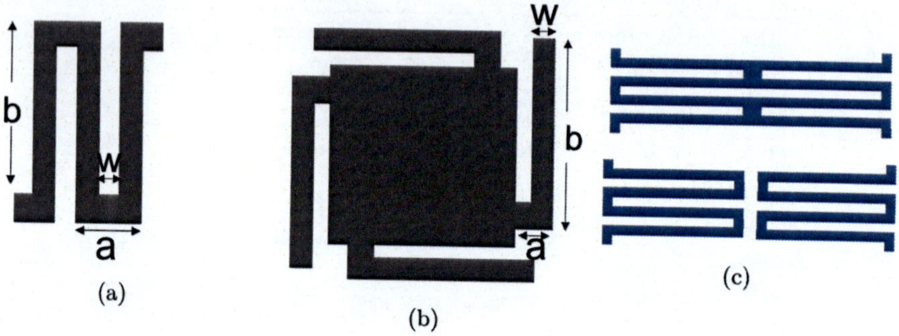

Fig. 4. (a) Meandering beams, (b) Modified crab leg configuration, and (c) Central connecting beams as a parallel combination of two meandering beams

Thus, stiffness of central connecting beam configuration (k_{cc}) and the coupling beam configuration (k_c) can be given by,

$$k_{cc} = 2k_m$$
$$k_c = \frac{k_m}{\sqrt{2}} \quad (2)$$

2.2 Roll and Pitch Modes

The roll and pitch modes are out of plane modes. The auxiliary proof mass that sense these modes are connected to the central flange by modified crab leg beam configuration as shown in Fig. 5.

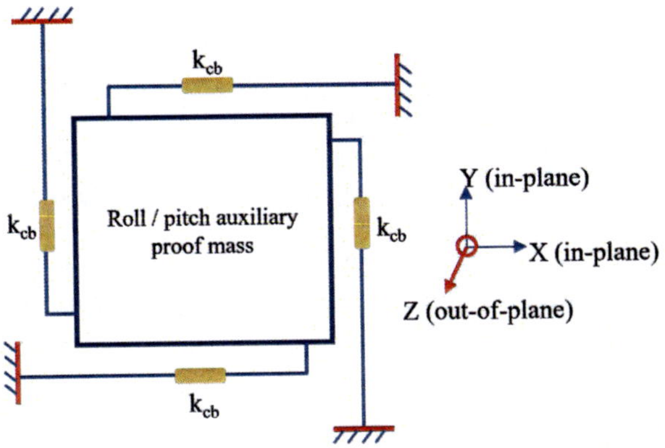

Fig. 5. Stiffness components for roll/pitch mode for out-of-plane motion

This configuration restricts the independent motion of the auxiliary proof masses in the in-plane directions and allows only out-of-plane movement. The expression for cab leg configuration as provided by Fedder [27] is given by

$$k_{cb} = \frac{P}{Q} \tag{3}$$

where

$$P = 48S_{ea}S_{eb}(S_{gb}L_a + S_{ea}L_b)(S_{eb}L_a + S_{ga}L_b)$$
$$Q = S_{eb}^2 S_{gb} L_a^5 + 4S_{ea}S_{eb}^2 L_a^4 L_b + S_{eb}S_{ga}S_{gb}L_a^4 L_b + 4S_{ea}S_{eb}S_{ga}L_a^3 L_b^2 + 4S_{ea}S_{eb}S_{gb}L_a^2 L_b^3 + 4S_{ea}^2 S_{eb} L_a L_b^4 + S_{ea}S_{ga}S_{gb}L_a L_b^4 + S_{ea}^2 S_{ga} L_b^5$$

Let E is the elastic Modulus and G be the shear modulus, then $S_e \equiv EI_x$, $S_g \equiv GJ$. Where, $I_x = \frac{tw^3}{12}$, L_a and L_b are the lengths of the thigh and shin, respectively, w_a and w_b are the widths of the thigh and shin, respectively.

2.3 Yaw Mode

An auxiliary proof mass capacitively senses the yaw motion with comb fingers that vibrate along an in-plane (Y axis) against a set of fixed comb fingers, as shown in Fig. 2. The meandering beams on both sides of this proof mass act in series to allow this in-plane motion, as shown in Fig. 6.

Fig. 6. Stiffness components for yaw mode

However, this beam configuration does not restrict the out-of-plane motion of the yaw proof mass. The geometric parameters of the beam and proof mass have been optimized such that the natural frequency of the out-of-plane mode is farther away than the gyroscopes drive and sense mode frequencies to avoid any unwanted vibrations. The expression for the meandering beam configuration provided in Eq. 1 can also be used here.

Further, the mass of the moving structure for each mode can be calculated from the geometric dimensions. As far as the effective modal mass of the beams is considered, the difference between the effective modal mass and the actual mass of the beams is negligible compared to the total mass of the structure. For simplicity, actual beam mass has been considered for analytical modelling. The natural frequency ω_n of a given mode with effective stiffness k and mass m is given by

$$\omega_n = \sqrt{\frac{k}{m}} \tag{4}$$

3 Numerical Studies

The geometry was built using a model builder in COMSOL Multiphysics. Boundary conditions were provided by fixing the anchors, and an eigenfrequency study was run to carry out the modal analysis. The mode shapes were studied to identify the drive, and sense modes and their natural frequencies were noted. Parametric analysis was performed by changing the geometric dimensions of beams and proof masses so that the structure is optimized to provide frequencies

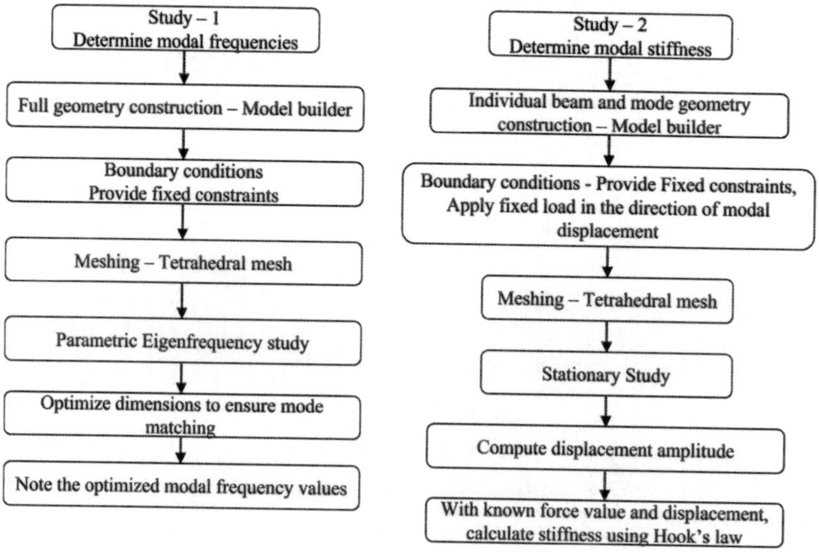

Fig. 7. Steps in numerical modelling

of drive and sense modes close-by with minimum intermediate modes between them. They were modelled individually to confirm the stiffness contributed by the beam configurations. A known force was applied to the beams in the direction of the mode shape, and the displacement was noted. Hook's law was then used to calculate the stiffness contributed by each beam. The resultant stiffness due to a combination of one or more beam configurations was then calculated for each mode.

4 Results and Discussion

In this study, we present a novel design of tri-axis MEMS gyroscope, with all modes decoupled and the geometric dimensions of the structure are tuned to bring the drive and sense mode frequencies close by. Analytical and numerical studies are carried out to evaluate the natural frequencies, and modeshapes of MEMS gyroscope. Figure 8 shows the mode shapes of tri-axis mems gyroscope in drive and sense directions.

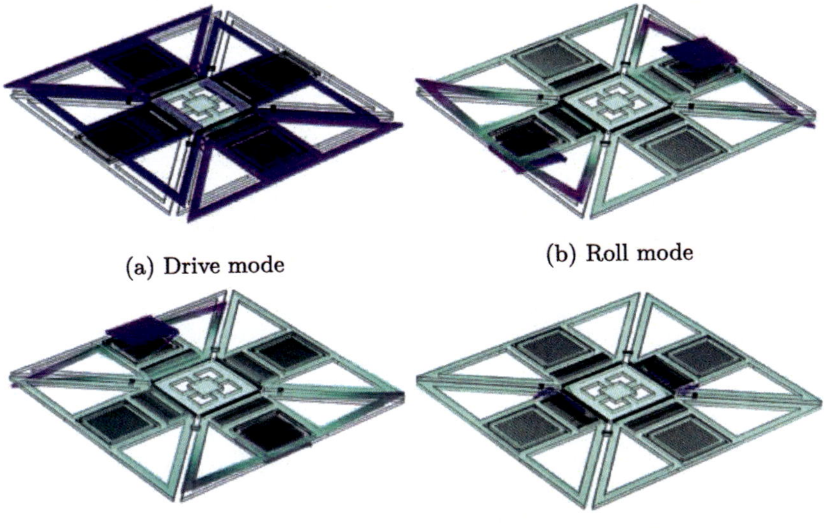

(a) Drive mode

(b) Roll mode

(c) Pitch mode

(d) Yaw mode

Fig. 8. Mode shapes of tri-axis MEMS gyroscope

From the figure, it is evident that the in-plane motions of the drive and sense modes are as expected, but for roll and pitch modes, along with the out-of-plane motion caused by the crab leg beam configuration, there is some unwanted out-of-plane motion of the external flange frames as well. Although this motion might not cause any functional obstruction, it still affects the modal stiffness and hence the natural frequency of vibration of the roll and pitch modes. This is demonstrated in the 'initial' column of Table 2 which summarizes individual numerical and analytical modal stiffness values. This unwanted flange frame movement is further modelled by considering each side frame as a combination of two guided cantilevers in out-of-plane motion, as shown in Fig. 9. Incorporating this assumption, the stiffness values were recalculated and summarized under the 'updated' columns of Table 2. It can be seen that the stiffness now reasonably matches the numerical value. Natural frequencies of the modes were then calculated and summarized in Table 3. The rest of the error is because we have considered the total mass of the beams and not the effective mass. If the effective mass of the beams is precisely modelled, the error can further be reduced.

Table 2. Comparison of numerical and analytical stiffness values (N/m)

	Initial			Updated		
	Drive	Roll, pitch	Yaw	Drive	Roll, pitch	Yaw
Analytical	2799	15.4	79.8	2799	248.2	79.8
Numerical	2875	212.6	78.8	2875	212.6	78.8

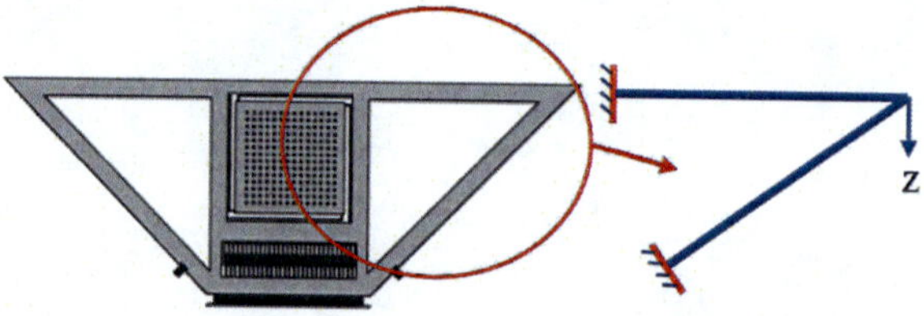

Fig. 9. Assumption of flange frames as cantilever beams

Table 3. Comparison of Numerical and Analytical natural frequencies

–	Drive (Hz)	Roll (Hz)	Pitch (Hz)	Yaw (Hz)
Numerical	18,899	18,532	19,265	18,959
Analytical	19,489	17,850	17,850	18,848
Error (%)	3	3.7	7.3	0.5

Table 3 shows comparison of frequency values obtained analytically and numerically. The roll and pitch geometries are precisely similar, and the structure is reasonably symmetric except that one pair of opposite flanges hosts the electrostatic comb drive while the other hosts the comb drive for yaw sensing. As observed earlier, since the frames and flanges have a parasitic influence on the roll and pitch modes, their frequencies differ by a few hundred Hertz. This is reflected in the numerical frequency results. However, modelling this analytically requires further detailed dynamics studies and will be explored in future work.

5 Conclusion

A novel design of a tri-axis MEMS gyroscope has been presented with in-plane drive mode. Roll and pitch mode displacement is out-of-plane, and yaw mode is sensed in-plane. Numerical simulations are carried out to extract modal frequencies, stiffness, and mass. An analytical model is proposed that closely matches the numerical results with an error of less than 10% for all the modes. The model accuracy can be further improved by considering the effective mass of the beams more precisely. This model provides direct relation between the structural dimensions and the corresponding modal frequencies.

References

1. Persson, A.: The Coriolis effect. History Meteorol. **2**, 1–24 (2005)
2. Greiff, P., Boxenhorn, B., King, T., Niles, L.: Silicon monolithic micromechanical gyroscope. In: TRANSDUCERS'91: 1991 International Conference on Solid-State Sensors and Actuators. Digest of Technical Papers, pp. 966–968. IEEE (1991)
3. Bernstein, J., Cho, S., King, A., Kourepenis, A., Maciel, P., Weinberg, M.: A micromachined comb-drive tuning fork rate gyroscope. In: Proceedings IEEE Micro Electro Mechanical Systems, pp. 143–148 (1993)
4. Usubamatov, R.: Mathematical models for principles of gyroscope theory. In: AIP Conference Proceedings, vol. 1798, no. 1, p. 020133. AIP Publishing LLC (2017)
5. Venkatesh, K., Patil, N., Pandey, A.K., Pratap, R.: Design and characterization of in-plane mems yaw rate sensor. Sadhana **34**(4), 633 (2009)
6. Mohite, S., Patil, N., Pratap, R.: Design, modelling and simulation of vibratory micromachined gyroscopes. J. Phys.: Conf. Ser. **34**(1), 125 (2006) (IOP Publishing)
7. Clark, W.A., Howe, R.T., Horowitz, R.: Surface micromachined z-axis vibratory rate gyroscope. In: Technical Digest Solid-State Sensor and Actuator Workshop, pp. 283–287 (1996)
8. Lutz, M., Golderer, W., Gerstenmeier, J., Marek, J., Maihofer, B., Mahler, S., Munzel, H., Bischof, U.: precision yaw rate sensor in silicon micromachining. in: Proceedings of International Solid State Sensors and Actuators Conference (Transducers'97), vol. 2, pp. 847–850. IEEE (1997)
9. Juneau, T., Pisano, A., Smith, J.H.: Dual axis operation of a micromachined rate gyroscope. In: Proceedings of International Solid State Sensors and Actuators Conference (Transducers'97), vol. 2, pp. 883–886. IEEE (1997)
10. Chang, H., Shen, Q., Zhou, Z., Xie, J., Jiang, Q., Yuan, W.: Design, fabrication, and testing of a bulk micromachined inertial measurement unit. Sensors **10**(4), 3835–3856 (2010)
11. Tanaka, K., Mochida, Y., Sugimoto, M., Moriya, K., Hasegawa, T., Atsuchi, K., Ohwada, K.: A micromachined vibrating gyroscope. Sens. Actuators A: Phys. **50**(1–2), 111–115 (1995)

12. Shcheglov, K., Evans, C., Gutierrez, R., Tang, T.K.: Temperature dependent characteristics of the JPL silicon mems gyroscope. In: 2000 IEEE Aerospace Conference. Proceedings (Cat. No. 00TH8484), vol. 1, , pp. 403–411. IEEE (2000)
13. Dussy, S., Durrant, D., Moy, A., Perriault, N., Celerier, B.: Mems gyros for space applications-overview of European activities. In: AIAA Guidance, Navigation, and Control Conference and Exhibit, p. 6466 (2005)
14. Qu, H., Fang, D., Sadat, A., Yuan, P., Xie, H.: High-resolution integrated microgyroscope for space applications. In: 41st Space Congress, pp. 27–30 (2004)
15. White, A.: A review of some current research in microelectromechanical systems (mems) with defence applications (2002)
16. Ge, H., Liu, Y., Chen, X., Wang, Y., Zhang, S., Jiang, Y., Fan, Z.: Performance degradation analysis of mems gyroscopes in typical weapon equipment under vibration environment. In: 2019 International Conference on Quality, Reliability, Risk, Maintenance, and Safety Engineering (QR2MSE), pp. 1–6. IEEE (2019)
17. Classen, J., Frey, J., Kuhlmann, B., Ernst, P., Bosch, R.: Mems gyroscopes for automotive applications. In: Advanced Microsystems for Automotive Applications, pp. 291–306. Springer (2007)
18. Acar, C., Schofield, A.R., Trusov, A.A., Costlow, L.E., Shkel, A.M.: Environmentally robust mems vibratory gyroscopes for automotive applications. IEEE Sens. J. **9**(12), 1895–1906 (2009)
19. Antonello, R., Oboe, R.: Mems gyroscopes for consumer and industrial applications. Microsensors, pp. 253–280. Intech (2011)
20. Li, Y., Georgy, J., Niu, X., Li, Q., El-Sheimy, N.: Autonomous calibration of mems gyros in consumer portable devices. IEEE Sens. J. **15**(7), 4062–4072 (2015)
21. Tanaka, M.: An industrial and applied review of new mems devices features. Microelectron. Eng. **84**(5–8), 1341–1344 (2007)
22. Tsai, N.-C., Sue, C.-Y.: Design and analysis of a tri-axis gyroscope micromachined by surface fabrication. IEEE Sens. J. **8**(12), 1933–1940 (2008)
23. Tsai, N.C., Sue, C.Y.: Fabrication and analysis of a micro-machined tri-axis gyroscope. J. Micromech. Microeng. **18**(11), 115014 (2008)
24. Tingkai, Z., Chaoyang, X., Ling, Z., Wei, W.: Study on a vibratory tri-axis mems gyroscope with single drive and multiple axes angular rate sense. Microsyst. Technol. **21**(10), 2145–2154 (2015)
25. Gabriele Cazzinga, L.C.: Integrated microelectromechanical gyroscope with improved driving structure. US Patent 8,459,110, 11 June 2013
26. Shah, M.A., Iqbal, F., Lee, B.-L.: Design and analysis of a single-structure three-axis mems gyroscope with improved coupling spring. In: 2016 IEEE 11th Annual International Conference on Nano/Micro Engineered and Molecular Systems (NEMS), , pp. 188–191. IEEE (2016)
27. Fedder, G.K.: Simulation of Microelectromechanical Systems. University of California, Berkeley (1994)

Analysis of the Equilibrium of a Magnetic Contactless Suspension

Pavel Udalov[1(✉)], Alexey Lukin[1], Ivan Popov[1], and Dmitriy Skubov[1,2]

[1] Peter the Great St. Petersburg Polytechnical University, Saint Petersburg, Russia
{udalov_pp,lukin_av,popov_ia}@spbstu.ru
[2] IPME RAS, Saint-Petersburg, Russia

Abstract. In this paper a one-dimensional model of magnetic suspension in the case of a variable magnetic field is considered. Analytical expressions for the averaged over a period of an external current source for the suspension equilibrium position are obtained. It is shown that the magnetic stiffness of the suspension has a sign-variable character in the general case. An expression for the averaged value of the magnetic stiffness is obtained and it is shown that this value is always positive.

Keywords: Levitation · Inertial sensor · Contactless suspension · Micro-electro-mechanical system

1 Introduction

Magnetic contactless suspension is a device that works on the principle of magnetic levitation [1, 2]. As a result, the moving part of the suspension becomes mechanically contactless, which leads to reduce the deterioration of the structural units due to the absence of mechanical friction, loss of mechanical energy, etc.

Magnetic levitation [3] of micro- and nano-suspensions [4, 5] is based on the law of electromagnetic induction [6]. The moving part of this device loses mechanical contact with the stationary part of the structure and gain a sufficiently large (theoretically infinite) sensitivity [7], which allows to use it as a sensor that registers the smallest deviations of motion relative to the initial equilibrium position of the device [7]. To demonstrate how the device works, let's assume that the suspension works as an acceleration sensor. In the simplest case, the accelerometer can be described by a transfer function of a linear oscillator [7]:

$$\frac{Y(p)}{A(p)} = \frac{1}{p^2 + \frac{\mu}{m}p + \frac{c}{m}}, \tag{1}$$

where $Y(p)$ is the mapping of the motion of the levitating part of the suspension, $A(p)$ is the effective transport acceleration, p is the Laplace transform variable, m is the inertial mass, μ is the viscous friction coefficient of the environment, c is the rigidity of the suspension.

A. K. Pandey et al. (Eds.): MAMM 2022, 126, pp. 183–190, 2023.
https://doi.org/10.1007/978-3-031-20353-4_14

The static sensitivity of the accelerometer is written as [7]

$$\frac{Y(p)}{A(p)} = \frac{m}{c}. \tag{2}$$

As the suspension stiffness $c \to 0$ decreases, the sensitivity of the accelerometer, defined by Eq. (2), increases indefinitely and becomes infinite [7]. The transfer function (1) in this case will take the following form

$$\frac{Y(p)}{A(p)} = \frac{1}{p\left(p + \frac{\mu}{m}\right)}. \tag{3}$$

It is noted in works [7, 8] that the integration of acceleration takes place in the absence of the stiffness parameter c. Thus, with the technological possibility of excluding the mechanical stiffness from the model, it is possible to obtain a device with infinite (very large) sensitivity and, consequently, the perspective of measuring arbitrarily small external accelerations. To exclude stiffness, a combination of magnetic and electric suspension is usually used to evaluate the total stiffness of the system to zero for certain parameters of the electric and magnetic parts of the structure. This approach is considered in work [7] when designing an electromagnetic suspension in the case of a disk-like mass, considering only the axial motions of the levitating mass. In works [8, 9] the dynamic equations of the levitating mass with additional consideration of angular displacements due to the appearance of magnetic moments are obtained. In works [10] general conditions for the stable motion of a levitating mass were considered.

The main purpose of this work is an analytical study of the magnetic part of the electromagnetic suspension and obtaining conditions for finding the equilibrium position and stability of vibrations of a levitating object near its equilibrium. The magnetic stiffness of the magnetic suspension is also evaluated. It is shown that in this formulation, it has a harmonic character, which should be considered when adjusting the electrical part of the system to eliminate the magnetic stiffness of the system.

2 Mathematical Model

A contactless magnetic suspension (see Fig. 1) in the simplest case is a system consisting of two basic elements - an inductor coil, to which an alternating current is applied, and a proof levitating mass (PM) made of a conductive material. Due to the law of electromagnetic induction, currents are induced in the PM when current flows in the coil, and an electromagnetic force arises between the parts of the structure. Due to this effect, it is possible to use the PM, for example, as a stand or support for engineering and other structures.

The energy of the magnetic field W of the coil of radius r_c and thickness t_c powered by alternating current $\imath_1$ and the disc-shaped proof mass (PM) of the thickness t_{pm}, radius r_{pm} with induced eddy current $\imath_2$ is

$$W = \frac{1}{2}L_1\imath_1^2 + M_{12}(y)\imath_1\imath_2 + \frac{1}{2}L_2\imath_2^2, \tag{4}$$

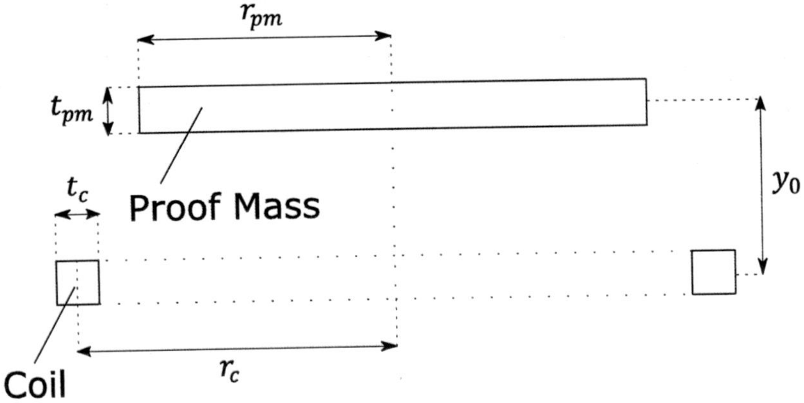

Fig. 1. Scheme of magnetic suspension

where L_1 and L_2 are the self-inductance of coil and PM, respectively [9].

$$L_1 = \mu_0 r_c w^2 \left(\ln \frac{8r_c}{t_c} - 2 \right), L_2 = \mu_0 r_{pm} \left(ln \frac{16r_{pm}}{t_{pm}} - 2 \right), \quad (5)$$

where w is the number of turns of coil, $M_{12}(y)$ is the mutual inductance, μ_0 is the magnetic permeability of free space, $(\dot{\cdot}) = \frac{d}{dt}$, t is the time, y is the distance between coil and PM.

In general case, the mutual inductance $M_{12}(y)$ is a complex non-analytic function. This represents the main difficulty for analytical investigation of the magnetic suspension model [7]. However, it is possible to consider some features of the micro-machine suspension device, which allow the application of the mutual inductance approximation formula [11]. These simplifications consist in the assumption that the linear dimensions of the coil and PM are much larger than the height of the levitation equilibrium position y_0. It is also assumed that the induced eddy current i_2 is distributed along the levitated control mass such that a contour corresponding to the maximum eddy current density can be identified [7].

The eddy current contour is geometrically defined as a circle having the same diameter as the reference mass [7]. Due to the above features of the device, the force interaction in the vertical direction is reduced to the interaction between the eddy current and the levitating coil current [12]. In the case of considering PM and the induced current as circles, the mutual inductance between the PM and the eddy current can be described by the Maxwell formula [11] as follows:

$$M_{12}(\kappa) = \mu_0 \sqrt{r_c r_{pm}} \left[\left(\frac{2}{\kappa} - \kappa \right) K(\kappa) - \frac{2}{\kappa} E(\kappa) \right], \kappa^2(y) = \frac{4 r_c r_{pm}}{(r_c + r_m)^2 + y^2}, \quad (6)$$

where $K(\kappa)$ and $E(\kappa)$ are complete eliptic integrals of the first and second kind, $\kappa(y)$ is the elliptic modulus.

In deriving the equations of dynamics of PM, we consider the dynamics of its center of gravity. From this assumptions potential Π and kinetic T energies of the PM are:

$$\Pi = \tilde{m} g y,\, T = \frac{1}{2}\tilde{m}\dot{y}^2, \tag{7}$$

where $\tilde{m}$ is the mass of the PM, g is the gravity acceleration.

The dissipation function of the system Ψ can be written as follow:

$$\Psi = \frac{1}{2}\mathrm{R}_2 \imath_2^2 + \frac{1}{2}\mu \dot{y}^2, \tag{8}$$

where R_2 is PM's electrical resistance, μ is the coefficient of mechanical friction between PM and external gas environment.

We assume that the current $\imath_1$ generated in the coil by the current generator has the following form

$$\imath_1 = \imath_\mathrm{a} \sin \omega t, \tag{9}$$

where $\imath_a$ and ω are the amplitude and the high frequency of the current $\imath_1$, correspondingly.

We use the Lagrange-Maxwell equations to write equation of motion of the upper ring. The displacement y and the current $\imath_2$ are taken as generalized coordinates:

$$\begin{aligned} &\frac{d}{dt}\frac{\partial W}{\partial \imath_2} + \frac{\partial \Psi}{\partial \imath_2} = 0, \\ &\frac{d}{dt}\frac{\partial T}{\partial \dot{y}} + \frac{\partial(\Pi - T)}{\partial y} + \frac{d\Psi}{d\dot{y}} = 0. \end{aligned} \tag{10}$$

Substituting Eqs. (5)–(9) into Eq. (10), we obtain

$$\begin{aligned} &L_2 \dot{i}_2 + \frac{dM_{12}(y)}{dy}\dot{y}\imath_a \sin \omega t + M_{12}(y)\imath_a \omega \cos \omega t + R_2 i_2 = 0, \\ &\tilde{m}\ddot{y} + \mu\dot{y} - \frac{dM_{12}(y)}{dy}\imath_2 \imath_a \sin \omega t + \tilde{m} g = 0. \end{aligned} \tag{11}$$

We introduce the nondimensional quantities as [13]

$$\begin{aligned} &\xi = \frac{y}{2r_c}, a = \frac{r_{pm}}{r_c}, \beta = \frac{\mu_0 r_c}{L_1}, \gamma = \frac{t_c}{t_{pm}},\ \ j_1 = \frac{\imath_1}{\imath_a}\ \ j_2 = \frac{\imath_2}{\imath_a}, r = \frac{R_2}{L_1\omega}, \\ &m_{12} = \frac{M_{12}}{L_1}, l = \frac{L_2}{L_1}, \tau = \omega t, \varepsilon = \frac{g}{2\omega^2 r_c}, \alpha = \frac{L_1 \imath_a^2}{2\tilde{m} g r_c}, \lambda = \frac{\mu}{\tilde{m}\omega}. \end{aligned} \tag{12}$$

and rewrite Eq. (11) in nondimensional form as

$$\begin{aligned} &l j_2' + r j_2 = -\frac{dm_{12}(\xi)}{d\xi}\xi' sin\tau - m_{12}(\xi)\mathrm{cos}\tau, \\ &\xi'' + \lambda\xi' = \varepsilon\left(\alpha\frac{dm_{12}(\xi)}{d\xi} j_2 \mathrm{sin}\tau - 1\right). \end{aligned} \tag{13}$$

where the prime indicates the derivative with respect to τ. The small nondimensional quantity $\varepsilon\alpha = \frac{L_1 i_a^2}{4r_c^2\omega^2}$ defines the ratio between magnetic and electrical energies.

For further study of the system (13) we find its equilibrium position $(j_2, \xi) = (j_{20}, \xi_0)$, where $j_{20} = j_{20}(\tau)$; ξ_0 is the constant.

3 Equilibrium

WE assume the condition under which the PM reaches an average value $\xi_0 = \frac{y_0}{2r_c}$ (y_0 is the dimensional average equilibrium point) over the period of change in coil current $\imath_1$. In other words, due to the harmonicity of the excitation current $\imath_1$, the magnetic force acting on the PM is also harmonic, which leads to a steady-state oscillatory process of the PM relative to some average value ξ_0. When the magnetic force changes, the equilibrium point changes, which corresponds to the equilibrium of the gravity force and the Ampere force also acting on the PM in the considered formulation. Moreover, in the general case, a variable magnetic field forms a variable magnetic stiffness, as will be shown later.

$$\begin{aligned} j_{20}' + rj_{20} &= -\frac{m_{12}(\xi_0)}{2} e^{i\tau} + c.c, \\ \frac{i\alpha}{2} \frac{dm_{12}(\xi_0)}{d\xi} j_{20} e^{i\tau} &+ \frac{1}{2} + c.c = 0, \end{aligned} \tag{14}$$

where $c.c$ is the complex conjugate value of equation [14].

Solving first equation in (14), we obtain the steady-state soultion for the current j_{20} as

$$j_{20} = \frac{im_{12}(\xi_0)}{2\sqrt{r^2 + l^2}} e^{i(\tau+\phi)} + c.c, \tag{15}$$

where $\cos\phi = \frac{l}{\sqrt{l^2+r^2}}$, $\sin\phi = \frac{r}{\sqrt{l^2+r^2}}$, $i^2 = -1$.

Substituting Eq. (15) in the second equation in (14), we integrate the current j_{20} within $\tau \in [0, \pi]$ and obtain the expression for the average equilibrium ξ_0.

$$m_{12}(\xi_0) \frac{dm_{12}(\xi_0)}{d\xi} = -\frac{2(r^2 + l^2)}{\alpha l}, \tag{16}$$

Figures 2 and 3 show the dependence of the value of the equilibrium position ξ_0 on the physical parameters of the system.

Figures 2 and 3 show that with increasing parameter α (which corresponds to increasing current $\imath_a$ the value ξ_0 increases, which is associated with an increase in electromagnetic force of the ring and, therefore, Ampere force. When the parameter r increases (which corresponds to a decrease in frequency ω), the equilibrium position ξ_0 takes smaller values, which is caused by an increase in reactance.

After finding the condition on the integral position of the suspension equilibrium, it is interesting to estimate the magnetic stiffness for its subsequent compensation by including an electric field, which is studied in the papers [7].

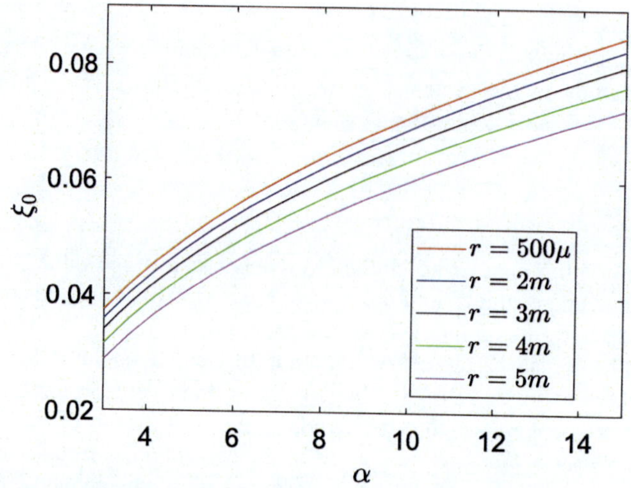

Fig. 2. Dependence of the integral position of the frame equilibrium ξ_0 on the parameter α with ($r = 500\,\mu$; 2 m; 4 m; 5 m)

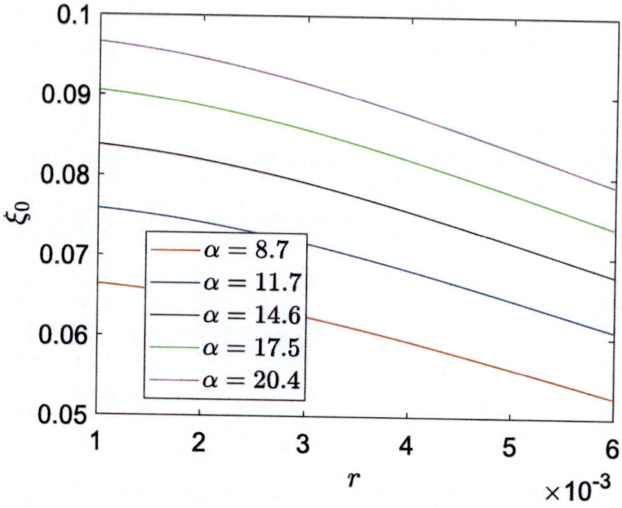

Fig. 3. Dependence of the integral equilibrium position ξ_0 on the parameter r ($\alpha = 8.7$; 11.7; 14.6; 17.5; 20.4)

4 Magnetic Spring Constant of the Suspension

Let us study the behavior of the PM near to the equilibrium point ξ_0. It is assumed that the linear displacement of the PM ξ is small in comparison with ξ_0, hence the following inequality

$$\frac{\xi}{\xi_0} \ll 1, \tag{17}$$

Because of (17), the function of the mutual inductance $m_{12}(\xi)$ can be extended by a Taylor series at the point ξ_0 in the form can be written as follow

$$m_{12}(\xi) = m_{12}(\xi_0) + m'_{12}(\xi_0)(\xi - \xi_0) + m''_{12}(\xi_0)(\xi - \xi_0)^2, \quad (18)$$

Substituting (18) into the last equation of set (10) and considering (16) the differential equation of the linear displacement of the PM near to the equilibrium point can be written as

$$\xi'' + \lambda\xi' + c_m\xi = F_\xi, \quad (19)$$

where F_ξ is generalized force acting on the PM, c_m is the spring constant of the magnetic suspension

$$c_m = \frac{\varepsilon\alpha}{l}\left(m'_{12}(\xi_0)\right)^2 \cos\phi\,\cos 2\phi \sin\tau\,\sin(\tau + \phi). \quad (20)$$

Equation (20) show that the magnetic stiffness has a periodic character and, in the general case, is sign-variable, which means an oscillatory mode of PM levitation near the integral equilibrium position. For a more detailed study of the dependence of the stiffness c_m on the system parameter, consider the average value of the stiffness $c_m = \frac{1}{\pi}\int_0^\pi c_m(\tau)d\tau$

$$c_m = \frac{\alpha\varepsilon l\left(l^2 - r^2\right)}{2\left(r^2 + l^2\right)^2}\left(m'_{12}(\xi_0)\right)^2. \quad (21)$$

The Eq. (21) shows that the average value of the magnetic stiffness is always positive. The only possible problem is related to the sign-variability of the initial stiffness (20), which should be considered in the modes of operation associated with the compensation of magnetic stiffness by adding electrodes to the system and creating a negative electric stiffness, which is done, for example, in the paper [7].

5 Conclusion

In this paper, a magnetic suspension consisting of an inductance coil and a levitating disk was considered. When the coil is energized by alternating current, it is shown that in this case the character of the equilibrium position has an alternating harmonic form. If the equilibrium position comes to the same point during the coil current period, an estimate of the integral mean value of this quantity was given. An expression for the magnetic stiffness of the suspension is obtained and it is shown that it also has a periodic form, which should be considered in the construction of devices of this type.

Funding. The research is funded by Russian Science Foundation grant № 21–71-10009, https://rscf.ru/en/project/21-71-10009/.

References

1. Post, R.F., Ryutov, D.D.: The Inductrack: a simpler approach to magnetic levitation. IEEE Trans. Appl. Supercond. **10**(1), 901–904 (2000)
2. Kordyuk, A.A.: Magnetic levitation for hard superconductors. J. Appl. Phys. **83**(1), 610–612 (1998)
3. Han, H.S., Kim, D.S.: Magnetic Levitation. Springer, Netherlands (2016)
4. Younis, M.I.: MEMS Linear and Nonlinear Statics and Dynamics. Springer, Germany (2011)
5. Gangele, A., Pandey, A.K.: Frequency analysis of carbon and silicon nanosheet with surface effects. Appl. Math. Model. **76**, 741–758 (2019)
6. Grover, F.: Inductance Calculations: Working Formulas and Tables. Dover Publication, Mineola, NewYork (2004)
7. Poletkin, K.V., Chernomorsky, A.I., Shearwood, C.: Proposal for micromachined accelerometer, based on a contactless suspension with zero spring constant. IEEE Sens. J. **12**(7), 2407–2413 (2012)
8. Poletkin, K.V.: Static pull-in behavior of hybrid levitation microactuators: simulation, modeling, and experimental study. IEEE/ASME Trans Mechatron **26**(2), 753–764 (2021)
9. Poletkin, K.V., Shalati, R., Korvink, J.G., Badilita, V.: Pull-in actuation in hybrid micromachined contactless suspension. IOP Publishing, Brislol (2018)
10. Poletkin, K.V., Lu, Z., Wallrabe, U., Korvink, J.G., Badilita, V.: Stable dynamics of micromachined contactless suspension. Int. J. Mech. Sci. **131**(132), 754–766 (2017)
11. Rosa, E.B., Frederick W.G.: Formulas and Tables for the Calculation of Mutual and Self-inductance. No. 169. US Government Printing Office, Washington, D.C. (1948)
12. Lu, Z., Poletkin, K.V., Wallrabe, U., Badilita, V.: Performance characterization of micromachined inductive suspensions based on 3D wire-bonded microcoils. Micromachines **5**, 1469–1484 (2014)
13. Dmitrii, S., Yu., Popov, I.A., Udalov, P.P.: Suspension of Conductive Microring and Its Gyroscopic Stabilization, 08 March 2021, PREPRINT (Version 1). Available at Research Square. https://doi.org/10.21203/rs.3.rs-171012/v1
14. Nayfeh, A.H.: Perturbation Methods. Wiley, New York (2008)

Frequency Analysis of Microbeam with Axial Pretension Using MSGT

Sai Kishore Jujjuvarapu(✉), Indrasena Reddy Erravelly, and Ashok Kumar Pandey

Mechanical and Aerospace Engineering, Indian Institute of Technology Hyderabad, Kandi, Sangareddy, Telangana 502285, India
me18resch11003@iith.ac.in, isreddyerravelly15@gmail.com, ashok@mae.iith.ac.in

Abstract. This paper presents the variation in natural frequencies of Euler-Bernoulli microbeams when subjected to axial pretension. The influence of size effects has been analysed using modified strain gradient theory (MSGT). The governing equation of motion has been derived using extended Hamilton's principle and variational calculus. The sixth order non-local governing differential equation is solved by analytical procedure and numerical differential quadrature method (DQM). The three end conditions of beams are considered: cantilever, simply supported, and clamped-clamped beams. It is found that MSGT accurately models the size effects compared to other theories. As the axial pretension increases from 0.0001 to 1 N, the natural frequency values for the beam with different boundary conditions increase. Subsequently, surface elasticity effects have been analysed for a silicon and aluminium-based nanobeams with the pretension of 0.0001 N for all boundary conditions. From the results of surface elasticity modeling, it has been concluded that the natural frequencies of the nanobeam get influenced either positive or negative based on the value of surface elastic modulus. The difference in natural frequency values with and without surface elasticity effects are approximately 5 and 2% for Si and Al nanobeams respectively. The methodology presented in this work can further be validated for nanoscale devices in which the higher-order strain gradient and surface elasticity effects subjected to pretension dominate.

Keywords: Microbeam · Natural frequency · Modified strain gradient theory · Differential qudrature method · Pretension · Material length scale · Size effects · Surface elasticity

1 Introduction

Microstructures like beams, bars and membranes are used in micro-electro-mechanical systems (MEMS). Generally, MEMS-based sensors or actuators consist of a microbeam as the sensing element [1]. Due to rapid advancements in

A. K. Pandey et al. (Eds.): MAMM 2022, 126, pp. 191–214, 2023.
https://doi.org/10.1007/978-3-031-20353-4_15

nanotechnology and fabrication methods, MEMS devices have various applications in the engineering, medical, and navigation sectors [2]. As many researchers have reported in their work [3–6], microbeams are designed for some basic requirements like natural frequencies of the beams and deflection. In the past, researchers [7–9] have studied the mechanical behaviour of microbeams within the context of classical continuum-based theories. These classical continuum-based theories are just an approximation due to the absence of material length scale parameters to capture the size effect phenomenon, which will influence the performance of the microbeam at the microlevel. Initially, [10–15] have conducted experiments to understand the influence of size effects on the mechanical behaviour of the structures at a microlevel .

Hence, size-dependent higher-order continuum-based theories have been developed to deal with the failure of the classical elasticity theory at small scales. These higher-order continuum-based theories can capture the effects of strain gradient (size effects) through material length scale parameters absent in classical continuum mechanics. Recently, Lam et al. [14] have proposed a modified strain gradient theory (MSGT), in which higher-order stresses and moments of couples are considered along with equilibrium equations. MSGT can be used effectively to predict the static and dynamic responses of microbeams. According to MSGT, the total strain energy density function contains the symmetric strain, dilatation, deviatoric stretch, and rotation gradient tensors. As a result, MSGT has three material length scale parameters to capture dilatation , deviatoric and rotation gradients of isotropic linear elastic materials, respectively and two classical elastic constants(Lame's constants).

Due to its superior characteristics of handling systems at the microlevel, MSGT has received more attention in recent times. Based on MSGT, researchers have done many developments in MEMS. Lu et al. [16] have used nonlocal elasticity theory [17] to investigate the dynamic behaviour of axially prestressed microbeams. Kong et al. [18] have formulated an analytical solution for static and dynamic response of epoxy-based Euler-Bernoulli cantilever beam using MSGT. Results have shown that MSGT predicts the higher natural frequency of microbeam when compared to other theories like classical theory and modified couple stress theory (MCST). Wang et al. [19] have found an analytical solution for static bending and free vibrations of the Timoshenko beam model for simply supported end conditions using MSGT. Akgöz and Civalek [20] have performed buckling analysis of axially loaded epoxy-based, simply-supported and cantilever Euler-Bernoulli microbeams with the help of MSGT. Vatankhah and Kahrobaiyan [21] have investigated vibrational characteristics of the clamped-clamped resonator with attached mass, subjected to axial load, based on MSGT, and results are compared with MCST and classical elasticity theory(CT). Sajal et al. [1] have performed free vibration analysis of epoxy microbeams based on MSGT using the differential quadrature method. Zhao et al. [22] have used MSGT to develop a new Euler-Bernoulli beam model.

In addition to strain gradients, surface elasticity has other effects that affect the dynamic response of the micro/nanostructure. Surface elasticity arises due to forming a surface layer with different mechanical properties than a bulk medium [23]. There are mainly two reasons for the formation surface layer one is surface relaxation [24], and the other is surface reconstruction [25]. Both surface residual stresses and surface elasticity effects are considered to model the surface-related effects in the nano/microstructures [26,27]. Earlier, Lagowski et al. [28] have performed frequency analysis on the microcantilever beam and found that the surface stress is the primary cause for the shift in the natural frequency of the structure. Later Gurtin et al. [29] stated that the surface elasticity is the only phenomenon that causes the shift in the natural frequencies of the structure and is independent of surface stresses. Zhang et al. [30] have measured the shift in resonant frequencies of micro/nanostructure to study surface related effects. Gangele and Pandey [31] have performed frequency analysis using a multi-scale finite element approach on silicon nanocomposites with surface effects. Recently, Fu et al. [32] have used SGT and surface elasticity theory to perform dynamic analysis on the Euler-Bernoullis nanobeam with size and surface-related effects.

Hence, from the above literature, it is found that there is a need to understand the dynamic response of microbeams with axial pretension by considering size and surface elasticity effects. This work presents a dynamic analysis of microbeams with pretension based on MSGT for different end conditions. The sixth order non-local governing differential equation is solved through the analytical procedure and then numerically using the differential quadrature method (DQM). Numerical examples are presented for the three(cantilevered, simply supported, and calmped-clamped) boundary conditions. We show that these two solution methods converge well. The method we developed here can model micro and nanostructures' size and surface elasticity effects.

This paper first presents the general formulation of MSGT and surface-related effects, using Hamilton's principle for the Euler-Bernoulli beam with cantilever, clamped-clamped and simply-supported boundary conditions at varying pretension. Subsequently, we described the analytical and DQM based numerical method and discussed their performance at different pretension.

2 Mathematical Formulation

This section described the modified strain gradient theory (MSGT) and applied it to obtain the governing differential equation.

2.1 Modified Strain Gradient Theory

According to the modified strain gradient theory (MSGT), in addition to the classical strain tensor (ϵ_{ij}),deviatoric stretch gradient tensor ($\tilde{\eta}_{ijk}$), dilatation tensor (γ_i), and symmetric rotation gradient tensor (χ^S_{ij}) are introduced. Therefore MSGT uses three-length scale parameters to consider the size effects.

For a linear isotropic elastic material with V as volume element and occupying region Ω, the total strain energy is defined as [14]

$$U_t = \int_{\Omega} \left(\sigma_{ij}\epsilon_{ij} + p_i\gamma_i + \tilde{\tau}_{ijk}\tilde{\eta}_{ijk} + m_{ij}^S\chi_{ij}^S\right) dV, \tag{1}$$

where, ϵ_{ij}, γ_i, χ_{ij}^S and $\tilde{\eta}_{ijk}$ are defined as

$$\epsilon_{ij} = \frac{1}{2}\left(\partial_i u_j + \partial_j u_i\right), \tag{2}$$

$$\gamma_i = \partial_i \epsilon_{nn}, \tag{3}$$

$$\chi_{ij}^S = \frac{1}{2}\left(e_{ipq}\partial_p\epsilon_{qj} + e_{jpq}\partial_p\epsilon_{qi}\right), \tag{4}$$

$$\begin{aligned}\tilde{\eta}_{ijk} = &\frac{1}{3}\left(\partial_i\epsilon_{jk} + \partial_j\epsilon_{ki} + \partial_k\epsilon_{ij}\right) - \frac{1}{15}\delta_{ij}\left(\partial_k\epsilon_{mm} + 2\partial_m\epsilon_{mk}\right)\\ &-\frac{1}{15}\left[\delta_{jk}\left(\partial_i\epsilon_{mm} + 2\partial_m\epsilon_{mi}\right) + \delta_{ki}\left(\partial_j\epsilon_{mm} + 2\partial_m\epsilon_{mj}\right)\right],\end{aligned} \tag{5}$$

where ∂_i is the differential operator, u_j, ϵ_{nn}, δ_{ij} and e_{ijk} are the displacement vector, dilation strain tensor, Kronecket delta, and permutation tensor respectively. Here, subscripts (i, j, k) are summed over 1–3.

The work-conjugates of the strain gradients γ_i,χ_{ij}^S , and $\tilde{\eta}_{ijk}$ are defined by the higher-order stresses p_i, m_{ij}^S, and $\tilde{\tau}_{ijk}$ respectively. Therefore the stress measures are related to the strains given by the following relationships

$$\sigma_{ij} = k\delta_{ij}\epsilon_{nn} + 2\mu\epsilon_{ij}', \tag{6}$$

$$p_i = 2\mu l_0^2\gamma_i, \tag{7}$$

$$\tilde{\tau}_{ijk} = 2\mu l_1^2\tilde{\eta}_{ijk}, \tag{8}$$

$$m_{ij}^S = 2\mu l_2^2\chi_{ij}^S, \tag{9}$$

where, k represents the bulk modulus and μ is shear modulus of the material. ϵ_{ij}' is given by $\epsilon_{ij} - \frac{1}{3}\epsilon_{nn}\delta_{ij}$ and known as deviatoric strain, ϵ_{ij} is the strain tensor. l_0, l_1 and l_2 are the three material length scale parameters related to the material's dilation, deviatoric and symmetric gradients, respectively.

2.2 Surface Elasticity

When the characteristic dimensions of the beam are in the order of micro or nanometer range, surface related effects often play an essential role in the design of MEMS/NEMS structures; this is due to an increase in the ratio between the surface area to volume of the structure. Due to surface effects, the layer on the structure's bulk will have different mechanical properties than the bulk medium.

Classical theory is just an approximation where the surface effects are not considered to perform frequency analysis. Hence classical theory would require proper corrections which can capture the small scale and surface-related effects. The drawback of the classical elasticity theory can be overcome by adequately incorporating the surface elasticity effects to determine its impact on the mechanical properties of micro/nanostructures.

From the basic assumption involved in the design of micro/nanostructure is that "Structure = Bulk + Surface" [23], the effective elastic modulus of micro/nanobeam as [31,33]

$$E^* = E\left[1 + 6\frac{h_0}{h} + 2\frac{h_0}{b}\right], \tag{10}$$

here, $h_0 = \frac{C_s}{E}$ is a material length parameter to capture the surface related effects at micro/nano level and C_s is the surface elastic modulus (Nm^{-1}). Equation (10) will be used in governing equation of motion to perform frequency analysis of micro/nanobeam by considering the surface elasticity effects.

2.3 Governing Equation of Motion

Let us consider a prismatic beam as shown in Fig. 1 of length L, having cross sectional area A, and the axial load N_0. The area of cross-section is specified by the x (longitudinal) and y (lateral) axes. The load q is applied in the transverse direction, i.e., the z axis.

Let the displacements in the x, y, and z directions are represented by u, v, w respectively. The displacement field for the beam from the Euler-Bernoulli beam model is defined as

$$u = -z\frac{\partial w(x)}{\partial x}, v = 0, w = w(x). \tag{11}$$

The non-local governing differential equation of motion and the corresponding boundary conditions are derived using extended Hamilton's and variational principle.

$$\delta\left[\int_{t_1}^{t_2} (T - U_t + W_{ext})\,dt\right] = 0, \tag{12}$$

here, δ indicates the first variation, T, U_t and W_{ext} are kinetic energy, strain energy and work done by the external forces respectively.

The governing differential equation the motion (GDE) of the microbeam by MSGT is found as

$$P\frac{\partial^4 w}{\partial x^4} - Q\frac{\partial^6 w}{\partial x^6} - N_0\frac{\partial^2 w}{\partial x^2} + \rho A\frac{\partial^2 w}{\partial t^2} + q(x) = 0, \tag{13}$$

and the corresponding boundary conditions are

$$\left[\left(V - Pw^{(3)} + Qw^{(5)} - N_0 w'\right)\delta w\right]_0^L = 0, \tag{14}$$

$$\left[\left(M + Pw^{(2)} - Qw^{(4)}\right)\delta w'\right]_0^L = 0, \tag{15}$$

$$\left[\left(M^h + Qw^{(3)}\right)\delta w''\right]_0^L = 0. \tag{16}$$

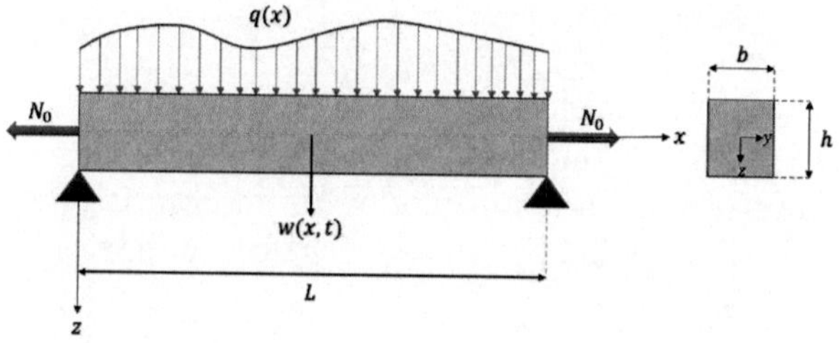

Fig. 1. Beam configuration when subjected to flexural and axial loads

Here, P and Q are,

$$P = EI + 2\mu A l_0^2 + \frac{8}{15}\mu A l_1^2 + \mu A l_2^2, Q = I\left(2\mu l_0^2 + \frac{4}{5}\mu l_1^2\right). \tag{17}$$

When the two material length scale parameters, $l_0 = l_1 = 0$, the sixth order non-local GDE (Eq. 13) is reduces to MCST. Hence, the equation of motion based on MCST is obtained as

$$\left(EI + \mu A l_2^2\right)\frac{\partial^4 w}{\partial x^4} - N_0\frac{\partial^2 w}{\partial x^2} + \rho A\frac{\partial^2 w}{\partial t^2} + q = 0, \tag{18}$$

and the boundary conditions are

$$\left[\left(V - \left(EI + \mu A l_2^2\right) w^{(3)} - N_0 w'\right)\delta w\right]_0^L = 0, \tag{19}$$

$$\left[\left(M + \left(EI + \mu A l_2^2\right) w^{(2)}\right)\delta w'\right]_0^L = 0. \tag{20}$$

If all the three material length scale parameters are assumed to zero, then the GDE (Eq. 13) is reduced to classical elasticity theory.

3 Solution Procedure

This section will discuss the analytical procedure and implementation of the DQM numerical method to solve the governing differential equation.

3.1 Analytical Solution

Now, we discuss the solution of the GDE (Eq. 13). The transverse load ($q(x) = 0$) is assumed to zero. Hence, the governing differential equation becomes

$$P\frac{\partial^4 w}{\partial x^4} - Q\frac{\partial^6 w}{\partial x^6} - N_0\frac{\partial^2 w}{\partial x^2} + m\frac{\partial^2 w}{\partial t^2} = 0, \tag{21}$$

where, $m = \rho A$.
Let us consider the solution to the above equation as

$$w(x,t) = \tilde{w}(x)e^{i\omega t}. \tag{22}$$

Substituting Eq. (22) into the Eq. (21), we get

$$P\tilde{w}^{(4)}(x) - Q\tilde{w}^{(6)}(x) - N_0\tilde{w}^{(2)}(x) - m\omega^2\tilde{w} = 0. \tag{23}$$

Equation (23) has the solution of the form

$$\tilde{w}(x) = \sum_{i=1}^{6} C_i e^{\lambda_i x}. \tag{24}$$

The exponents, $\lambda_i (i = 1, 2, \ldots, 6)$ are complex in nature and C_i are the constants of integration and found by using the boundary conditions.

$$P\lambda^4 - Q\lambda^6 - N_0\lambda^2 - m\omega^2 = 0, \tag{25}$$

The auxiliary equation (25) has six roots (λ_i) which are complex in nature and can be found as,
let $\alpha' = -36N_0PQ - 108m\omega^2Q^2 + 8P^3$ and

$$\alpha_1 = \left(\alpha' + 12\sqrt{3}\sqrt{4N_0^3Q - N_0^2P^2 + 18N_0PQm\omega^2 + 27m^2\omega^4Q^2 - 4m\omega^2P^3K}\right)^{\frac{1}{3}},$$

$$\lambda_1 = \frac{-1}{\sqrt{6Q\alpha_1}}\left(\alpha_1^2 - 12N_0Q + 4P^2 + 2P\alpha_1\right)^{\frac{1}{2}},$$

$$\lambda_2 = \frac{1}{\sqrt{6Q\alpha_1}}\left(\alpha_1^2 - 12N_0Q + 4P^2 + 2P\alpha_1\right)^{\frac{1}{2}},$$

$$\lambda_3 = \frac{-1}{2\sqrt{3Q\alpha_1}}\left(-\alpha_1^2 + 12N_0Q - 4P^2 + 4P\alpha_1 - I\sqrt{3}\alpha_1^2 - 12I\sqrt{3}N_0Q + 4I\sqrt{3}P^2\right)^{\frac{1}{2}},$$

$$\lambda_4 = \frac{1}{2\sqrt{3Q\alpha_1}}\left(-\alpha_1^2 + 12N_0Q - 4P^2 + 4P\alpha_1 - I\sqrt{3}\alpha_1^2 - 12I\sqrt{3}N_0Q + 4I\sqrt{3}P^2\right)^{\frac{1}{2}},$$

$$\lambda_5 = \frac{-1}{2\sqrt{3Q\alpha_1}}\left(-\alpha_1^2 + 12N_0Q - 4P^2 + 4P\alpha_1 + I\sqrt{3}\alpha_1^2 + 12I\sqrt{3}N_0Q - 4I\sqrt{3}P^2\right)^{\frac{1}{2}},$$

$$\lambda_6 = \frac{1}{2\sqrt{3Q\alpha_1}}\left(-\alpha_1^2 + 12N_0Q - 4P^2 + 4P\alpha_1 + I\sqrt{3}\alpha_1^2 + 12I\sqrt{3}N_0Q - 4I\sqrt{3}P^2\right)^{\frac{1}{2}}.$$

Here, I represents the complex nature of the roots.

Now, we discuss the application of the above described analytical procedure to find the natural frequencies of the microbeam for different boundary conditions.

Simply Supported Beam Let us consider the case of simply supported beam with length L, the classical boundary conditions in terms of displacement and moments are

$$\tilde{w}(0) = \tilde{w}(L) = 0, \tag{26}$$

$$M(0) = M(L) = 0, \tag{27}$$

and the non-classical boundary conditions are

$$\tilde{w}^{(4)}(0) = \tilde{w}^{(4)}(L) = 0. \tag{28}$$

By substituting the Eq. (24) into the above equations, we get the matrix form of the equation as

$$[B(\omega)]\{C\} = \{0\}. \tag{29}$$

In the Eq. (29), $\{C\} = \{C_i\}^T, i = 1, 2, \ldots, 6$ and the matrix $[B(\omega)]$ is found as

$$B_{1i} = 1, B_{2i} = e^{\lambda_i L}, B_{3i} = \lambda_i^2, B_{4i} = \lambda_i^2 e^{\lambda_i L}$$
$$B_{5i} = \lambda_i^4, B_{6i} = \lambda_i^4 e^{\lambda_i L}, \tag{30}$$

where $\lambda_i = \lambda_i(\omega)$ are the six roots of the Eq. (31) and $i = 1, 2, \ldots, 6$.
In order to get the non-zero trivial solution of the Eq. (29), the following condition has to be satisfied,

$$\mathbf{det}[B(\omega)] = 0. \tag{31}$$

The above Eq. (31) is called the "frequency equation". The analytical solution of the above system gives all the microbeam natural frequencies that can be found using a complex arithmetic procedure [1].

Clamped-Clamped Beam Let us consider a clamped-clamped beam with length L, having fixed at both the ends, the classical boundary conditions in terms of displacement and slope are

$$\tilde{w}(0) = \tilde{w}(L) = 0, \tag{32}$$

$$\tilde{w}'(0) = \tilde{w}'(L) = 0, \tag{33}$$

and the non-classic boundary conditions are

$$\tilde{w}^{(4)}(0) = \tilde{w}^{(4)}(L) = 0. \tag{34}$$

By substituting the Eq. (24) into the above equations, we get the matrix form of the equation as

$$[B(\omega)]\{C\} = \{0\}. \tag{35}$$

In the Eq. (35), $\{C\} = \{C_i\}^T, i = 1, 2, \ldots, 6$ and the matrix $[B(\omega)]$ is found as

$$B_{1i} = 1, B_{2i} = e^{\lambda_i L}, B_{3i} = \lambda_i, B_{4i} = \lambda_i e^{\lambda_i L}$$
$$B_{5i} = \lambda_i^4, B_{6i} = \lambda_i^4 e^{\lambda_i L}. \tag{36}$$

By following the procedure mentioned above, we can find the natural frequencies of the clamped-clamped beam with axial pretension.

Cantilever Beam Consider a cantilever microbeam with length L, one end is clamped and other end is free to move, the classical boundary conditions are

$$\tilde{w}(0) = \tilde{w}'(0) = 0, \tag{37}$$

$$Q\tilde{w}^{(5)}(L) - P\tilde{w}^{(3)}(L) + N_0\tilde{w}'(L) = 0, P\tilde{w}''(L) - Q\tilde{w}^{(4)}(L) = 0, \tag{38}$$

and the non-classical boundary conditions are

$$\tilde{w}''(0) = \tilde{w}^{(3)}(0) = 0. \tag{39}$$

By substituting the Eq. (24) into the above equations, we get the matrix form of the equation as

$$[B(\omega)]\{C\} = \{0\}. \tag{40}$$

In the Eq. (40), $\{C\} = \{C_i\}^T, i = 1, 2, \ldots, 6$ and the matrix $[B(\omega)]$ is found as

$$\begin{aligned} B_{1i} = 1, B_{2i} = \lambda_i, B_{3i} = \lambda_i^2, B_{4i} = \lambda_i^3 e^{\lambda_i L} \\ B_{5i} = \left(\lambda_i^5 Q - \lambda_i^3 P + N_0\lambda_i\right) e^{\lambda_i L}, B_{6i} = \left(\lambda_i^2 P - \lambda_i^4 Q\right) e^{\lambda_i L}. \end{aligned} \tag{41}$$

Following the procedure mentioned earlier, we can find the natural frequencies of the cantilevered microbeam subjected to axial pretension.

Now, we discuss the theoretical framework and implementation of the differential quadrature method (DQM) to solve the sixth order governing differential equation of motion.

3.2 Differential Quadrature Method (DQM)

The differential quadrature method(DQM) was proposed by Bellman et al. [34]. The DQM is a numerical method that approximates a function's derivative at a point by taking the sum of weighted functional values at other grid points in the discretized domain. The first derivative of the function $f(x)$ is expressed as [35]

$$f_x(x_i) = \frac{df}{dx}|_{x_i} = \sum_{j=1}^{N} a_{ij} \cdot f(x_j), \tag{42}$$

where $f(x_j)$ is the functional value at the jth sampling point, a_{ij} are the weighting coefficients and N_s is the number of grid points in the domain. It reduces the differential equations into a set of algebraic equations.
The Chebyshev-Gauss-Lobatto distribution will discretise the domain into N_s sampling points.

$$X_i = \frac{1}{2}\left[1 - cos\left(\frac{i-1}{N_s - 1}\pi\right)\right], \tag{43}$$

The weighting coefficients are defined as

$$a_{ii}^{(1)} = -\sum_{j=1, j\neq i}^{N_s} a_{ij}^{(1)}, \quad i = j, \tag{44}$$

$$a_{ij}^{(1)} = \frac{L^{(1)}(x_i)}{(x_i - x_j)L^{(1)}(x_j)}, \quad i \neq j, \tag{45}$$

here, $L^{(1)}(x_i)$ is the Lagrange interpolating polynomials' first derivative, and it is written as

$$L^{(1)}(x_i) = \prod_{k=1,k\neq i}^{N_s} (x_i - x_k), \tag{46}$$

The higher order weighted coefficient matrices can be obtained using

$$a_{ii}^{(n)} = - \sum_{j=1,j\neq i}^{N_s} a_{ij}^{(n)}, \quad i = j, n = 2, 3, \ldots, N_s - 1. \tag{47}$$

$$a_{ij}^{(n)} = n\left(a_{ij}^{(n-1)} a_{ij}^{(1)} - \frac{a_{ij}^{(n-1)}}{x_i - x_j}\right), \quad i \neq j, n = 2, 3, \ldots, N_s - 1, \tag{48}$$

here, note that the subscripts i, j are repeated indices from 1 to N_s.
Hence, from the DQM, the differential equation of motion Eq. (13), can be written as [41]

$$P\sum_{n=1}^{N_s} a_{mn}^{(4)}\tilde{w}_n - Q\sum_{n=1}^{N_s} a_{mn}^{(6)}\tilde{w}_n - N_0\sum_{n=1}^{N_s} a_{mn}^{(2)}\tilde{w}_n - \rho A\omega^2\tilde{w}_n = 0, \quad m = 1, 2, \ldots, N_s. \tag{49}$$

Various approaches are proposed to apply the boundary conditions for solving the Eq. (49). In DQM method, the vibrational problems are solved with the SBCGE (Substitution of Boundary Conditions into Governing Equations) technique [36] and δ-technique [37], used for clamped-clamped end simply-supported and conditions. The complete details of DQM, implementation of boundary conditions, and solution method are available in the literature [1].

4 Results and Discussion

This section discusses the implementation of MSGT for the different sets of beams and compares both analytical and numerical DQM results. For the frequency analysis, we considered epoxy-based microbeam with strain gradient effects.

4.1 Strain Gradient Effects

Now we discuss the influence of the strain gradient effects on the microbeams at varying axial pretension. To study the effect of pretension, the following properties are considered for the numerical purpose, the beam properties are taken as that of epoxy [14], and the material properties are, Young's Modulus $(E) = 1.44$

Table 1. Set of Epoxy microbeams with different geometries

Sets	1	2	3	4	5	6	7	8	9	10
Thickness, h (μm)	10	20	30	40	50	60	70	100	150	200
Width, b = 2h (μm)	20	40	60	80	100	120	140	200	300	400
Length, L = 20h (μm)	200	400	600	800	1000	1200	1400	2000	3000	4000

GPa, Poisson's ratio (ν) = 0.38 and density (ρ) = 1000 kg/m^3 [18] and corresponding geometrical information has provided in Table 1. From MSGT material length scales for epoxy micro beam are considered as $l_0 = l_1 = l_2 = l = 17.6 \times 10^{-6}$ μm.

Table 2 shows the three fundamental modal frequencies of the simply-supported (SS) microbeam with a thickness of 20 μm using the analytical method, differential quadrature method (DQM). The obtained results are compared with Zhao et al. [38] and matched with each other. Hence, MSGT can be effectively used to model and design micro/nanobeams with axial pretension.

Table 2. Comparison of the first three natural frequencies of simply supported beam ($h = 20$ μm) with axial pretension

	Axial force	ω_1 (MHz)	ω_2 (MHz)	ω_3 (MHz)
Zhao et al. [38]	$N_0 = 1$	8.902774	18.519468	29.50064
Analytical		8.896462	18.613040	29.853069
DQM		8.914291	18.610294	29.788022
Zhao et al. [38]	$N_0 = 0.01$	1.708429	6.13419	13.53744
Analytical		1.772355	6.402038	14.148805
DQM		1.767054	6.403434	14.153800
Zhao et al. [38]	$N_0 = 0.0001$	1.46974	5.88052	13.28097
Analytical		1.535676	6.159607	13.906361
DQM		1.536444	6.159937	13.908860

Simply-Supported Microbeam Figures 2, 3 and 4 show the first three natural frequencies for the simply supported (SS) microbeam, subjected to different axial loads ($N_0 = 1, 0.01, 0.0001$ N). The analytical method and the numerical technique (DQM) give closer results in the case of the simply supported beam, with the error being less than 0.2%. One important observation found is that when the thickness of the microbeam is reduced from $200 \mu m$ to $10 \mu m$, there was an appreciable difference in different theories, and strain gradients effects have prevailed. The natural frequencies found in the different theories are separated by reducing the axial force. With the reduction in the axial force, the natural frequencies decrease, and the size effects are turned up. The MSGT gives a higher value of natural frequencies than the MCST and classical theory because the formulation of MSGT considers the extra equilibrium equations, thereby accounting for the additional length scale effects. The present results show the same trend, with MSGT giving higher values of natural frequencies than MCST and classical theory. The effect of material length scale parameters reduces as the beam's size increases, and all the theories' results converge.

Clamped-Clamped Microbeam In Figs. 5, 6 and 7, the first three fundamental natural frequency values of the clamped-clamped beams with three different axial loads ($N_0 = 1, 0.01, 0.0001\,\mathrm{N}$) are compared. The MSGT gives high values of natural frequencies, and the classical theory gives lower frequency values. In this case, both analytical and differential quadrature methods give closer values with an error less than 2%. The same trend is observed here; with the reduction in the axial force, the microbeam becomes softened. Hence, the natural frequencies decrease, and the size effects are turned up. The effect of material length scale parameters reduces as the beam's size increases, and the results from all the theories converge. The clamped-clamped beam is slightly stiffer for the same geometrical configuration as the SS beam. Hence, the first three natural frequencies of the clamped-clamped beam are higher than the SS beam for the same axial pretension. Hence, the results show that thin beams need to be modeled accurately by considering strain gradients at low pretension values.

Cantilever Microbeam Figures 8, 9 and 10 show the first three natural frequencies of the cantilever beam obtained from MSGT, MCST and classical theories. The natural frequencies obtained from MSGT are higher than those obtained from MCST and classical theory. The analytical and DQ methods give closer results, with the error being less than 9.71%. In the case of the cantilever beam, the error is slightly higher than the SS beam and FF beams due to the application of boundary conditions in the differential quadrature method. As we are applying the boundary conditions at a distance of δ away from the boundaries, this gives a slight variation in natural frequencies [1]. The difference in natural frequency from different theories increases by reducing the axial pretension value. For lower values of pretension, MSGT is capable enough to capture size effects quite accurately compared to other theories. As the thickness of the beam increases, all the theories converge well, and size effects are reduced. A cantilever microbeam has lower natural frequencies than a simply supported and clamped-clamped beam for the same geometrical configuration.

4.2 Strain Gradient with Surface Elasticity

This section discusses the effect of strain gradients and surface elasticity effects in Silicon and Aluminium-based nanobeams using MSGT theory. The first three natural frequencies of the nanobeam with a pretension of 0.0001 N have been computed for different boundary conditions using the analytical method.

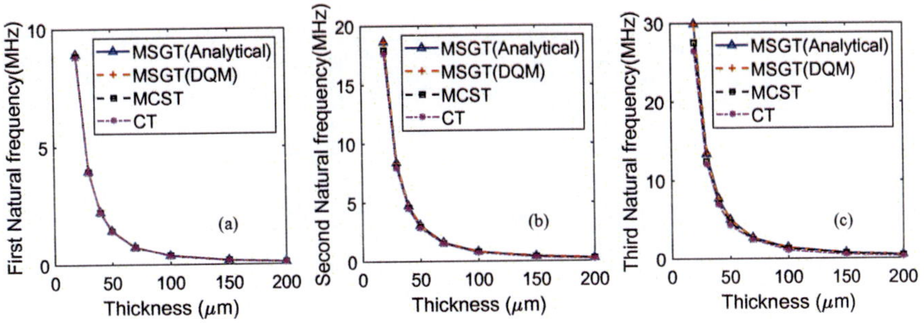

Fig. 2. Comparison of the first three natural frequencies ((a), (b) and (c)) of SSB with axial load $N_0 = 1\,\mathrm{N}$

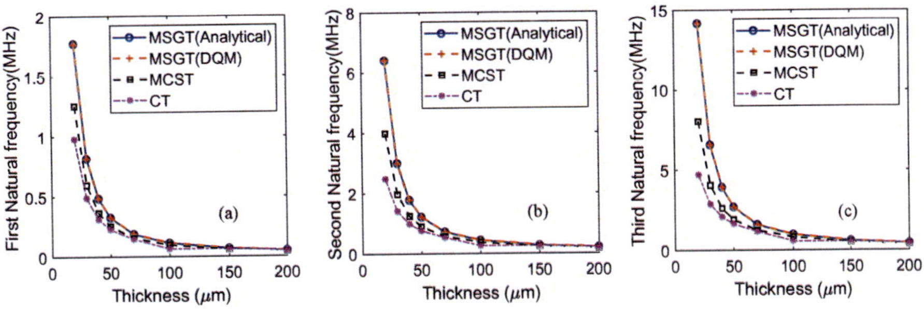

Fig. 3. Comparison of the first three natural frequencies ((a), (b) and (c)) of SSB with axial load $N_0 = 0.01\,\mathrm{N}$

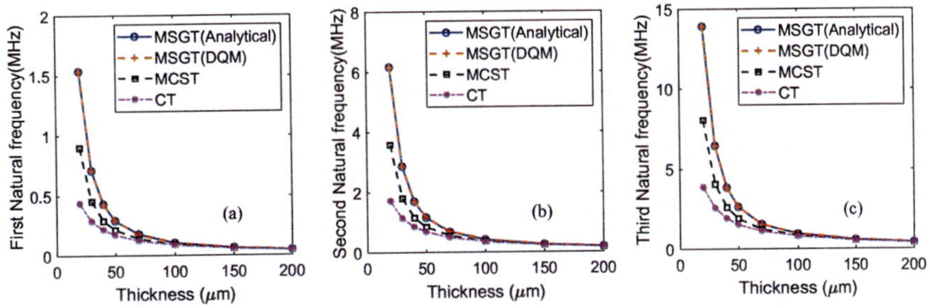

Fig. 4. Comparison of the first three natural frequencies ((a), (b) and (c)) of SSB with axial load $N_0 = 0.0001\,\mathrm{N}$

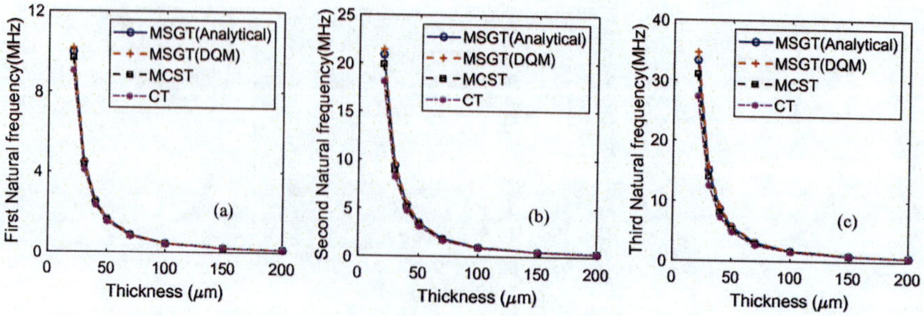

Fig. 5. Comparison of the first three natural frequencies ((a), (b) and (c)) of clamped-clamped beam with axial load $N_0 = 1\,\mathrm{N}$

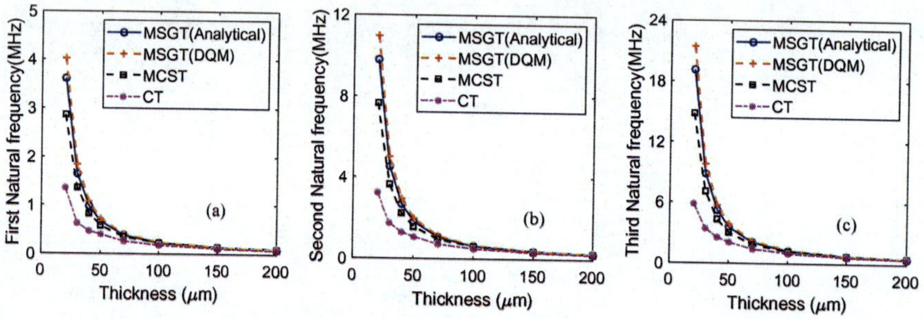

Fig. 6. Comparison of the first three natural frequencies ((a), (b) and (c)) of clamped-clamped beam with axial load $N_0 = 0.01\,\mathrm{N}$

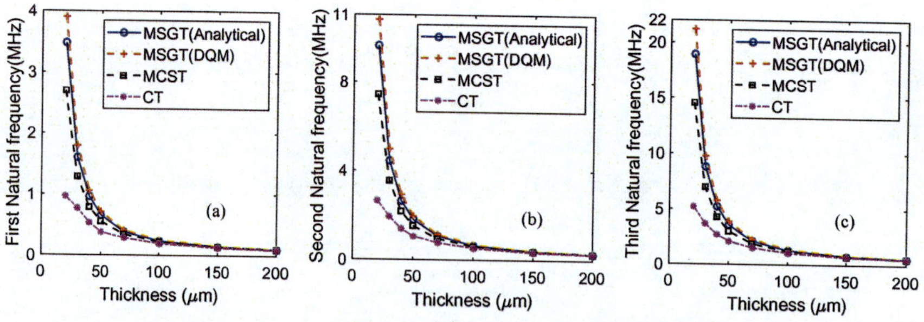

Fig. 7. Comparison of the first three natural frequencies ((a), (b) and (c)) of clamped-clamped beam with axial load $N_0 = 0.0001\,\mathrm{N}$

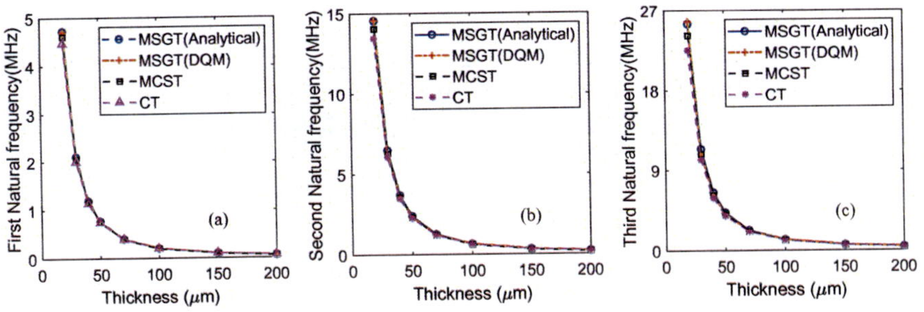

Fig. 8. Comparison of the first three natural frequencies ((a), (b) and (c)) of Cantilever beam with axial load $N_0 = 1\,\text{N}$

Table 3. Set of nanobeams with different geometries (From Fu et al. [32])

Sets	1	2	3	4	5	6	7	8	9	10
Thickness, h (nm)	1	2	3	4	5	6	7	8	9	10
Width, b = h (nm)	1	2	3	4	5	6	7	8	8	10
Length, L = 10h (nm)	10	20	30	40	50	60	70	80	90	100

Table 4. Mechanical properties of Si and Al-based nanobeams

	E(Gpa)	ν	ρ (kg/m^3)	C_S (Nm^{-1})	$l_0 = l_1 = l_2$(μm)	$h_0 = C_S/E$ (nm)
Silicon [32]	107	0.33	2330	−11.5	0.2961 [32]	−0.1074
Aluminum [39]	70	0.30	2700	5.1882	0.3500 [40]	0.0741

Silicon Nanobeam For frequency analysis on silicon-based nanobeam with different configurations, Table 3 and the corresponding mechanical properties are presented in Table 4. The analytical method computed the first three natural frequencies for different end conditions using MSGT with and without considering the surface elasticity effects with axial pretension of 0.0001 N. Figures 11, 12 and 13 show the first fundamental three natural frequencies for SSB, clamped-clamped and cantilever beam respectively. In Silicon nanobeams, surface-related effects negatively influence natural frequencies because of Silicon's negative surface elastic modulus (C_S), making the structure soften. Considering surface elasticity effects, the maximum difference in natural frequencies is approximately 5%. Hence, the effect of surface elasticity will have an appreciable influence on the frequency response of the nanobeams.

Aluminum Nanobeam Subsequently, we also performed frequency analysis on Aluminum based nanobeam with considering the surface elasticity effects and the corresponding geometrical configurations and mechanical properties are shown in Tables 3 and 4, respectively. The first three natural frequencies have been computed for different end conditions using analytical method of MSGT

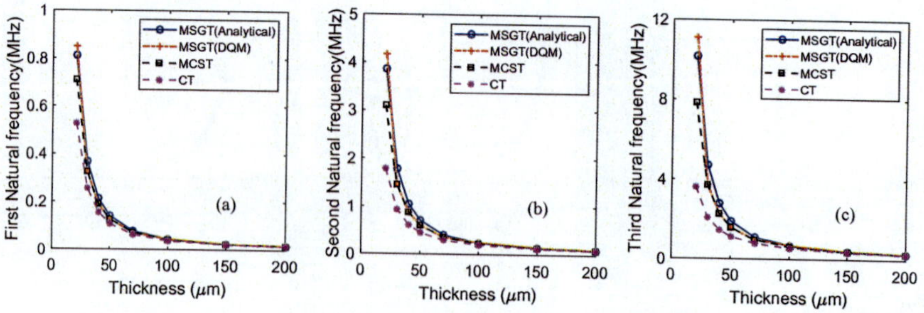

Fig. 9. Comparison of the first three natural frequencies ((a), (b) and (c)) of Cantilever beam with axial load $N_0 = 0.01\,\mathrm{N}$

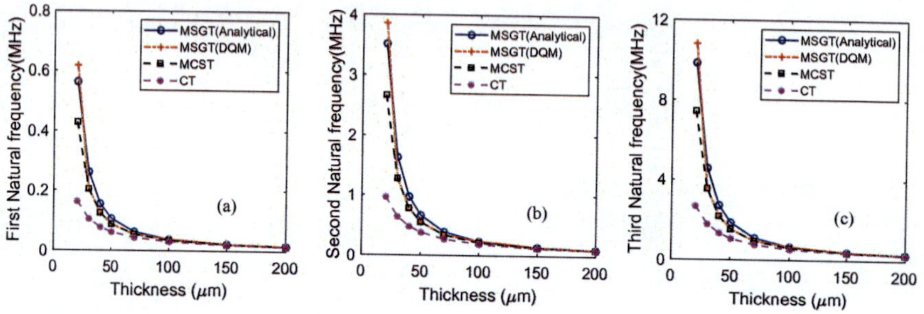

Fig. 10. Comparison of the first three natural frequencies ((a), (b) and (c)) of Cantilever beam with axial load $N_0 = 0.0001\,\mathrm{N}$

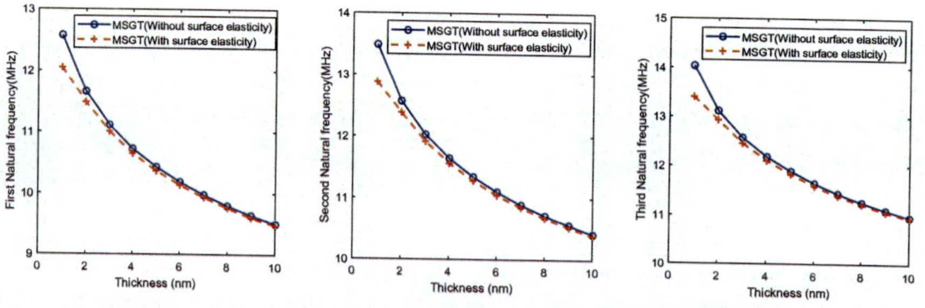

Fig. 11. Comparison of the first three natural frequencies ((a), (b) and (c)) of Silicon SSB including surface elasticity effects and subjected to axial load of $N_0 = 0.0001$

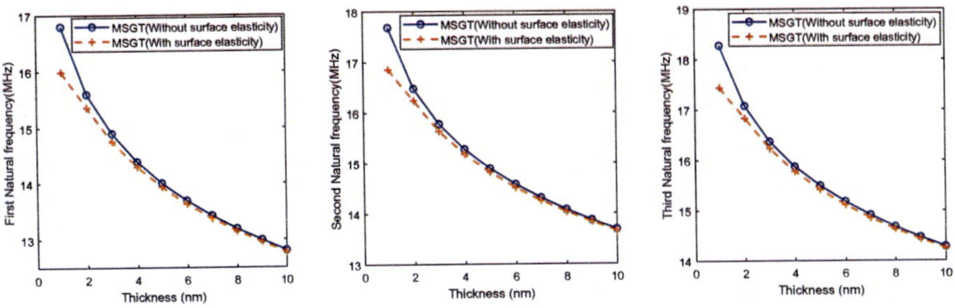

Fig. 12. Comparison of the first three natural frequencies ((a), (b) and (c)) of Silicon clamped-clamped beam including surface elasticity effects and subjected to axial load of $N_0 = 0.0001$

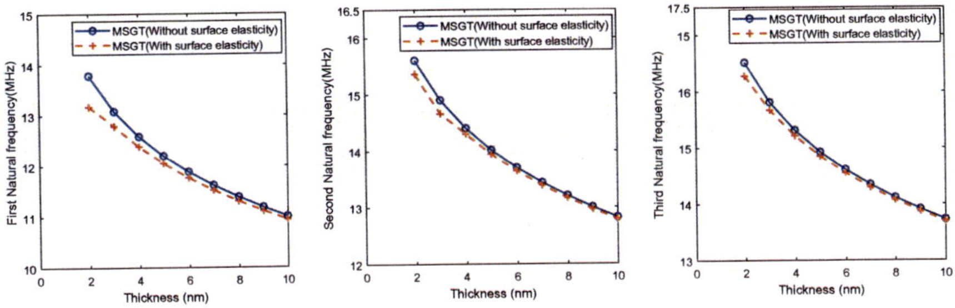

Fig. 13. Comparison of the first three natural frequencies ((a), (b) and (c)) of Silicon cantilever beam including surface elasticity effects and subjected to axial load of $N_0 = 0.0001$

with and without considering surface elasticity effects when they are subjected to axial loading of 0.0001 N. Figures 14, 15, and 16 show the first three fundamental natural frequencies of Aluminium-based SSB, clamped-clamped and cantilever beam respectively. From Figs. 14, 15 and 16, the natural frequencies corresponding to surface elasticity effects are higher than those without surface elasticity effects. In the case of Aluminum nanobeam, surface-related effects positively influence natural frequencies because of positive surface elastic modulus (C_S), making the structure stiffer. The natural frequency values with surface elasticity are approximately 2% higher than those without surface effects. When the dimensions increase further, the difference in frequencies is reduced. This difference is small because the choice of geometrical configuration (length, width and thickness) greatly influences the structure's frequency response. The surface elasticity effects in Silicon and Aluminum nanobeams become significant at lower thickness and length of the beam. These effects are accurately captured by introducing an effective elastic modulus [31]. Hence, the effect of surface elasticity on the structure's natural frequency can't be neglected to have the optimum design of nanobeams.

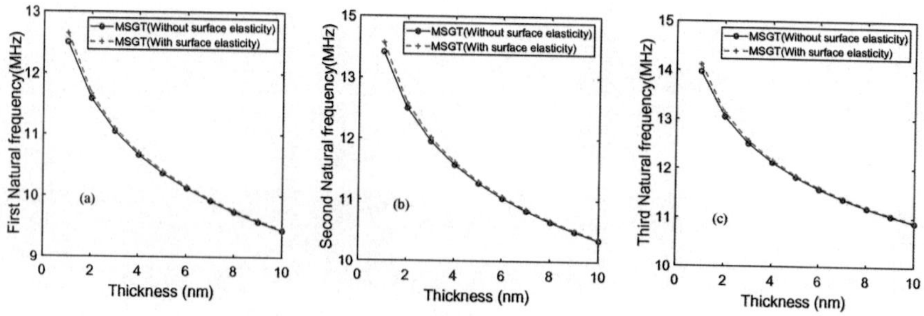

Fig. 14. Comparison of the first three natural frequencies ((a), (b) and (c)) of Al SSB including surface elasticity effects and subjected to axial load of $N_0 = 0.0001\,\mathrm{N}$

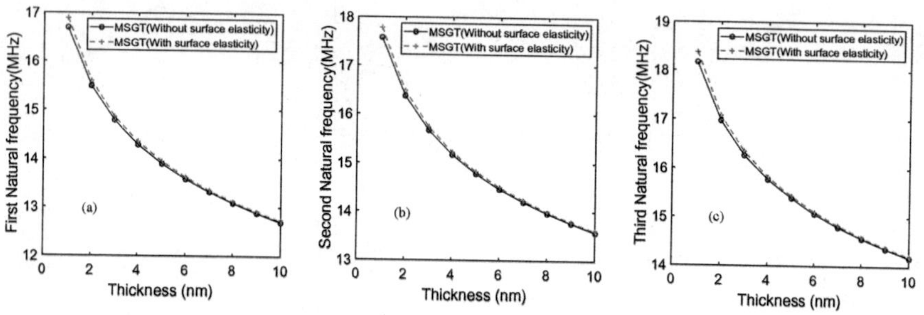

Fig. 15. Comparison of the first three natural frequencies ((a), (b) and (c)) of Al clamped-clamped beam including surface elasticity effects and subjected to axial load of $N_0 = 0.0001\,\mathrm{N}$

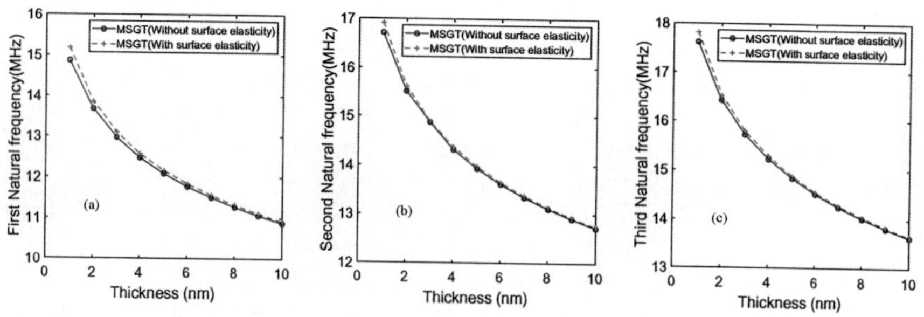

Fig. 16. Comparison of the first three natural frequencies ((a), (b) and (c)) of Al cantilever beam including surface elasticity effects and subjected to axial load of $N_0 = 0.0001\,\mathrm{N}$

Further, from the above discussion, we can state that the surface elasticity effects will affect the system's frequency response at the nanolevel either positively or negatively, depending upon the surface elastic modulus of the beam [30]. Therefore it is critical to consider surface elasticity effects in addition to strain gradients while modeling the systems at the micro/nano level.

5 Conclusions

This work presented the frequency analysis of epoxy-based microbeams with axial pretension using modified strain gradient theory. Theoretical formulation and corresponding governing equation of motion has been derived by using extended Hamilton's principle. Three end conditions are considered (SSB, Clamped-Clamped, and Cantilever). The governing sixth order differential equation has been solved using the standard analytical method and numerical differential quadrature technique. The first three natural frequencies are found for different beam configurations by varying the axial pretension. The obtained results are validated with existing literature and showed good agreement. The results show that MSGT can capture size effects accurately compared to other theories like MCST and classical theory. As we increase the axial pretension from 0.0001N to 1N, the microbeam's stiffness increases, increasing natural frequency for all end conditions. The results have shown that the frequency response of the micro/nanostructure is affected by size effects when the characteristic dimensions are in the orders of material length scales. Subsequently, surface elasticity effects have been analysed for Silicon and Aluminium-based nanobeams with the pretension of 0.0001 N for different boundary conditions. Results of surface elasticity modeling have shown that the frequency response of the nanobeam is either positive or negative depending on the value of the material length scale, h_0. The maximum difference in natural frequencies is 5% and 2% for Silicon and Aluminum based nanobeams, respectively. Hence, the present methodology can effectively model the strain gradients and surface elasticity effects in micro/nanostructures.

Acknowledgments. The first author would like to acknowledge the fellowship provided by the Ministry of Education, Government of India.

Appendix

In this section, we derived the governing equation of motion and the corresponding boundary conditions using extended Hamilton's principle.

Consider a prismatic beam of length L, having cross sectional area A, is subjected to the distributed load $q(x)$ and the axial load N_0 as shown in Fig. 1. The cross-section is specified by the x (longitudinal) and y (lateral) axes. The load q is applied in the transverse direction, i.e., the z axis.

Let u, v, w are the displacements in the x, y, and z directions, respectively. From the Euler-Bernoulli theory, the displacements can be written as :

$$u = -z\frac{\partial w(x)}{\partial x}, v = 0, w = w(x). \tag{50}$$

On substitution of the above expression in the Eqs. (2) and (3),

$$\epsilon_{xx} = -z\frac{\partial^2 w}{\partial x^2}, \gamma_x = -z\frac{\partial^3 w}{\partial x^3}, \gamma_y = 0, \gamma_z = -\frac{\partial^2 w}{\partial x^2}, \tag{51}$$

and from the Eqs. (4) and (5) we get

$$\chi^S_{xy} = \chi^S_{yx} = -\frac{1}{2}\frac{\partial^2 w}{\partial x^2} \tag{52}$$

$$\begin{aligned}
\tilde{\eta}_{111} &= -\frac{2}{5}\left(z\frac{\partial^3 w}{\partial x^3}\right), \tilde{\eta}_{113} = -\frac{4}{15}\left(\frac{\partial^2 w}{\partial x^2}\right), \tilde{\eta}_{122} = \frac{1}{5}\left(z\frac{\partial^3 w}{\partial x^3}\right)\\
\tilde{\eta}_{133} &= \frac{1}{5}\left(z\frac{\partial^3 w}{\partial x^3}\right), \tilde{\eta}_{212} = \frac{1}{5}\left(z\frac{\partial^3 w}{\partial x^3}\right), \tilde{\eta}_{221} = \frac{1}{5}\left(z\frac{\partial^3 w}{\partial x^3}\right)\\
\tilde{\eta}_{223} &= \frac{1}{15}\left(\frac{\partial^2 w}{\partial x^2}\right), \tilde{\eta}_{232} = \frac{1}{15}\left(\frac{\partial^2 w}{\partial x^2}\right), \tilde{\eta}_{311} = -\frac{4}{15}\left(\frac{\partial^2 w}{\partial x^2}\right)\\
\tilde{\eta}_{313} &= \frac{1}{5}\left(z\frac{\partial^3 w}{\partial x^3}\right), \tilde{\eta}_{322} = \frac{1}{15}\left(\frac{\partial^2 w}{\partial x^2}\right), \tilde{\eta}_{331} = \frac{1}{5}\left(z\frac{\partial^3 w}{\partial x^3}\right)\\
\tilde{\eta}_{333} &= \frac{1}{5}\left(z\frac{\partial^3 w}{\partial x^3}\right), \tilde{\eta}_{131} = -\frac{4}{15}\left(\frac{\partial^2 w}{\partial x^2}\right)
\end{aligned} \tag{53}$$

On substitution of Eqs. (51)–(53) in Eq. (1) and after simplification, we get the following equation for the total strain energy U_t of linear isotropic elastic material is found as,

$$U_t = \frac{1}{2}\int_0^L \left[P \cdot (w'')^2 + Q \cdot \left(w^{(3)}\right)^2\right] dx, \tag{54}$$

where, w is the transverse deflection, $w'' = \frac{\partial^2 w}{\partial x^2}$, $w^{(3)} = \frac{\partial^3 w}{\partial x^3}$ and P, Q are defined as,

$$P = EI + 2\mu A l_0^2 + \frac{8}{15}\mu A l_1^2 + \mu A l_2^2, Q = I\left(2\mu l_0^2 + \frac{4}{5}\mu l_1^2\right). \tag{55}$$

and A is the area of cross section and I is the moment of inertia of the beam. μ is the shear modulus of the material(Lame's constant) and is given by $\mu = \frac{E}{2(1+\nu)}$. The first variation of integral of the kind $H = \int_0^L F\left(w'', w^{(3)}\right) dx$

can be defined as,

$$\delta H = \int_0^L \left[\frac{d^2}{dx^2}\left(\frac{\partial F}{\partial w''}\right) - \frac{d^3}{dx^3}\left(\frac{\partial F}{\partial w^{(3)}}\right)\right]\delta w dx$$
$$+ \left[\left[-\frac{d}{dx}\left(\frac{\partial F}{\partial w''}\right) + \frac{d^2}{dx^2}\left(\frac{\partial F}{\partial w^{(3)}}\right)\right]\delta w\right]_0^L \quad (56)$$
$$+ \left[\left[\left(\frac{\partial F}{\partial w''}\right) - \frac{d}{dx}\left(\frac{\partial F}{\partial w^{(3)}}\right)\right]\delta w'\right]_0^L + \left[\frac{\partial F}{\partial w^{(3)}}\delta w''\right]_0^L.$$

In the present method ,consider $H = U_t$ of Eq. (54) and the Lagrangian function F is defined by,

$$F = \frac{1}{2}\left[P\cdot(w'')^2 + Q\cdot\left(w^{(3)}\right)^2\right]. \quad (57)$$

With the use of above relations one can write the first variation (Eqn.(56)) of the total strain energy as,

$$\delta U_t = \int_0^L \left[Pw^{(4)} - Qw^{(6)}\right]\delta w dx + \left[-Pw^{(3)} + Qw^{(5)}\right]\delta w|_0^L$$
$$+ \left[Pw'' - Qw^{(4)}\right]\delta w'|_0^L + Qw^{(3)}\delta w''|_0^L, \quad (58)$$

where, $w' = \frac{\partial w}{\partial x}, w^{(3)} = \frac{\partial^3 w}{\partial x^3}, w^{(4)} = \frac{\partial^4 w}{\partial x^4}, w^{(5)} = \frac{\partial^5 w}{\partial x^5}, \ \& \ w^{(6)} = \frac{\partial^6 w}{\partial x^6}$.

The variational kinetic energy and the work due to external forces can be written as :

$$\delta T = \left[\int_0^L \rho A \frac{\partial^2 w}{\partial t^2}\right]\delta w dx, \quad (59)$$

$$\delta W_{ext} = \int_0^L \left(q + N_0\frac{\partial^2 w}{\partial x^2}\right)\delta w dx + \left[V\delta w\right]_0^L + \left[M\delta w'\right]_0^L + \left[M^h\delta w''\right]_0^L. \quad (60)$$

Substituting the Eqs. (58)–(60), into the extended Hamilton's principle Eq. (12), we get the following governing differential equation (GDE) the motion of the beam by modified strain gradient theory (MSGT) as,

$$P\frac{\partial^4 w}{\partial x^4} - Q\frac{\partial^6 w}{\partial x^6} - N_0\frac{\partial^2 w}{\partial x^2} + \rho A\frac{\partial^2 w}{\partial t^2} + q(x) = 0, \quad (61)$$

and the corresponding boundary conditions which satisfy the equations are,

$$\left[\left(V - Pw^{(3)} + Qw^{(5)} - N_0 w'\right)\delta w\right]_0^L = 0, \quad (62)$$

$$\left[\left(M + Pw^{(2)} - Qw^{(4)}\right)\delta w'\right]_0^L = 0, \tag{63}$$

$$\left[\left(M^h + Qw^{(3)}\right)\delta w''\right]_0^L = 0. \tag{64}$$

When the two material length scale parameters, i.e., related with dilatation gradient and deviatoric stretch gradient becomes zero ($l_0 = l_1 = 0$) then the governing differential equation reduces to that of modified couple stress theory (MCST). Hence, the governing differential equation of motion based on MCST is,

$$\left(EI + \mu A l_2^2\right)\frac{\partial^4 w}{\partial x^4} - N_0\frac{\partial^2 w}{\partial x^2} + \rho A\frac{\partial^2 w}{\partial t^2} + q = 0 \tag{65}$$

and the boundary conditions become

$$\left[\left(V - \left(EI + \mu A l_2^2\right) w^{(3)} - N_0 w'\right)\delta w\right]_0^L = 0, \tag{66}$$

$$\left[\left(M + \left(EI + \mu A l_2^2\right) w^{(2)}\right)\delta w'\right]_0^L = 0. \tag{67}$$

When all the three material length scale parameters re assumed to zero ($l_0 = l_1 = l_2 = 0$), then the governing differential equation of motion is reduces to that of classical elasticity theory.

References

1. Singh, S.S., Nair, D.K., Rajagopal, A., Pal, P., Pandey, A.K.: Dynamic analysis of microbeams based on modified strain gradient theory using differential quadrature method. Eur. J. Comput. Mech. **27**(3), 187–203 (2018)
2. Fu, Y., Du, H., Huang, W., Zhang, S., Hu, M.: TiNi-based thin films in MEMS applications: a review. Sens. Actuators A: Phys. **112**, 395–408 (2004)
3. Stelmashenko, N.A., Walls, M.G., Brown, L.M., Milman, Y.V.: Microindentations on W and Mo oriented single crystals: an STM study. Acta Metallurgica et Materialia **41**, 2855–2865 (1993)
4. Poole, W.J., Ashby, M.F., Fleck, N.A.: Micro-hardness of annealed and work-hardened copper polycrystals. Scripta Materialia **34**, 559–564 (1996)
5. Park, S.K., Gao, X.-L.: Bernoulli-Euler beam model based on a modified couple stress theory. J. Micromech. Microeng. **16**, 2355–2359 (2006)
6. Asghari, M., Ahmadian, M.T., Kahrobaiyan, M.H., Rahaeifard, M.: On the size-dependent behavior of functionally graded micro beams. Mater. Des. **31**, 2324–2329 (2010)
7. Wang, L.: Size-dependent vibration characteristics of microtubes conveying fluid. J. Fluids Struct. **26**, 675–684 (2010)
8. Nayfeh, A.H., Mohammad, I.Y., Abdel-Rahman, E.M.: Dynamic pull-in phenomenon in MEMS resonators. Nonlinear Dyn. **48**, 153–163 (2007)
9. Wang, L.: Dynamical behaviors of double-walled carbon nanotubes conveying fluid accounting for the role of small length scale. Comput. Mater. Sci. **45**, 584–588 (2009)

10. Fleck, N.A., Muller, G.M., Ashby, M.F., Hutchinson, J.W.: Strain gradient plasticity: theory and experiment, Acta Metallurgica et materialia **42**(2), 475–487 (1994)
11. Ma, Q., Clarke, D.R.: Size dependent hardness of silver single crystals. J. Mater. Res. **10**(4), 853–863 (1995)
12. Stolken, J.S., Evans, A.G.: A microbend test method for measuring the plasticity length scale. Acta Materialia **46**(14), 5109–5115 (1998)
13. Chong, A.C.M., Lam, D.C.C.: Strain gradient plasticity effect in indentation hardness of polymers. J. Mater. Res. **14**(10), 4103–4110 (1999)
14. Lam, D.C.C., Yang, F., Chong, A.C.M., Wang, J., Tong, P.: Experiments and theory in strain gradient elasticity. J. Mech. Phys. Solids **51**(8), 1477–1508 (2003)
15. McFarland, A.W., Colton, J.S.: Role of material microstructure in plate stiffness with relevance to microcantilever sensors. J. Micromech. Microeng. **15**(5), 1060 (2005)
16. Lu, P.: Dynamic analysis of axially prestressed micro/nanobeam structures based on nonlocal beam theory. J. Appl. Phys. **101**, 073504 (2007)
17. Eringen, A.C., Edelen, D.G.B.: On nonlocal elasticity. Int. J. Eng. Sci. **10**(3), 233–248 (1972)
18. Kong, S., Zhou, S., Nie, Z., Wang, K.: Static and dynamic analysis of micro beams based on strain gradient elasticity theory. Int. J. Eng. Sci. **47**, 487–498 (2009)
19. Wang, B., Zhao, J., Zhou, S.: A micro scale Timoshenko beam model based on strain gradient elasticity theory. Eur. J. Mech. -A/Solids **29**(4), 591–599 (2010)
20. Akgöz, B., Civalek, Ö.: Strain gradient elasticity and modified couple stress models for buckling analysis of axially loaded micro-scaled beams. Int. J. Eng. Sci. **49**(11), 1268–1280 (2011)
21. Vatankhah, R., Kahrobaiyan, M.H., Investigation of size-dependency in free-vibration of micro-resonators based on the strain gradient theory. Latin Am. J. Solids Struct. 13(3), 498–515(2016)
22. Zhao, B., Liu, T., Chen, J., Peng, X., Song, Z.: A new Bernoulli-Euler beam model based on modified gradient elasticity. Arch. Appl. Mech. **89**(2), 277–289 (2019)
23. Shenoy, V.B.: Atomic calculations of elastic properties of metallic fcc crystal surfaces. Phys. Rev. B **71**, 094104 (2005)
24. Guo, J., Zhao, Y.P.: The size-dependent bending elastic properties of nanobeams with surface effects. Nanotechnology **18**, 295701 (2007)
25. Ibach, H.: The role of surface stress in reconstruction, epitaxial growth and stabilization of mesoscopic structures. Surf. Sci. Rep. **29**, 195–263 (1999)
26. Lu, P., Lee, P., Lu, C., O'Shea, S.: Surface stress effects on the resonance properties of cantilever sensors. Phys. Rev. B **72**, 085405 (2005)
27. He, J., Lilley, C.M.: Surface stress effect on bending resonance of nanowires with different boundary conditions. Appl. Phys. Lett. **93**, 263108 (2008)
28. Lagowski, J., Gatos, H.C., Sproles, E.S.: Surface stress and normal mode of vibration of thin crystals: GaAs. Appl. Phys. Lett. **26**, 493–495 (1975)
29. Gurtin, M.E., Markenscoff, X., Thurston, R.N.: Effect of surface stress on the natural frequency of thin crystals. Appl. Phys. Lett. **29**, 529–530 (1976)
30. Zhang Y., Zhuo L.J., Zhao, H.S.: Determining the effects of surface elasticity and surface stress by measuring the shifts of resonant frequencies. Proc. R. Soc. A **469** (2013)
31. Gangele, A., Pandey, A.K.: Frequency analysis of carbon and silicon nanosheet with surface effects. Appl. Math. Model. **76**, 741–758 (2019)
32. Fu, G., Zhou, S., Qi, L.: A size-dependent Bernoulli–Euler beam model based on strain gradient elasticity theory incorporating surface effects. ZAMM-J. Appl.

Math. Mech./Zeitschrift für Angewandte Mathematik und Mechanik, e201800048 (2019)

33. Wang, G.F., Feng, X.Q.: Effects of surface elasticity and residual surface tension on the natural frequency of microbeams. Appl. Phys. Lett. **90**, 231904 (2007)
34. Bellman, R., Kashef, B.G., Casti, J.: Differential quadrature: a technique for the rapid solution of nonlinear partial differential equations. J. Comput. Phys. 10(1), 40–52 (1972)
35. Shu, C.: Differential Quadrature and Its Application in Engineering. Springer (2012)
36. Shu, C., Du, H.: Implementation of clamped and simply supported boundary conditions in the GDQ free vibration analysis of beams and plates. Int. J. Solids Struct. **34**(7), 819–835 (1997)
37. Bert, C.W., Jang, S.K., Striz, A.G.: Two new approximate methods for analyzing free vibration of structural components. AIAA J. **26**(5), 612–618 (1988)
38. Zhao, J., Zhou, S., Wang, B., Wang, X.: Nonlinear microbeam model based on strain gradient theory. Appl. Math. Model. **36**(6), 2674–2686 (2012)
39. Nazemnezhad, R., Hosseini-Hashemi, S., Rokni, H., Nonlocal nonlinear free vibration of nanobeams with surface effects. Eur. J. Mech./A Solids (2015)
40. Son, D., Jeong, J.-H., Kwon, D.: Film-thickness considerations in microcantilever-beam test in measuring mechanical properties of metal thin film. Thin Solid Films **437**, 182–187 (2003)
41. Erravelly, I.R., Pandey, A.K.: Frequency analysis of microbeams with axial. Mtech thesis, Indian Institute of Technology, Hyderabad, 2019 pretension

Nonlocal Nonlinear Analysis of Functionally Graded Nano Plates Used in MEMs Devices

Dhaladhuli Pranavi(✉) and Amirtham Rajagopal

Department of Civil Engineering, Indian Institute of Technology Hyderabad, Hyderabad, India
ce18m20p000001@iith.ac.in, rajagopal@ce.iith.ac.in

Abstract. In this work, an extension of Eringens model for analysis of functionally graded nano plates used in MEMS (micro-electromechanical systems) devices will be presented. Rule of mixtures and Mori-Tanaka methods are adopted for through thickness homogenization of material properties. The role of nonlocal parameter together with the choice of homogenization method is studied in FGM plates. Reddy's third order shear deformation theory [1] is adopted for the analysis of the plates. The limitations in use of Eringen's model are also discussed in detail.

Keywords: Functionally graded · Nonlocal · Homogenization · Rule of mixtures · Mori-Tanaka

1 Introduction

In functionally graded materials (FGMs), the volume fraction of different materials are varied regularly along the structure to achieve a desired purpose. Applications of these FGMs include thermal barrier coatings [2], bio materials [3], and dental implants [4] and wear resistant coatings [5]. Therefore, understanding the mechanical response of such systems is necessary. Modeling of FGMs is complex as it involve small scale effects due to its applications in nano components, heterogenity and nonlinearity. Nonlocal theories helps in modeling them by incorporating a length scale parameter.

The structures in MEMS devices are idealized as nano beams [6] and plates [7,8]. The material possesses discreteness and inhomogeneity at these lower length scales and is considered to be functionally graded. Aifantis et al., have demonstrated introduced higher order gradients in the constitutive relations to capture the heterogeneity of deformation occurring at the smaller length scale [9,10]. Eringen's nonlocal model has been extended to heterogeneous or functionally graded materials [11,12]. In [12], an equivalent strain approach is adopted

Indian Institute of Technology Hyderabad.

A. K. Pandey et al. (Eds.): MAMM 2022, 126, pp. 215–222, 2023.
https://doi.org/10.1007/978-3-031-20353-4_16

to incorporate the varying kernel function. Nonlocal strain tensor is defined by considering the material to be functionally graded only along the thickness of the plate. Three dimensional functionally graded plate is solved by Kashtalyan et al. [16].

The main objectives of the present work are to (a) study the response of FGM plates under different loading conditions by homogenizing the material properties using Rule of mixtures and Mori-Tanaka techniques, (b) understand the effect of in-plane dimension to thickness (a/h) ratio, and nonlocal parameter μ on the load-deflection response of the plates. The obtained numerical results are compared with the results from the existing literature. The applications and limitations of Eringen's nonlocal model to heterogeneous materials are summarized.

2 Eringens Nonlocal Model

Nonlocal elasticity theory assumes the dependence of stress at any particular point on strain field at every possible point in the body [17, 18]. Using certain class of kernels, the partial differential equations can be developed which will be easier to solve for obtaining displacements when compared to linear local theories. The equilibrium equation can be written as

$$\boldsymbol{\sigma}^{nl}_{kl,k} + \rho(\boldsymbol{f}_l - \ddot{\boldsymbol{u}}_l) = 0 \tag{1}$$

The relation between nonlocal stress tensor $\boldsymbol{\sigma}^{nl}_{kl}$ can be written in terms of the classical stress tensor as

$$\boldsymbol{\sigma}^{nl}_{kl}(\boldsymbol{x}) = \int_V \alpha(|\boldsymbol{x}' - \boldsymbol{x}|, \tau)\, \boldsymbol{\sigma}_{kl}(\boldsymbol{x}') d\mathrm{V}(\boldsymbol{x}') \tag{2}$$

$$\boldsymbol{\sigma}_{kl}(\boldsymbol{x}') = \lambda \boldsymbol{e}_{rr}\boldsymbol{x}'\delta_{kl} + 2\mu \boldsymbol{e}_{kl}\boldsymbol{x}' \tag{3}$$

The conditions that are to be satisfied for choosing the kernel function and the properties of it are given by Eringen in [18]. It mainly depends on ratio of internal to external characteristic lengths. When the limit $\tau \to 0$, the nonlocal model reduces to classical elasticity theory.

The differential form of the Eringen's nonlocal stress tensor can be written as follows [24]

$$(1 - \mu\nabla^2)\, \sigma^{nl}_{ij} = \sigma_{ij} \tag{4}$$

where the nonlocal parameter, $\mu = (e_o a)^2$, e_o is a material constant and a is the internal characteristic length. An equivalent strain of the form

$$\varepsilon^{nl}_{ij} = \int_V \alpha(|\boldsymbol{x}' - \boldsymbol{x}|, \tau)\, \varepsilon_{ij}(\boldsymbol{x}') d\mathrm{V}(\boldsymbol{x}') \tag{5}$$

that results in a nonlocal equilibrium equation

$$\sigma^{nl}_{ij} = \mathrm{C}_{ijkl}\varepsilon^{nl}_{kl} \tag{6}$$

C_{ijkl} is the fourth order elasticity tensor.

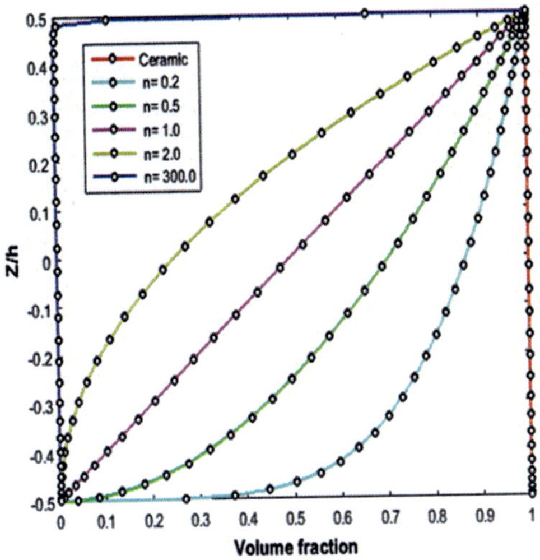

Fig. 1. Volume fraction across the thickness z/h using power law, $V_{Ce} = (\frac{1}{2} + \frac{z}{h})^n, V_{Me} = 1 - V_{Ce}, n =$ power index, V_{Ce} and V_{Me} are the volume fractions of Ceramic and Metal phases in a composite material.

3 Equivalent Properties of Functionally Graded Materials

Exponential law, power law or sigmoidal variation can be used to account for the variation of properties across the thickness for a continuous graded FGM. For the present work, power law is adopted which is presented in Fig. 1.

In order to predict the response of a functionally graded material, the effective properties need to be evaluated. For this purpose, various micro-mechanical models like Rule of mixtures [19,20], and Mori-Tanaka method [21,22] have been developed. For lower volume fraction of inhomogenities, Rule of mixtures method can be adopted, whereas, for higher volume fraction of inhomogenities, Mori-Tanaka method gives the most accurate results as it includes the effects of multiple inhomogeneities by taking the average strain ($\bar{\epsilon}_o$) in the matrix phase, refer Fig. 2. The elastic modulus through the thickness at different values of power index n using Rule of mixtures(RM) and Mori-Tanaka(MT) schemes are plotted in Fig. 3.

4 Governing Equations

$$\nabla \cdot \boldsymbol{N} = 0 \tag{7}$$

$$\nabla \cdot \boldsymbol{Q} + \nabla \cdot (\boldsymbol{N} \nabla w) + \nabla^2 \boldsymbol{P} + q = 0 \tag{8}$$

$$\nabla \cdot \boldsymbol{M} - \boldsymbol{Q} = 0 \tag{9}$$

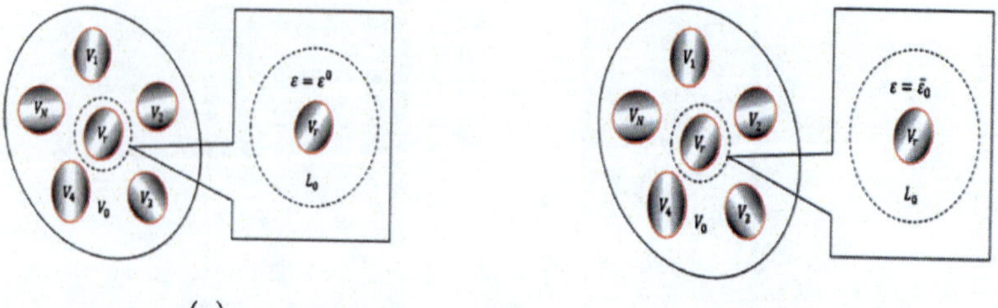

Fig. 2. Schematic representation of (a) Rule of mixtures, and (b) Mori-Tanaka methods.

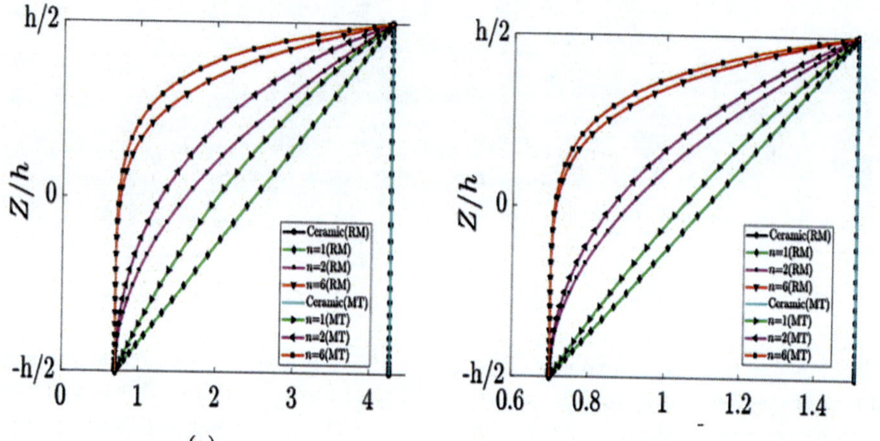

Fig. 3. Comparison of elastic modulus through the thickness using RM and MT for different values of power index n for Al/SiC FGM plate.

Where $\boldsymbol{N}$, $\boldsymbol{Q}$ and $\boldsymbol{M}$ are the stress resultants of inplane force, shear force and bending moment, q is the applied load, and w is the displacement field along thickness direction. Interesting readers can refer [24] for finite element approximations and implementation in detail.

5 Numerical Examples

A simply supported square plate of dimension $a \times a$ is considered which is subjected to SS1 boundary conditions (see Fig. 4). $a/h = 10$ is considered for the analysis. Following material properties are considered, $E_{Ce} = 145$ GPa, $E_{Me} = 60$ GPa and $\nu = 0.28$. The non-dimensionalized deflection is calculated which is given by the relation $w' = \frac{w(0,0)}{h}$.

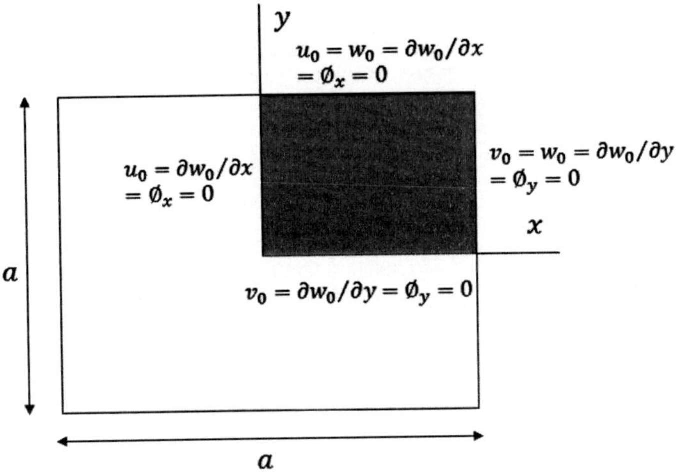

Fig. 4. Geometry and boundar conditions of square plate of dimension $a \times a$, where u_o, v_o, w_o are in-plane displacements, ϕ_x and ϕ_y are the rotations of a transverse normal line at the mid-plane ($z = 0$).

Table 1. Comparison of w' obtained for various values of power index(n).

a/h	n	w'			
		Rule of mixtures	Ferreira [25]	Mori-Tanaka	Ferreira [25]
20	Ceramic	0.0209	0.0205	0.0209	0.0205
	0.2	0.0236	–	0.0243	–
	0.5	0.0267	0.0262	0.0276	0.0276
	1	0.0299	0.0294	0.0307	0.0305
	2	0.0327	0.0323	0.0333	0.033
	Metal	0.0452	0.0443	0.0452	0.0443

For Al/Zr FGM plate, the values of normalized central deflections w' obtained using (a) Rule of mixtures, and (b) Mori-tanaka schemes, are compared with the results from Ferreira [25] in Table 1. A difference of less than 2% is observed. The values of w' for different values a/h, nonlocal parameter μ subjected to sinusoidal load are tabulated in Table 2. w' versus a/h is plotted for different values of μ in Fig. 5(a) and w' versus μ is plotted for various values of n in Fig. 5(b) under sinusoidal load.

6 Conclusions

Reddy's third order shear deformation theory(TSDT) [1] can evaluate the transverse shear strains as quadratic whereas the classical plate theory neglects the shear deformation and Mindlin's first order shear deformation theory(FSDT)

Table 2. Effect μ on w' for different values of a/h.

a/h	n	w'			
		$\mu = 0$	$\mu = 1$	$\mu = 3$	$\mu = 5$
10	Ceramic	0.0141	0.0453	0.1193	0.2276
	0.2	0.0158	0.0510	0.1344	0.2565
	0.5	0.0179	0.0576	0.1519	0.2897
	1	0.0201	0.0645	0.1701	0.3245
	2	0.0220	0.0708	0.1866	0.3560
	Metal	0.0304	0.0977	0.2574	0.4910
5	Ceramic	0.0169	0.0538	0.1406	0.2674
	0.2	0.0189	0.0603	0.1577	0.2999
	0.5	0.0213	0.0679	0.1776	0.3378
	1	0.0239	0.0762	0.1993	0.3790
	2	0.0265	0.0844	0.2206	0.4195
	Metal	0.0365	0.1161	0.3034	0.5769

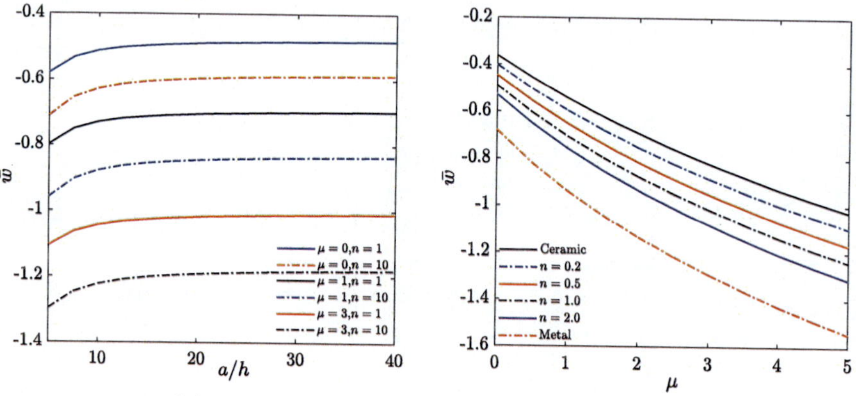

(a) (b)

Fig. 5. Plots of (a) w' Vs a/h for various values of μ and n, (b) w' Vs μ for various values n.

predicts a constant variation of shear strain and a shear correction factor is to be used. Therefore, TSDT theory is adopted for the analysis of the plates in the present work. Eringen's nonlocal model together with homogenization techniques (Rule of mixtures [19] and Mori-Tanaka schemes [21]) are used to evaluate the response of FGM plate. The effect of nonlocality is not considered along thickness direction of the plate. Though Eringen's nonlocal model cannot be applied directly for 3D structures or structures with FGM material where all dimensions are necessary to study, it can be applicable to flat structures where thickness is significantly less compared to other dimensions as the differential nonlocal operator is always applied with respect to x, y for flat structures as effect on z direction may not be significant. Therefore, if the nonlocal effects are not considered in thickness direction and thickness integrated resultant forces and moments are evaluated in thickness direction, then application of Eringen's nonlocal model can still be valid to study the deformations in functionally graded plates.

References

1. Reddy, J.N., Lee, S.J.: Nonlinear deflection control of laminated plates using third-order shear deformation theory. Mech. Mater. Des. **2**(5), 99–110 (2016)
2. Koizumi, M.: FGM activities in Japan. Compos. B: Eng. **28**(1–2), 1–4 (1997)
3. Pompe, W., Worch, H., Epple, M., Friess, W., Gelinsky, M., Greil, P., Hempel, U., Scharnweber, D., Schulte, K.: Functionally graded materials for biomedical applications. Mater. Sci. Eng. A **362**(12), 4060 (2003)
4. Watari, F., Yokoyama, A., Saso, F., Kawasaki, T: Fabrication and properties of functionally graded dental implant. Compos. B **28**(12), 5–11 (1997)
5. Schulz, U., Peters, M., Bach, F.W., Tegeder, G.: Graded coatings for thermal, wear and corrosion barriers. Mater. Sci. Eng. A **362**(12), 6180 (2003)
6. Reddy, J.N., El-Borgi, S., Romanoff, J.: Non-linear analysis of functionally graded microbeams using eringens non-local differential model. Int. J. NonLinear Mech. **67**, 308–318 (2014)
7. Rahmani, O., Pedram, O.: Analysis and modeling the size effect on vibration of functionally graded nanobeams based on nonlocal timoshenko beam theory. Int. J. Eng. Sci. **77**, 55–70 (2014)
8. Simsek, M., Yurtcu, H.H.: Analytical solutions for bending and buckling of functionally graded nano beams based on the nonlocal Timoshenko beam theory. Compos. Struct. **97**, 378–386 (2013)
9. Aifantis, E.C.: Gradient effects at macro, micro, and nano scales. J. Mech. Behav. Mater. **5**(3), 355–375 (1994)
10. Aifantis, E.C.: Strain gradient interpretation of size effects. Int. J. Fract. **95**, 299–314 (1999)
11. Kunin, I.A.: Inhomogeneous elastic medium with nonlocal interaction. J. Appl. Mech. Tech. Phys. **8**, 60–66 (1967)
12. Evgrafov, A., Bellido, J.C.: From non-local eringen's model to fractional elasticity. Math Mech. Solids **24** (2019)
13. Salehipour, H., Shahidi, A.R., Nahvi, H.: Modified nonlocal elasticity theory for functionally graded materials. Int. J. Eng. Sci. **90**, 44–57 (2015)
14. Reddy, J.N., Cheng, Z.Q.: Three dimensional thermomechanical deformations of functionally graded rectangular plates. Eur. J. Mech. A/Solids **20**, 841–855 (2001)
15. Vel, S.S., Batra, R.C.: Exact solution for thermoelastic deformations of functionally graded thick rectangular plates. AIAA Journal **40**(7), 14211433 (2002)
16. Kashtalyan, M.: Three-dimensional elasticity solution for bending of functionally graded rectangular plates. Eur. J. Mech. A/Solids **23**, 853864 (2004)
17. Eringen, A.C.: Linear theory of nonlocal elasticity and dispersion of plane waves. Int. J. Eng. Sci. **10**(5), 425–435 (1972)
18. Eringen, A.C.: On differential equations of nonlocal elasticity and solutions of screw dislocation and surface waves. J. Appl. Phys. **54**, 4703–4710 (1983)

19. Praveen, G.N., Reddy, J.N.: Nonlinear transient thermoelastic analysis of functionally graded ceramic-metal plates. J. Solids. Struct. **35** (33), 4 4574476 (1998)
20. Reddy, J.N.: Analysis of functionally graded plates. Int. J. Numer. Meth. Eng. **47**(13), 663–684 (2000)
21. Mori, T., Tanaka, K.: Average stress in matrix and average elastic energy of materials with misfitting inclusions. Acta Metall. **21**, 571–574 (1973)
22. Zuiker, J.R.: Functionally graded materials choice of micro mechanics model and limitations in property variation. Compos. Eng. **5**(7), 807–819 (1995)
23. Benveniste, Y.: A new approach to the application of Mori-Tanaka's theory in composite materials. Mech. Mater. **6**, 147–157 (1987)
24. Srividhya, S., Raghu, P., Rajagopal, A., Reddy, J.N.: Nonlocal nonlinear analysis of functionally graded plates using third-order shear deformation theory. Int. J. Eng. Sci. **125**, 1–22 (2018)
25. Ferreira, A.J.M., Batra, R.C, Rouque, C.M.C., Qian, L.F., Martins, P.A.L.S.: Static analysis of functionally graded plates using third-order shear deformation theory and a Meshless method. Compos. Struct. **69** , 449–457 (2005)

Finite Element Analysis of Squeezed Film Damping on Trapezoidal Microcantilever Resonators at Different Pressure Levels

Varun P. Sharma[1,2] and Rahul Shukla[1,2](✉)

[1] Accelerator Physics and Synchrotrons Utilization Division, Raja Ramanna Centre for Advanced Technology, Indore, Madhya Pradesh, India
rahulshukla@rrcat.gov.in

[2] Homi Bhabha National Institute, Anushaktinagar, Mumbai, Maharashtra, India

Abstract. Squeezed film damping (SFD) becomes a dominant damping mechanism in micro-electro-mechanical system (MEMS). Depending on the pressure variation in MEMS, SFD governs the dynamic parameters like quality factors (Q factors) and damping ratio. In the present paper, we calculate the Q factor and damping ratio of the trapezoidal shaped microcantilever beam by eigenfrequency analysis using finite element method (FEM) in COMSOL Multiphysics. The effective viscosity method is used to calculate the SFD in FEM for the continuum, slip, transition and molecular flow regimes as described by the Knudsen no. (K_n). K_n is varied by altering the operating pressure, keeping the thickness of air gap constant. It is observed that the Q factor and damping ratio of trapezoidal microcantilever beam varies by an order of million for different flow regimes with the change of operating pressure from atmospheric to 0.1 Pa.

Keywords: Squeezed film damping · MEMS · Microcantilever · Knudsen number · Eigen frequency · Quality factor · Damping ratio · COMSOL multiphysics

1 Introduction

Microcantilever based microresonators are very demanding and widely used in MEMS for varieties of applications such as, Atomic Force Microscope (AFM) [1], temperature and pressure measurement sensors, [2], biomedical sensors [3], vertical comb-drives [4, 5] and accelerometers [6, 7] due to its capability of tuning the resonance frequencies, sensitivity and, excellent resolution in dynamic operations. Besides, microcantilevers are easy to fabricate by bulk micromachining, surface micromachining and SoI-based processes. In addition, Metal-assisted Chemical Etching (MaCEtch) is explored for the fabrication of microstructures in recent works [8, 9] and can be extended for the development of the microcantilevers and other microstructures in MEMS.

Several authors reported the prismatic microcantilever beam for the study of quality factor (Q factor), damping ratios and other dynamic performance parameter of MEMS resonators. The performance of the microresonators is greatly affected by the damping

A. K. Pandey et al. (Eds.): MAMM 2022, 126, pp. 223–231, 2023.
https://doi.org/10.1007/978-3-031-20353-4_17

mechanisms named as; squeezed film damping (SFD), thermoelastic damping (TED), anchor loss and surface loss. Out of the aforesaid damping mechanisms, SFD is found to be responsible for the losses associated with damping in majority of cases in MEMS [10]. SFD is affected by the several parameters such as, operating pressure, temperature, relative humidity, squeezing media (air, or any other gas) and degree of rarefaction. Bao and Yang [10] have contributed in the SFD analysis of MEMS and proposed the damping models with perforation and without perforation. Veijola [11] have developed the SFD model with effective viscosity method and is widely accepted for the damping analysis in MEMS. COMSOL Multiphysics also uses Veijola's effective viscosity method for SFD analysis. The effects of environmental factors i.e. temperature, humidity and pressure on Q factors is studied by Nguyen et al. [12]. Martin et al. [13] have studied the different geometries of MEMS resonators and corrected the existing models of SFD by calculating the damping factors theoretically and experimentally. SFD coefficients and dynamic response of higher flexural mode is reported by Pandey and Pratap [14] for the MEMS cantilever resonators. They have compared the analytical response of damping at higher modes for different MEMS cantilever beams with the experimental results. Ashok et al. [15] have studied recently the tuning of wideband frequency and Q factors by coupling the non-uniform microbeam arrays. They reported the eigenfrequency and Q factor variation with tapering parameter for converging and diverging microbeams in atmosphere (1.01325 bar) and vacuum (8×10^{-3} mbar) operating conditions. From the best of our knowledge, Q factor and eigenfrequency analysis with wide pressure variation (10^5 Pa to 0.1 Pa) for the trapezoidal microcantilever beam is not yet reported in the literature.

In the present study, we have focused on the trapezoidal microcantilever beam to broaden the applicability of microresonators with variable cross sections in MEMS. The usefulness of the variable cross section of trapezoidal beam lies with the uniform distribution of stress along its length. The methodology developed in the present study using finite element method (FEM) for SFD analysis will be extended for the estimation of Q factor, damping ratio and other related dynamic parameters of MEMS-based vertical-comb drive microactuator in torsion at different pressure levels.

2 Squeezed Film Damping

In the MEMS, air remains trapped between the moving and fixed microstructures and acts as a thin film. On actuation, microstructures squeeze the thin film of air (or fluid) between the moving and fixed microstructures and exerts a back pressure or force on the walls of the microstructures. The back pressure dampens the movement and results in higher dissipation of energy per cycle. This phenomenon is known as squeezed film damping or SFD [10]. SFD depends on the various environmental factors i.e. pressure, temperature and humidity along with the boundary conditions and mode of actuation [16]. Pressure is one of the dominating factors in estimating the damping characteristics of the microresonators. The operating pressure flow regimes i.e. continuum, slip, transition and molecular flow, are defined by the Knudsen no. ($K_n = \lambda/h_0$) where λ is the mean free path of surrounding media and h_0 is the thickness of air gap between the moving and fixed microstructures and given in Table 1. The atmospheric pressure is considered as 10^5 Pa instead of 1.013×10^5 Pa for the calculations in the current study.

Table 1. Flow regimes based on K_n and operating pressure

S. No	Knudsen No. (K_n)	Pressure (Pa)	Type of flow regime
1	$K_n < 0.01$	$P \geq 10^5$	Continuum flow regime
2	$0.01 < K_n < 0.1$	$10^4 < P < 10^5$	Slip flow regime
3	$0.1 < K_n < 10$	$10^4 < P < 10^5$	Transition flow regime
4	$K_n > 10$	$10^3 < P < 10^4$	Molecular flow regime

The modified Reynolds equation to calculate the pressure distribution in the SFD analysis for different flow regime is given by Eq. (1).

$$\frac{h_0^3}{12\mu_{eff}}\left(\frac{\partial^2 p}{\partial x^2}+\frac{\partial^2 p}{\partial y^2}\right)=\frac{h_0}{p_a}\frac{\partial p}{\partial t}+\frac{\partial w}{\partial t} \tag{1}$$

where, p is the pressure distribution due to SFD, p_a is the ambient pressure, w is the transverse deflection and μ_{eff} is the effective viscosity of the air and given by Eq. (2):

$$\mu_{eff}=\frac{\mu}{1+9.638K_n^{1.159}} \tag{2}$$

here, μ is the dynamic viscosity of air.

Effective viscosity method [11] is used to solve the modified Reynolds equation (refer Eq. (1)) for the continuum, slip, transition and molecular flow regimes.

Total Q factor (Q_{total}) and damping ratio (ζ) in the SFD analysis are given by Eqs. (4) and (5), respectively.

$$\frac{1}{Q_{total}}=\frac{1}{Q_{SFD}}+\frac{1}{Q_{TED}}+\frac{1}{Q_{anchor}} \tag{3}$$

$$\zeta=\frac{1}{2Q_{total}} \tag{4}$$

In microcantilever based resonators, Q_{total} is governed by the SFD, thermoelastic damping (TED), and anchor loss. In the present study, only SFD is considered to calculate the Q factor and damping ratio for the microcantilever beam. TED and anchor loss are neglected due to the absence of temperature and simpler clamping at the fixed end of the trapezoidal microcantilever beam.

3 FEM Analysis

Schematic diagram of trapezoidal microcantilever beam is shown in Fig. 1. The width at distance x from the fixed end can be given by Eq. (5). Length and thickness of microcantilever beam are 150 μm and 5 μm, respectively. The width of microcantilever beam at fixed end (w_0) is 50 μm whereas at free end (w_l) is varying from 50 to 20 μm.

Solid mechanics physics with eigenfrequency study is used in COMSOL Multiphysics to evaluate the SFD of microcantilever. Tetrahedral element is used to discretize

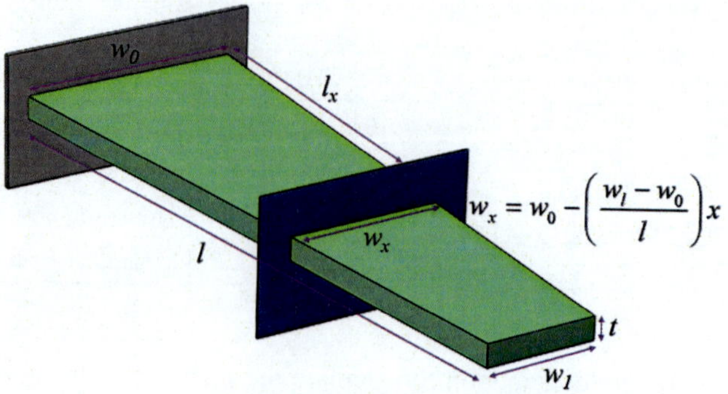

Fig. 1. Schematic diagram of trapezoidal microcantilever beam showing the dimensional parameters and variation of the width along the length

the domain and serendipity approach is used to define the shape functions. Mesh convergence is done in order to determine the optimum size of the mesh element. The verification of eigen frequency and damping ratio is done with the literature [14] to validate the formulation in FEM.

$$w_x = w_0 - \left(\frac{w_l - w_0}{l}\right)x \tag{5}$$

3.1 Verification of FEM Model

The verification of FEM formulation is done using COMSOL Multiphysics for a known problem from the literature [14]. The microcantilever with different configurations are simulated at atmospheric pressure and the results are compared. From Table 2, it can be seen that FEM results of the current work are in good agreement with the simulation and experimental results of Pandey and Pratap [14].

Table 2. Verification and comparison of damping ratio and eigen frequencies with the literature [14] and current work in COMSOL Multiphysics 6.0

Length (μm)	Damping ratio (ζ)			Eigen frequency (kHz)	
	Pandey and Pratap [14]		Current work	Pandey and Pratap [14]	Current work
	Experimental	Simulation			
150	0.071 ± 0.002	0.074	0.077	240	239
200	0.125 ± 0.001	0.135	0.145	133	132
250	0.201 ± 0.005	0.22	0.234	80	84
300	0.320 ± 0.002	0.33	0.344	50	56
350	0.415 ± 0.003	0.45	0.475	43	38.7

Besides the verification of FEM model of SFD for microcantilevers at atmospheric pressure, the accuracy of the developed model is further verified using microcantilever beam of uniform length (l_b) 150 μm, width (w_b) 50 μm and thickness (t_b) 5 μm for varying pressures (10^5–0.1 Pa). For verification, the analytical solution of damping ratio (ζ) (refer Eq. (6)) for the first mode of microcantilever beam of uniform cross section is used from the literature [14]. Various coefficients σ, χ, φ, ρ, ω_n and b can be found from literature [14]. It is found that the analytical and FEM results of Q factor and damping ratio (ζ) are in good agreement for the pressures varying from 10^5 to 0.1 Pa and given in Table 3.

$$\zeta_d = \frac{384\mu_{eff} w_b^2}{\phi t_b \rho \omega_n h_0^3 \pi^6}\left(\frac{(4+\chi^2)b^2}{(4+\chi^2)+(\sigma^2/4\pi^4)}\right) \quad (6)$$

Table 3. Verification of FEM model with analytical model of Q factors and damping ratio of microcantilever beam of uniform cross section at varying pressures (10^5–0.1 Pa)

Pressure (Pa)	$Q_{analytical}$	$Q_{numerical}$	$\zeta_{analytical}$	$\zeta_{numerical}$
0.1	4.4×10^7	3.69×10^7	1.23×10^{-8}	1.35×10^{-8}
1	2.81×10^6	2.6×10^6	1.77×10^{-7}	1.94×10^{-7}
10	1.97×10^5	1.8×10^5	2.53×10^{-6}	2.76×10^{-6}
10^2	1.4×10^4	1.28×10^4	3.56×10^{-5}	3.88×10^{-5}
10^3	1.07×10^3	1.0×10^3	4.66×10^{-4}	4.98×10^{-4}
10^4	137.67	148	0.0036	0.0034
10^5	72.21	88	0.0069	0.0056

4 Results and Discussions

The variation of Q factor, from atmospheric pressure (10^5 Pa) to vacuum (0.1 Pa) is shown in Fig. 2 for variable width (50 μm to 20 μm) of free end of the trapezoidal beam. The Q factor decreases sharply from 0.1 to 10^3 Pa and changes its slope at 10^3 and 10^4 Pa. This is due to the fact that at lower pressure, mean free path of the gas molecules increase and energy losses gets minimized in successive collisions whereas at higher pressure mean free path of the gas molecules decreases and dissipates more energy in each collision, results in decrease in Q factor. Q factors for the proposed trapezoidal microcantilever beam are ranging from 3.7×10^7 at 0.1 Pa to 88.8 at 10^5 Pa and 1.2×10^8 at 0.1 Pa to 287.92 at 10^5 Pa for 50 and 20 μm width at free end, respectively. It can be observed that during the operation from atmospheric pressure to vacuum (0.1 Pa) the Q factor changes by nearly an order of 10^6 for different configurations of the microcantilever beam which signifies the huge loss of energy in operating the MEMS at atmospheric pressure in comparison to vacuum.

Figure 3 shows the variation of damping ratio with pressure. The damping ratio is inversely proportional to Q factor as shown in Eq. (4) and increases with the increase in pressure. The damping ratio varies from 1.35×10^{-8} at 0.1 Pa to 0.0056 at 10^5 Pa and 4.18×10^{-9} at 0.1 Pa to 0.0017 at 10^5 Pa for 50 μm and 20 μm width at free end, respectively.

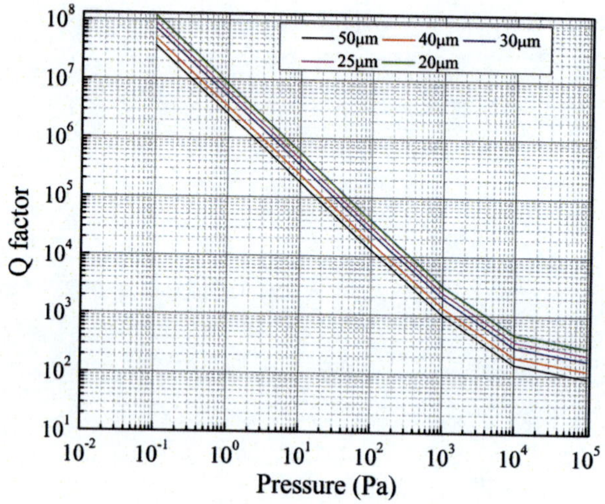

Fig. 2. Q factors variation with operating pressure for trapezoidal microcantilever beam with varying cross section width 50–20 μm at free end

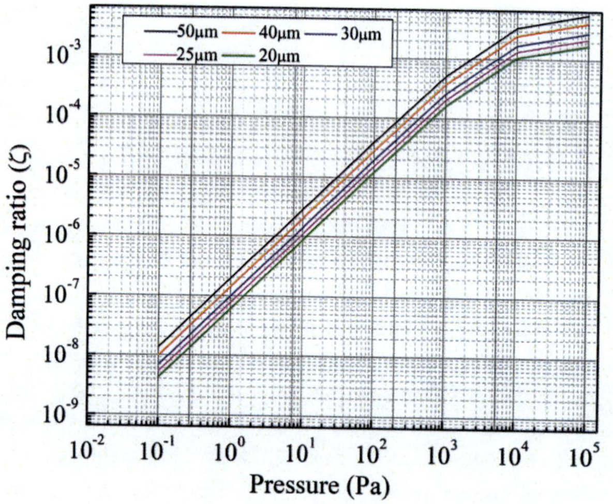

Fig. 3. Damping ratio variation with operating pressure for trapezoidal microcantilever beam with varying cross section width 50–20 μm at free end

Figure 4 shows the behavior of Q factor with varying width of free end of the trapezoidal beam at pressures 0.1, 10, 10^3 and 10^5 Pa. It can be seen that with the increase in width and pressure, Q factor is decreasing. The Q factor is minimum for

50 μm width at 10^5 Pa and maximum for 20 μm width at 0.1 Pa. The increase of width results in enhancement of losses due to SFD. It can be concluded that the losses due to damping can be minimized by the critical consideration of the design parameters in MEMS.

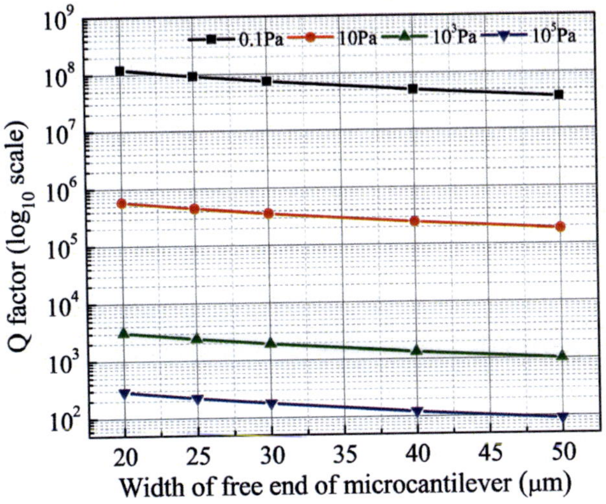

Fig. 4. Q factor variation with the variable width at free end of the trapezoidal microcantilever beam

5 Conclusions

In the present paper, a methodology for the SFD analysis in COMSOL Multiphysics for trapezoidal microcantilever based resonator is developed to calculate the dynamic performance parameters i.e., Q factors and damping ratio for pressure varying from atmospheric pressure (10^5 Pa) to vacuum (0.1 Pa). The verification of FEM formulation for eigen frequencies and damping ratio is done at atmospheric pressure (10^5 Pa) in COMSOL Multiphysics and the simulation results are in good agreement with the literature. The proposed methodology for Q factor and damping ratio (ς) is further verified with the analytical model at varying pressures from 10^5 to 0.1 Pa for the uniform microcantilever beam. In this way the correctness of the methodology for FEM simulation is authenticated and extended to analyze the trapezoidal microcantilever beam. It is observed that Q factors and damping ratio of trapezoidal microcantilever based resonator changes by an order of 10^6 when pressure varies from atmospheric (10^5 Pa) to vacuum (0.1 Pa).

The effects of SFD on Q factor and damping ratio are analyzed in the present study and will be extended to the other MEMS-based devices such as vertical comb-drives, accelerometers and micromotors.

Acknowledgements. Authors are thankful of Dr. Tapas Ganguli for technical discussion and necessary support at each stage of the work. Authors also acknowledge the support of computer division of RRCAT for providing the access of Kshitij-5 (High Performance Computing Cluster) where the COMSOL Multiphysics 6.0 is installed. One of the authors VPS is thankful of HBNI-RRCAT for providing the financial assistance.

References

1. Binnig, G., Quate, C.F., Gerber, C.: Atomic force microscope. Phys. Rev. Lett. **56**(9), 930–933 (1986). https://doi.org/10.1103/PhysRevLett.56.930
2. Lang, H.P., Hegner, M., Gerber, C.: Cantilever array sensors. Mater. Today **8**(4), 30–36 (2005). https://doi.org/10.1016/S1369-7021(05)00792-3
3. Gupta, A., Akin, D., Bashir, R.: Single virus particle mass detection using microresonators with nanoscale thickness. Appl. Phys. Lett. **84**(11), 1976–1978 (2004). https://doi.org/10.1063/1.1667011
4. Pham, P.H., Hoang, K.T., Nguyen, D.Q.: Trapezoidal-shaped electrostatic comb-drive actuator with large displacement and high driving force density. Microsyst. Technol. **25**(8), 3111–3118 (2019). https://doi.org/10.1007/s00542-019-04315-4
5. Engelen, J.B.C., Lantz, M.A., Rothuizen, H.E., Abelmann, L., Elwenspoek, M.C.: Improved performance of large stroke comb-drive actuators by using a stepped finger shape. In: TRANSDUCERS 2009—15th International Conference on Solid-State Sensors, Actuators and Microsystems, pp. 1762–1765. https://doi.org/10.1109/SENSOR.2009.5285744
6. Huang, W., Kwon, S.R., Zhang, S., Yuan, F.G., Jiang, X.: A trapezoidal flexoelectric accelerometer. J. Intell. Mater. Syst. Struct. **25**(3), 271–277 (2014). https://doi.org/10.1177/1045389X13491021
7. Hu, B., Liu, Y., Lin, B., Wu, G., Liu, W., Sun, C.: A novel trapezoidal ScAlN/AlN-based MEMS piezoelectric accelerometer. IEEE Sens. J. **21**(19), 21277–21284 (2021). https://doi.org/10.1109/JSEN.2021.3101219
8. Sharma, V.P., Shukla, R., Mukherjee, C., Tiwari, P., Sinha, A.K.: Study of metal-assisted chemical etching of silicon as an alternative to dry etching for the development of vertical comb-drives. J. Micromanuf. **6**, 1–8 (2021). https://doi.org/10.1177/25165984211033422
9. Miao, B., et al.: Improved metal assisted chemical etching method for uniform, vertical and deep silicon structure. J. Micromech. Microeng. **27**(5). https://doi.org/10.1088/1361-6439/aa6872
10. Bao, M., Yang, H.: Squeeze film air damping in MEMS. Sens. Actuators A **136**(1), 3–27 (2007). https://doi.org/10.1016/j.sna.2007.01.008
11. Veijola, T.: End effects of rare gas flow in short channels and in squeezed-film dampers. In: 2002 International Conference on Modeling and Simulation of Microsystems—MSM 2002, No. 1, pp. 104–107
12. Phan, M.T., Trinh, X.T., Le, Q.C., Ngo, V.K.T., Nguyen, C.C.: Effect of environmental conditions on quality factors of MEMS cantilever beam resonator in gas rarefaction. Sens Imaging **22**(1), 1–27 (2021). https://doi.org/10.1007/s11220-020-00329-9
13. Martin, M.J., Houston, B.H., Baldwin, J.W., Zalalutdinov, M.K.: Damping models for microcantilevers, bridges, and torsional resonators in the free-molecular-flow regime. J. Microelectromech Syst. **17**(2), 503–511 (2008). https://doi.org/10.1109/JMEMS.2008.916321
14. Pandey, A.K., Pratap, R.: Effect of flexural modes on squeeze film damping in MEMS cantilever resonators. J. Micromech. Microeng. **17**(12), 2475–2484 (2007). https://doi.org/10.1088/0960-1317/17/12/013

15. Ashok, A., Manoj Kumar, P., Singh, S.S., Raju, P. Pal, P. Pandey, A.K.: Achieving wideband micromechanical system using coupled non-uniform beams array. Sens. Actuators A Phys. **273**, 12–18 (2018). https://doi.org/10.1016/j.sna.2018.02.008
16. Nguyen, C.C., Li, W.L.: Effect of gas rarefaction on the quality factors of micro-beam resonators. Microsyst. Technol. **23**(8), 3185–3199 (2016). https://doi.org/10.1007/s00542-016-3068-z

Two-Dimensional Hydrodynamic Forces in an Array of Shape-Morphed Cantilever Beams

Lalsingh Devsoth(✉) and Ashok Kumar Pandey

Indian Institute of Technology Hyderabad, Kandi, Sangareddy, TS 502284, India
me17resch11006@iith.ac.in, ashok@mae.iith.ac.in

Abstract. In this paper, we estimated the hydrodynamic force in an array of cantilever beams separated by a distance $\bar{s}$ oscillating in a viscous fluid. The beam is assumed to be sufficiently long to consider 2D flow and has symmetric as well as asymmetric shape morphing curvature while oscillating in a fluid. The fluid-structure interaction problem is modelled by considering the unsteady Stokes equation. The resulting 1D boundary integral problem is solved by the boundary element method (BEM) numerically in MATLAB to obtain the desired pressure distribution on the beam. It is found that as the frequency oscillation of the rigid beam is increased, both the damping as well as added mass effects are decreased at different rates due to the gradual decrease in unsteady viscous layer. Finally, the hydrodynamic coupling effect on the beam is demonstrated at $\beta = 0.1$. However, for increase in the symmetric and asymmetric shape morphing parameters, the hydrodynamic decoupling appears lower than the gap ratio 5. The cantilever beam with optimal shape morphing parameter can be useful for the optimal designs of atomic force microscopy (AFM).

Keywords: Drag force · Added mass coefficient · Damping coefficient · Arrays

1 Introduction

Computing hydrodynamic drag force on an array of cantilever-based resonators is essential while operating in a viscous fluid. The array of shape-morphed cantilever beam widely used in new emerging technologies, such as atomic force microscopy (AFM) [1,2], robotic propulsion systems [3,4], energy harvester [5,6], data storage millipede [7,8], biological, pressure and viscosity sensors [9–12]. These devices are required to have a high signal-to-noise ratio which is achieved by improving the quality factor (Q). However, the quality factor is decreased due to fluid damping when the cantilever resonator operates in a fluid. Its resonance frequency shifts due to the added mass effect of the fluid on the structure.

A. K. Pandey et al. (Eds.): MAMM 2022, 126, pp. 232–243, 2023.
https://doi.org/10.1007/978-3-031-20353-4_18

Therefore, it is important to improve the quality factor of the cantilever beam array by improving its design. In this paper, we focus on minimizing the added mass and damping effects by introducing the shape-morphing strategy over its width during oscillation.

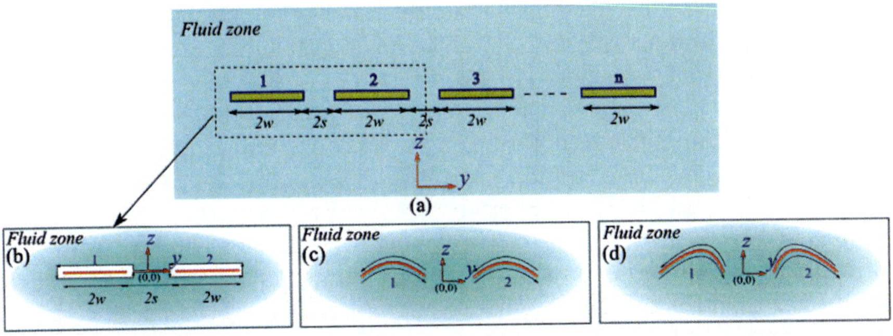

Fig. 1. A schematic representation of n array of beam arranged with a side gap of $2s$ and each beam has a width of $2w$, (a) sectional view of two dimensional array of beams (b) two rigid beam array oscillation in yz-plane, (c) two symmetric shape-morphing beam array oscillation in yz-plane and (d) two asymmetric shape-morphing beam array oscillation in yz-plane at a given longitudinal distance.

The two-dimensional unsteady Stokes equation is analytically solved in terms of stream function for a long arbitrary shape beam oscillating in a fluid by Tuck [13]. He then applied the solution to a limiting case of a thin ribbon to evaluate the in-phase and out-of-phase components of hydrodynamic functions with respect to the velocity. The resulting complex hydrodynamic function is expressed in terms of frequency parameter β. The real and imaginary parts of the hydrodynamic function represent the added mass and damping coefficients, respectively. Later, the theoretical and experimental approaches are employed to estimate the hydrodynamic coupling effect in an array of beams concerning the gap between two beams, the amplitude of oscillation, and the relative phase between the beams [7,10]. However, their analysis is limited to two-dimensional flow in a thin and long beam with a rectangular section. It is found that the hydrodynamic damping and added mass are directly dependent on the unsteady viscous layer [14]. For the beam oscillating with higher frequency, the fluid damping is drastically decreased as compared with added mass coefficient. Recently, the semi-analytical approach is used to compute the transverse hydrodynamic force, and velocity profiles on arrays of five to fifty beams [8]. From the analysis, they found that the hydrodynamic effects in the small and large arrays are different. Consequently, the existing hydrodynamic information obtained from the small-size array of coupled oscillators is not sufficient to predict the dynamics of a large-size array of beams. To control the drag force and resonance frequency shift, the shape-morphing of the oscillating cantilever has been used [3–5,15,16]. The beam vibrating with shape-morphing creates a smaller projected area than that during the rigid oscillation. Therefore, the optimal shape morphing parameter is estimated to minimize the energy loss and damping. It is also found that the hydrodynamic effects increase significantly if the beam oscillates near the bottom surface due to an increase in convective inertial effects compared to vortex phenomena near the beam edges [10,12,17,18]. Various other theoretical and experimental studies have been done to estimate the hydrodynamic force in oscillating single and array of cantilever beams by including the influence of the side wall and bottom gap effect [19,20].

Based on the above studies, it is found that the hydrodynamic function is used to compute forces in cantilever beam arrays considering the two-dimensional flow subjected to the rigid motion of cantilever beam of rectangular section as shown in Fig. 1a. In the present study, we are imposing the rigid and shape morphing strategy on the array of two beams with sections shown in Fig. 1b–c. Moreover, we have also introduced the asymmetric shape morphing shown in Fig. 1d to control the hydrodynamic damping. We are considering the two beams oscillating in phase with a maximum amplitude of one in the presence of viscous fluid. After validating the results for rigid motion of the beam arrays with literature, we study the effect of the gap between the beams with different shape morphing configurations of beam array. To perform the study, we present a brief mathematical model, solution methodology, and convergence criteria. Finally, the results are compared for different shape morphing cases, and a conclusion is drawn at the end.

2 Governing Equations

Consider the infinitely long array of thin cantilever beams oscillating in an incompressible, viscous Newtonian fluid with a prescribed shape morphing parameter. The physical problem is modeled by considering the beam length l to be much greater than the width $2w$, and thickness is negligibly small compared to width and length. The amplitude of oscillation is kept much smaller than the width of beams, and the gravity force is neglected. To study the influence of the interbeam gap in hydrodynamic forces of two beam arrays, we denote the gap between two cantilever beams as $2s$. In general, the flow is modeled using the complete Navier Stokes equation as

$$\nabla.\bar{V} = 0, \rho\left(\frac{\partial \bar{V}}{\partial t} + (\bar{V}.\nabla)\bar{V}\right) = -\nabla p + \mu\nabla^2\bar{V}. \tag{1}$$

Here, $\bar{V}(v_x,\ v_y,\ v_z)$ is the velocity domain, p is pressure, ρ is density, μ is viscosity of the fluid. Inertia force to viscous force is defined as the Reynolds number(Re). For low Re <1 (i.e., viscous force $\gg$ Inertia force), the nonlinear convective forces are negligible due to the small amplitude of oscillation of the beam [21], the linearized Navier Stokes equation is reduced from Eq. 1 to,

$$\rho\frac{\partial \bar{V}}{\partial t} = -\nabla p + \mu\nabla^2\bar{V}. \tag{2}$$

The deflection of the beam at free end is assumed to be unity. Since the length of the beam is much longer than the other two dimension, the velocity filed is assumed to the in YZ-direction i.e., (v_y, v_z). The linearized two-dimensional Navier Stokes equation is solved by using the Green's function method. The procedure is clearly explained by the Tuck [13] for a cylinder of arbitrary cross section in a viscous fluid. Later, Basak and Raman [7] obtained the stream function for an array of two beams in the frequency domain in terms of vorticity and pressure jumps as,

$$\psi(y', z') = \int_{-(2w+s)}^{-s} [\Delta\omega_1 \Psi_z - \frac{1}{\mu}\Delta p_1 \Psi_y]dy + \int_{s}^{(2w+s)} [\Delta\omega_2 \Psi_z - \frac{1}{\mu}\Delta p_2 \Psi_y]dy, \quad (3)$$

where, $\Psi = -\frac{1}{\alpha^2}(\Omega - G)$, $G = \frac{1}{2\pi}\log(\alpha R)$, $\Omega = -\frac{1}{2\pi}K_0(\alpha R)$ and $\Psi = -\frac{1}{2\pi\alpha^2}(logR + K_0(\alpha R)))$, $\alpha = (i\rho\omega)/\mu$, $i = \sqrt{-1}$, ρ is density of the fluid, ω is the oscillation frequency, μ is dynamic viscosity of the fluid, K_0 is the modified Bessel function of third kind of zero order [22] and $R = \sqrt{(y-y')^2 + (z-z')^2}$ with y, z and y', z' are the local points of a fluid and beam, respectively. Here, G and Ω have logarithmic singularity at $R = 0$ but $\Psi = 0$. G and Ψ have the logarithmic singular at infinity, but Ω is exponentially small. Also, $\Delta\omega_i$ and Δp_i represent vorticity and pressure jumps of a respective i^{th} beam. The subscript denotes the partial derivative respective to the spatial coordinates. The Green functions mainly depend on the dynamic flow condition α and spatial variables. The main unknowns of the above expressions are pressure, and vorticity jumps across the thin beam. The partial derivative of the Eq. (3) concerning y' and z' denote the velocity in transverse and lateral directions, respectively. Since the beams are considered very thin, i.e., $z = z' = 0$, the R is considered only as a function of y. The resulting transverse velocity can be written as,

$$v_z = -\psi_{y'}(y', 0) = \int_{-(2w+s)}^{-s} [\frac{1}{\mu}\Delta p_1 \Psi_{y'y}]dy + \int_{s}^{(2w+s)} [\frac{1}{\mu}\Delta p_2 \Psi_{y'y}]dy. \quad (4)$$

2.1 Transverse Velocity of a Shape-Morphed Cantilevers

An array of two long cantilever beams oscillates with a unit amplitude at the free end and simultaneous shape-morphing deflection of its section. An instantaneous deflection of the shape-morphing oscillation of a section of beams in the YZ-plane is shown in Fig. 1(b), (c). To induce the symmetric and asymmetric shape morphing curvature on both beams oscillating with the same amplitude, the displacement function at the free end location can be represented in terms of the shape morphing parameters $\bar{\kappa}$ and $\bar{y}'$,

$$[y, z] = \left[\frac{\sin(\bar{y}\bar{\kappa}\sin\omega t)}{\bar{\kappa}\sin\omega t}, \left(\delta\sin\omega t + \frac{\cos((\bar{y} - \bar{y}')\bar{\kappa}\sin\omega t) - 1}{\bar{\kappa}\sin\omega t}\right)\right], \quad (5)$$

where, δ is the amplitude of oscillation of the beam, ω is the frequency of oscillation, and t is the time of oscillation. Expanding the trigonometric function of the above equation and successive approximation, Eq. (5) can be written in a phasor form as,

$$[y, z] = \left[\bar{y}, \left(1 - \frac{\bar{\kappa}(\bar{y} - \bar{y'})^2}{2}\right)\delta \sin \omega t\right]. \tag{6}$$

To obtain the local transverse velocity of the structure, we differentiate local transverse displacement with respect to time. Hence, the velocity in Z-direction can be written as $v_z = \left(1 - \frac{\bar{\kappa}(\bar{y}-\bar{y'})^2}{2}\right)\delta\omega \cos \omega t$. This velocity condition is applied to both the beams with a no-slip condition in an array of two beams, thus, influencing the fluid film and, finally, the pressure jump around the beams which is described in the following section.

2.2 Non-dimensionalization and Numerical Procedure

Let us consider the pressure jumps $\Delta p_1, \Delta p_2$ and transverse velocities v_{z1}, v_{z2} as a function of y in an array of two beams. Assuming the velocity of the first beam as $v_{z1} = \left(1 - \frac{\bar{\kappa}(\bar{y}-\bar{y'})^2}{2}\right)V_{01}e^{i\omega t}$ and the second beam velocity with a phase difference of θ as $v_{z2} = \left(1 - \frac{\bar{\kappa}(\bar{y}-\bar{y'})^2}{2}\right)V_{02}e^{i(\omega t+\theta)}$, here, $V_{0j} = \delta_j\omega$ with $i = 1, 2$.

Using the non-dimensional parameter, $\xi_j = y_j/w$, the gap between two beams is $\bar{s} = s/w$, non-dimensional pressure jump with respect to the first beam velocity is $P(\xi_j') = w\Delta p_j(y')/\mu V_{01}$, amplitude ratio is $\eta = V_{02}/V_{01}$, frequency parameter is $\beta = \rho\omega w^2/2\pi\mu$, shape-morphing parameter $\kappa = 2w\bar{\kappa}$, and the skew-symmetric parameter is $\epsilon = \bar{y'}/w$. The transverse velocities on the each beam in the dimensionless form in terms of the non-dimensional pressure can be written as

$$\int_{-(2+\bar{s})}^{-\bar{s}} [P_1(\xi')\Psi_{\xi'\xi}]d\xi + \int_{\bar{s}}^{(2+\bar{s})} [P_2(\xi')\Psi_{\xi'\xi}]d\xi = \left(1 - \frac{\kappa(\xi - \epsilon)^2}{2}\right), \tag{7}$$

$$\int_{-(2+\bar{s})}^{-\bar{s}} [P_1(\xi')\Psi_{\xi'\xi}]d\xi + \int_{\bar{s}}^{(2+\bar{s})} [P_2(\xi')\Psi_{\xi'\xi}]d\xi = \alpha\left(1 - \frac{\kappa(\xi - \epsilon)^2}{2}\right)e^{i\theta}. \tag{8}$$

We solve the above equation using the boundary element method (BEM) based numerical scheme by Reimann summation. Assuming that the unknown pressure on each of the beams considers being constant, each set of kernel square matrix of size $M \times M$ convert into a global matrix with a size of $4(M \times M)$. The unknown column vector of pressure on each beam is re-framed with a size of $4(M \times 1)$. The right-hand side of the above equations is presented as a known column vector of size $4(M \times 1)$. We are inverting the global matrix at a given time numerically to obtain the desired output non-dimensional pressure jump. Finally, integrating the pressure on each beam, the complex hydrodynamic force is obtained. Its real and imaginary parts denote the damping force (in-phase with a velocity of the beam) and inertial or added mass force (in-phase with an acceleration of the beam). The hydrodynamic force F and hydrodynamic function Γ in the frequency domain are found as,

$$F(\omega) = \int_{d\xi'} P(\xi')d\xi' = \frac{\pi}{4}\rho\omega w^2\Gamma(\beta,\kappa,\epsilon), \tag{9}$$

$$\Gamma(\beta,\kappa,\epsilon) = \frac{2}{\pi^2\beta}\int_{\xi'} P(\xi')d\xi'. \tag{10}$$

The complex hydrodynamic function $\Gamma(\beta,\kappa,\epsilon)$ has a real and imaginary part. The real part of the hydrodynamic function denotes the added mass coefficient. Its imaginary part represents the damping coefficient which influences the frequency shift and lowers the oscillation amplitude while operating in the fluid medium. Most of the resonators operate at high frequencies with low amplitude. Thus, we perform the numerical study for the variation of different parameters such that frequency parameter $\beta = [0.1, 1000]$, shape morphing parameter $\kappa = [0, 25]$, asymmetric shape morphing parameter $\epsilon = [0, 0.5]$, amplitude ratio $\eta = [0, 1]$ and the phase change of the two beam oscillating between in-phase and out of phase motion, respectively, i.e., $\theta = [0, \pi]$. Considering the above non-dimensional parameters, we have performed a convergence study to obtain a grid-independent result. The grid independence study is mentioned in terms of hydrodynamic forces in the Table 1 for an extreme values of parameters $\beta = 1000$, $\theta = \pi$, $\eta = 1$ and the gap between two beams considered to be very small, i.e., $\bar{s} = 0.1$, at different shape morphing parameters.

Table 1. The grid independence study on first beam with extreme parameters

M	$\kappa = 0$		$\kappa = 10$		$\kappa = 25$	
	Re(F) $\times 10^3$	Im(F) $\times 10^2$	Re(F) $\times 10^3$	Im(F) $\times 10^1$	Re(F) $\times 10^3$	Im(F) $\times 10^2$
80	4.5767	2.6863	1.6031	2.5757	2.8574	4.6733
100	4.5771	2.6847	1.6030	2.6005	2.8581	4.6771
120	4.5773	2.6840	1.6029	2.6117	2.8586	4.6789
140	4.5774	2.6836	1.6029	2.6177	2.8588	4.6798
160	4.5774	2.6834	1.6029	2.6212	2.8590	4.6804

The real and imaginary parts of the hydrodynamic force are tabulated at various shape morphing parameters, $\kappa = 0, 10, 25$. It is found that the 140 elements on each beam is found to be sufficient for considering the extreme conditions. The effects of parameters such as $\beta, \bar{s}, \theta, \kappa, \epsilon$ are examined on quality factor (Q) and resonator frequency shift in the subsequent section.

3 Results and Discussions

In this section, we present the influence of different parameters on the variation of hydrodynamic forces and hydrodynamic coefficients.

3.1 Hydrodynamic Force

The hydrodynamic force per unit length is computed on an array of two beams in the frequency domain. To compute the forces, both the beams are assumed to oscillate at a maximum amplitude of oscillation in the same phase. Consequently, we estimate both the damping and inertial forces on rigid beam 1 with symmetric and asymmetric shape morphing oscillation with varying non-dimensional gap $\bar{s}$ as shown in Fig. 2.

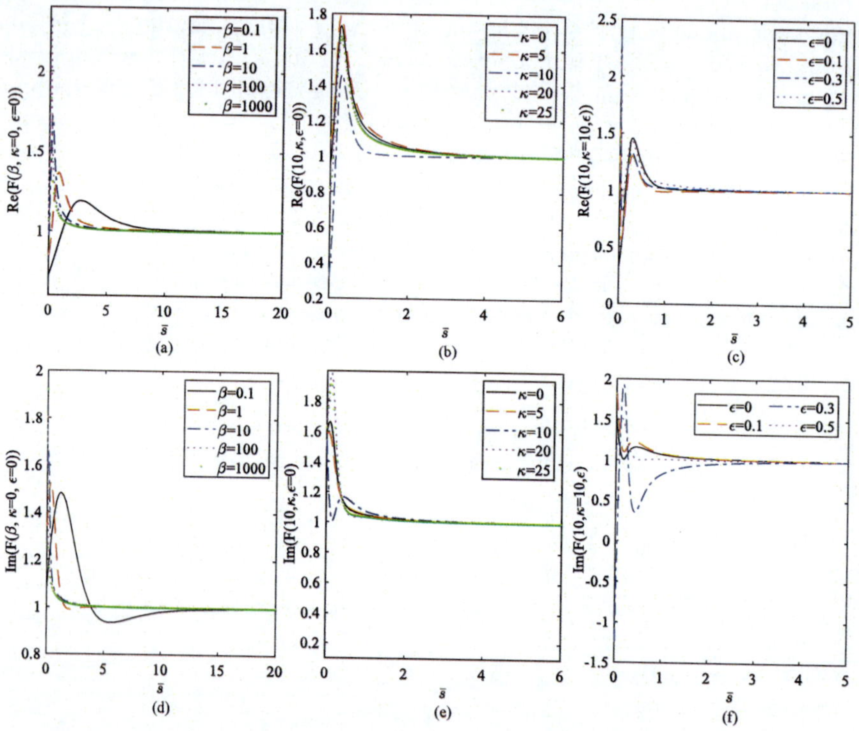

Fig. 2. The hydrodynamic force per unit length estimated with a rigid oscillation of the first beam with symmetric and asymmetric shape morphing oscillation of first beam of the array with respective to the gap between the beams, $\bar{s}$. (a) and (d) represent the real and imaginary part of the hydrodynamic force on 1st beam with rigid oscillation at different frequency parameter β. (b) and (e) represent the damping force and added mass force on the beam 1 with a varying κ at $\beta = 10$. (c) and (f) represent damping and added mass forces on beam at $\kappa = 10$, $\beta = 10$ with varying ϵ.

The fundamental effect can be described through the unsteady stokes layer $\delta_s = \sqrt{\nu/\omega}$ with ν as kinematic viscosity, and is inversely proportional to the frequency parameter β. The unsteady stokes layer of the fluid around the cantilever creates the fluid damping. It lowers the amplitude of oscillating beams. The fluid with more viscosity generates a very thick layer of δ_s. Thus, the more viscous fluid causes more damping. Furthermore, it is also true that the fluid damping happens due to oscillation of the beam at a lower frequency, i.e., β. As the frequency parameter increases, the unsteady viscous layer becomes smaller around the beam. Therefore, the fluid does not get sufficient time to move around the beam at high β. Thus, the inertial effects dominate, which is also a function of fluid viscosity and frequency of oscillation.

To quantify the effect, the drag force per unit length is computed using Eq.(9). Figure 2 illustrate both the damping and inertial force per unit length on beam 1. These forces are normalized with respect to the force on the beam oscillating in isolation (i.e., when the beams are far apart). Figures. 2 (a) and (d) represent normalised damping and added mass force per unit length with $\bar{s}$ at specified frequency parameter $\beta = 0.1, 1, 10, 100, 1000$ for a rigid oscillation of beam such that the shape morphing parameter $\kappa = 0$ and $\epsilon = 0$. At a small frequency of oscillation, the viscous layer is highly dominated. As the gap between the two beams increases from 0.01 to 20, the viscous layer diminishes at larger $\bar{s} = 12$ for $\beta = 0.1$. As β increases, the effect of the viscous layer gradually decreases, and it results in high drag force due to the vorticity formation near the edges of the beam. The vorticity significantly affects the added mass force as well.

Similarly, the damping and inertial force per unit length are also computed on a symmetric shape morphing beam at $\beta = 10$ as shown in Fig. 2b, e. As the magnitude of the shape morphing parameter increases from small curvature $\kappa = 5$ to $\kappa = 25$, the viscous layer gradually decreases. Thus, the gap between two beams **is** required to be less due to the shape morphing. The hydrodynamic forces, such as damping and added mass forces at lower gap $\bar{s}$ under symmetric shape morphing oscillation are compared with those of the rigid beam oscillation. At $\beta = 10$, the gap effect of rigid oscillation of array of beam $\kappa = 0$ occurs at $\bar{s} < 2.5$. However, as κ increases from 5 to 25, the gap between two beams influences below $\bar{s} = 2.5$ to 1.8. Thus, the higher shape morphing parameter creates a smaller projected surface area as compared to that with the rigid oscillation of the beam.

We have also analyzed the effect of asymmetric shape morphing of the array of two beams. The asymmetric case shows the peak shift towards the right side of the first beam and the left side of the second beam, as shown in Fig. 1d. These two beams moving with the same amplitude are taken under in-phase conditions. First, we perform different studies by varying β and then κ for various asymmetric shape morphing parameters ϵ. Figure 2c, f show that the hydrodynamic damping force and added mass force per unit length at $\beta = 10, \kappa = 10$ for different ϵ. It is found that the asymmetric shape morphing further reduces the effective surface area and, thus, reduces damping and added mass force per unit length than the cases in rigid and symmetric shape morphing beams. The hydrodynamic forces are found to be dominant at smaller values of $\bar{s}$. Therefore, the beams oscillating at high frequency with small amplitude may adopt the shape morphing strategy to create a small viscous layer around the beam and create the vortex shedding near the edges of the shape morphed beam.

3.2 Hydrodynamic Coefficient

The hydrodynamic function per unit length, which is inversely proportional to the frequency parameter, β, is estimated using the Eq. (10). For smaller β, both the real and imaginary parts of the hydrodynamic function are having larger magnitudes. The magnitude of the hydrodynamic coefficient gradually decreases as β increases. The real and imaginary parts of the hydrodynamic function denote the added mass and damping coefficients, respectively. The coefficients are dependent on the acceleration and velocity of the array of beam. Figure 3 represents the variation of hydrodynamic coefficient over β at different values of $\bar{s}$ for rigid, shape morphing and asymmetric shape morphing beams. The rigid oscillations of the two cantilever beams oscillating at same amplitude and in same phase at different gap ratios are illustrated in Fig. 3a, d for beam 1. As the frequency parameter increases from 0.1 to 1000, the added mass coefficient approaches to unity and damping coefficient approaches to zero. Thus, for higher β, the formation of the unsteady viscous layer thickness is much smaller. As a result, lowering of the amplitude of oscillation of the beam at higher β is significantly small. Similarly, the resonance frequency shift is also shows less shift. Considering the effect of the gap ratio $\bar{s}$, we found that for the smaller gap $\bar{s} = 0.1$, the effect of viscous layer of beam 2 on beam 1 is vivid at lower β. The hydrodynamic coupling appear at $\bar{s} = 5$. When the gap ratio $\bar{s} \longrightarrow \infty$, both the beams oscillate individually without any hydrodynamic coupling effect.

Similarly, the hydrodynamic coefficients are computed on the shape morphing array of the beams. We present the hydrodynamic coefficients on a cantilever beam 1 with a symmetric shape morphing parameter κ . The added mass and damping coefficients on beam 1 are shown in Fig. 3b, e at $\kappa = 10$ over a range of β for different $\bar{s}$. Thus, it is found that the shape morphing parameter plays a significant role in reducing the damping and added mass effect. As the κ increases, the projected surface area of array of micro beam becomes gradually decreases. Also, as the gap of the beam increases, the hydrodynamic coupling effect significantly diminishes.

Finally, we have described the variation of added mass and damping coefficients on the beam 1 with asymmetric shape morphing parameter $\epsilon = 0.3$ with respect to the gap variation when **varying** β, keeping $\kappa = 10$ as shown in Figs. 3c, f. The effect of one beam on the other beam become dominant at lower $\bar{s}$. Thus, the effect of the viscous layer on the asymmetric shape morphing gradually disappear when the gap ratio increases.

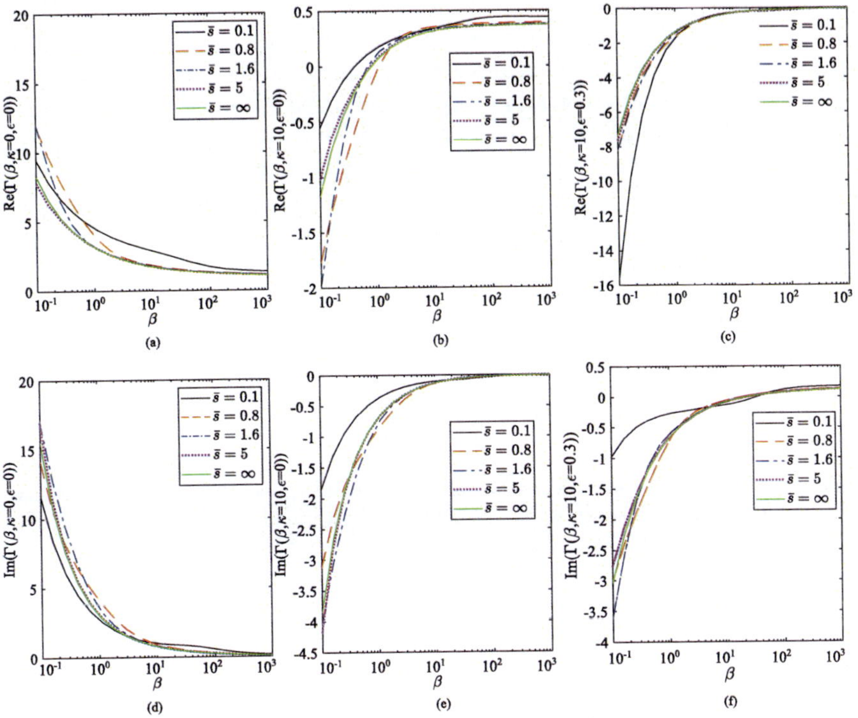

Fig. 3. The hydrodynamic coefficient per unit length is estimated with a rigid oscillation of the first beam, shape morphing oscillation of first beam and asymmetric shape morphing on the array of the first beam with respect to the $\bar{s}$ over β. (a) and (d) represents the real and imaginary part of the hydrodynamic function on 1st beam with rigid oscillation at different β. (b) and (e) represented the damping force and added mass force on beam 1 with given $\kappa = 10$, β. (c) and (f) represents damping and added mass force on beam at $\kappa = 10$, β with a given $\epsilon = 0.3$.

4 Conclusion

The hydrodynamic damping on the oscillating beam array in a viscous fluid is modelled considering a very long beam compared with the other dimensions. Therefore, we consider a fluid-structure interaction (FSI) problem in 2D domain. The integral boundary formulation of a stream function gives exposure to finding the velocity boundary condition. We considered the rigid, symmetric and asymmetric shape morphing parameters on the oscillation of an array of beams. Considering two beams, both oscillate with high frequency and the same amplitude in the same phase. The desired output pressure jump is estimated and is used to evaluate the hydrodynamic force and coefficient per unit length.

Subsequently, the hydrodynamic force and hydrodynamic function on beam 1 are estimated to study the effect of hydrodynamic coupling due to the gap ratio $\bar{s}$. As the frequency parameter β increases from 0.1 to 1000, the hydrodynamic coupling effect significantly reduces. Thus, the unsteady viscous layer formation near the beam surface becomes smaller and produces a lower value of drag force. Moreover, the vortex formation near the edges of the beam creates a more added mass effect than the damping. The increased symmetric and asymmetric shape morphing parameters of an array of beams significantly reduce damping and added mass. Such shape morphing effects can be obtained by using a layer of piezoelectric patches which can be useful in energy harvesting and robotic propulsion systems.

Acknowledgements. The author would like to acknowledge the fellowship provided by the Ministry of Education, New Delhi, the Government of India.

References

1. Sader, J.E.: Frequency response of cantilever beams immersed in viscous fluids with applications to the atomic force microscope. J. Appl. Phys. **84**(1), 64–76 (1998)
2. Sader, J.E., Chon, J.W., Mulvaney, P.: Calibration of rectangular atomic force microscope cantilevers. Rev. Sci. Instrum. **70**(10), 3967–3969 (1999)
3. Ahsan, S.N., Aureli, M.: Nonlinear oscillations of shape-morphing submerged structures: Control of hydrodynamic forces and power dissipation via active flexibility. J. Fluids Struct. **74**, 35–52 (2017)
4. Ahsan, S.N., Aureli, M.: Three-dimensional analysis of hydrodynamic forces and power dissipation in shape-morphing cantilevers oscillating in viscous fluids. Int. J. Mech. Sci. **149**, 436–451 (2018)
5. Ahsan, S.N., Aureli, M.: Finite amplitude oscillations of flanged laminas in viscous flows: vortex-structure interactions for hydrodynamic damping control. J. Fluids Struct. **59**, 297–315 (2015)
6. Hu, L., Yan, H., Zhang, W.M., Zou, H.X., Peng, Z.K., Meng, G.: Theoretical and experimental study on dynamic characteristics of V-shaped beams immersed in viscous fluids: from small to finite amplitude. J. Fluids Struct. **82**, 215–244 (2018)
7. Basak, S., Raman, A.: Hydrodynamic coupling between micromechanical beams oscillating in viscous fluids. Phys. Fluids **19**(1), 017105 (2007)
8. Ande, R., Gutschmidt, S., Sellier, M.: Fluid dynamics investigation of a large array. Phys. Fluids **33**(7), 073608 (2021)
9. Li, C., Ma, X., Guan, Y., Tang, J., Zhang, B.: Microcantilever array biosensor for simultaneous detection of carcinoembryonic antigens and α-fetoprotein based on real-time monitoring of the profile of cantilever. ACS Sens. **4**(11), 3034–3041 (2019)
10. Manickavasagam, A.K.: Hydrodynamic coupling of arrays in fluids (2020)
11. Hosaka, H., Itao, K.: Coupled vibration of microcantilever array induced by airflow force. J. Vib. Acoust. **124**(1), 26–32 (2002)
12. Green, C.P., Sader, J.E. Small amplitude oscillations of a thin beam immersed in a viscous fluid near a solid surface. Phys. Fluids **17**(7), 073102 (2005)
13. Tuck, E.O.: Calculation of unsteady flows due to small motions of cylinders in a viscous fluid. J. Eng. Math. **3**(1), 29–44 (1969)
14. Landau, L.D., Lifshitz, E.M.: Fluid Mechanics: Landau and Lifshitz: Course of Theoretical Physics, vol. 6. Elsevier (2013)
15. Lang, H.P., Hegner, M., Gerber, C.: Cantilever array sensors. Mater. Today **8**(4), 30–36 (2005)

16. Kimber, M., Garimella, S.V., Raman, A.: An experimental study of fluidic coupling between multiple piezoelectric fans. In: Thermal and Thermomechanical Proceedings 10th Intersociety Conference on Phenomena in Electronics Systems, 2006. ITHERM 2006, pp. 333–340. IEEE (2006)
17. Lang, H.P., Hegner, M., Gerber, C.: Nanomechanical cantilever array sensors. In: Springer Handbook of Nanotechnology, pp. 457–485. Springer, Berlin, Heidelberg (2017)
18. Amiri, I.S., Addanki, S.: Simulation fabrication and characterization of micro-cantilever array based ozone sensor. Results Phys. **10**, 923–933 (2018)
19. Akarapu, A., Nighot, R.P., Devsoth, L., Yadav, M., Pal, P., Pandey, A.K.: Experimental and theoretical analysis of drag forces in micromechanical-beam arrays. Phys. Rev. Appl. **13**(3), 034003 (2020)
20. Ashok, A., Kumar, P.M., Singh, S.S., Raju, P., Pal, P., Pandey, A.K.: Achieving wideband micromechanical system using coupled non-uniform beams array. Sens. Actuators A: Phys. **273**, 12–18 (2018)
21. Meirovitch, L., Parker, R.G.: Fundam. Vibr. Appl. Mech. Rev. **54**(6), B100–B101 (2001)
22. Kreyszig, E., Stroud, K., Stephenson, G.: Advanced engineering mathematics. Integration, **9**(4) (2008)

Design and Performance Analysis of TiO_x Based MEMS Bolometer Pixel

Isha Yadav[1,2](✉), Sudha Gupta[1], A. Vishwakarma[1], S. Dutta[1], and Ratnamala Chatterjee[2]

[1] Solid State Physics Laboratory, DRDO, Lucknow Road, Timarpur, Delhi 110054, India
isha.yadav.sspl@gov.in
[2] Department of Physics, IIT Delhi, Hauz Khas, New Delhi 110016, India

Abstract. In this study the electro-thermo-mechanical behavior and analysis of TiO_X microbolometer is presented. Different thermal, electrical and coupled electromechanical simulations were performed to achieve the target Noise Equivalent Temperature Difference (NETD) and thermal time constant. The influence of various device dimensions and film thickness on device thermal conductance, NETD and thermal time constant has been reported.

Keywords: Microbolometer · TiO_X · Thermal conductance · Thermal time constant · NETD

1 Introduction

Uncooled infrared focal plane arrays (IRFPAs) are the low-cost thermal imagers for the LWIR (8–14 μm) range [1]. Compared to cooled detectors, they offer the advantages of smaller size and lower power dissipation [2]. These uncooled thermal imagers are widely used in various defence and civilian applications like thermal sights for night conditions, smoke or fog environment, handheld thermal imagers for inspection of buildings, firefighting applications etc.

Microbolometer consists of an array of pixels fabricated over the read-out circuit. The read-out circuit provides voltage or current bias to each pixel. The pixel is a bridge-like structure suspended over the read-out circuit for thermal isolation [3]. It absorbs the incoming IR radiation and causes an increase in temperature, which changes the resistance of the bridge-like structure. The resistance change can be sensed as the electrical signal by using monolithically integrated readout circuitry with the sensor element.

Commercially, thin film of vanadium-oxide (VO_X) or amorphous-silicon are actively used as bolometer material. Though a-Si offers good TCR and mechanical strength, but it has high resistivity and hence high 1/f noise. VOx has issues of hysteresis with temperature and non-standard material for CMOS foundry. As an alternative to these materials, nonstoichiometric titanium-oxide (TiO_X) received considerable attention as the bolometer material for its lower resistivity, higher TCR, lower 1/f noise and better CMOS compatibility.

A. K. Pandey et al. (Eds.): MAMM 2022, 126, pp. 244–249, 2023.
https://doi.org/10.1007/978-3-031-20353-4_19

The technology of this device consists of material selection based on the compatibility of a thermal detector with the CMOS process, detector design, ROIC design and detector fabrication over the ROIC. In order to carry out the device design, the structure is designed using AutoCAD software and thermoelectric characteristics of the bolometer structure is analyzed by close-form equations [4, 5] and finite element method (FEM) simulation [6, 7].

Initially researchers had demonstrated the performance of MEMS bolometer with pixel pitch (separation between the centers of two consecutive pixels) was typically ~50 μm [1, 2]. However, smaller pixel size is needed to achieve improved spatial resolution and smaller imaging systems. To double the pixel density, the pixel pitch has to be reduced by $\sqrt{2}$ times for a given FPA area. Hence the pitch has been reducing in a sequence of 35 μm, 28 μm and 17 μm. The present work discusses about the design of 17 μm pitch TiO_X based bolometer pixel for development of large format infra-red focal plane array (IRFPA). The measured materials characteristics of the TiO_X films are utilized in the simulation database and performance of the bolometer pixel are analyzed.

2 Proposed Device Structure

The 17 μm pitch microbolometer pixel structure consists of the suspended stack structure of 100 nm-thick structural layer of Silicon Nitride (SiN), 100 nm-thick sensor layer of TiO_X and 10 nm-thick Titanium Nitride absorber layer, as shown in Fig. 1. The stack is supported by two L-shaped legs (consists of Ti and SiN stack) from the two opposite corners on a prefabricated CMOS ROIC wafer. To enhance the absorbent thermo-efficiency, a quarter-wave cavity resonator is formed by creating an Al reflector layer beneath the suspended stack. The temperature induced change in resistance in the bolometer stack due to the incoming IR flux is converted into electrical signal (voltage/current) by the underneath ROIC.

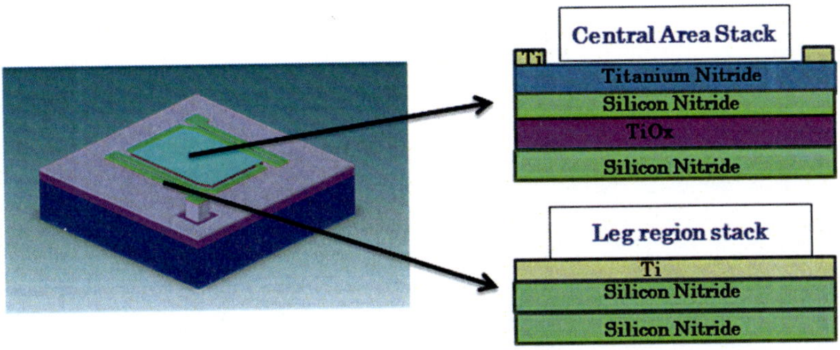

Fig. 1. Schematic structure of TiO_X based MEMS bolometer pixel

The characteristics of the pixel structure is simulated by using FEM simulation software (Coventorware). The pixel geometry is dissected into tiny Tetrahedron mesh (size: 0.3 μm) for the FEM simulation. Measured thermal and electrical characteristics of the previously optimized TiO_X films [7] are utilized in the present simulation study. The steady-state thermal analysis of the pixel structure is done at a fixed $2000 \times 10^{-12}\ \text{W}/\mu\text{m}^2$ heat-flux input (calculated as per the Planck's law for a distance of 1 m). Detailed FEM simulations of the bolometer structure are carried out to evaluate the thermal capacitance, thermal conductance, temperature increase, time-constant, voltage responsivity, Figure-of-merit (TCR/G), and Noise Equivalent Temperature Difference (NETD) of the bolometer pixel (Fig. 2).

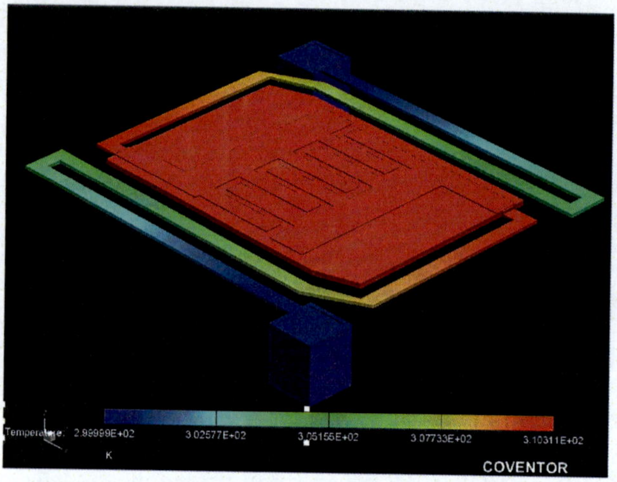

Fig. 2. Simulated temperature-rise in the TiO_x based MEMS bolometer pixel due to incoming heat (IR) flux

3 Results and Discussions

The pixel NETD and voltage responsivity is calculated by using the standard equations [7, 8]. The mathematical equations are given below:

$$R_{\lambda}^{V} = \alpha\eta R_{eff}\frac{I_b}{G_{th} \times \sqrt{(1+\omega^2\tau^2)}} \tag{1}$$

where, α represent the TCR of the thermistor material; R_{eff} be the pixel resistance; the thermal conductance of the pixel represented by G_{th}; I_b is the bias current; the absorbance of the absorber layer is shown by the symbol η; τ = thermal time constant and $\omega = 2\pi \times$ frame rate (f_i).

$$NETD = \sqrt{NETD_{1/f}^2 + NETD_{Jhonson}^2 + NETD_{Thermal}^2 + NETD_{ROIC}^2} \tag{2}$$

The contributing terms are as follows:

$$NETD_{\frac{1}{f}} = \left(4\frac{F^2}{\emptyset\left(\frac{\Delta P}{\Delta T}\right)}\right)\frac{G_{th}}{\beta A\eta}\frac{\sqrt{\frac{K}{\nu}\ln\left(\frac{x_l f_i}{f_s}\right)}}{\alpha}\sqrt{1+\omega^2\tau^2} \tag{3}$$

$$NETD_{Johnson} = \left(4\frac{F^2}{\emptyset\left(\frac{\Delta P}{\Delta T}\right)}\right) * 2\sqrt{k} * \frac{G_{th}}{\beta A\eta}\frac{\sqrt{T_1 R_{bol} x_i f_i}}{\alpha V_{bias}}\sqrt{1+\omega^2\tau^2} \tag{4}$$

$$NETD_{Thermal} = \left(4\frac{F^2}{\emptyset\left(\frac{\Delta P}{\Delta T}\right)}\right)\sqrt{k}\frac{G_{th}}{\beta A\eta}\frac{T_1}{\sqrt{C}}\sqrt{1+\omega^2\tau^2} \tag{5}$$

$$NETD_{ROIC} = \left(4\frac{F^2}{\emptyset\left(\frac{\Delta P}{\Delta T}\right)}\right)\frac{G_{th}}{\beta A\eta}\frac{(R_{ROIC}+R_{bol})}{\alpha V_{bias} R_{ROIC}}\sqrt{1+\omega^2\tau^2} \tag{6}$$

where, F-number of the optics of the IR imaging system is presented as F; Φ be the transmission of the IR optics in 8–12 μm wavelength (LWIR) range; the 1/f noise constant of the microbolometer thermistor material is marked as K; ν = the thermistor volume; R_{bol} be is the bolometer resistance at temperature T_1; the pixel fill factor is shown as β; x_i is number of bolometer pixels in each column for rolling integration.

The simulated values of G, τ and NETD for various film thickness (SiN and Ti), TCR values of the TiOx films and leg width are as shown in Fig. 3(a)–(d). Thickness of SiN film has been varied from 70 nm to 140 nm. The relationship between SiN film thickness and the G and τ, is given by the following equations:

$$G_{cond} = \frac{g_{leg} A_{leg}}{L_{leg}} \quad \tau = \frac{C_{total}}{G_{total}}$$

For achieving low NETD, the thermal conductance has to be minimized while for faster response, the time constant is also to be minimized. The G and τ values have opposite dependence on film thickness. So, the G and τ values have been plotted with different film thickness values. From the point of intersection of the two curves, the optimum value of SiN thickness is found to be 100 nm for the present design. Keeping the thickness of SiN as 100 nm, thickness of the Ti layer is varied and the optimum value of Ti film thickness is found to be ~30 nm. (28 nm precisely)

The resultant pixel NETD for these optimized values of SiN and Ti film thicknesses is calculated by varying the leg width. The leg width depends on the fabrication capabilities of the foundry in which the device is fabricated. In present study, leg width is varied from 0.3 to 1 μm and for a target NETD ≤ 50mK, the required value of leg width is found to be <0.6 μm. The sensor material TCR requirement is obtained by plotting the graph in Fig. 3(d). It can be seen that in order to get the desired NETD, TCR of the sensor film has to be higher than 2.8%/K.

The optimum device parameters for the target NETD of ≤ 50mK have been summarised Table 1 below:

4 Conclusions

In this study the detailed electro-thermo-mechanical behavior and analysis of TiO_x-microbolometer is presented. Pixel design for the target NETD ≤ 50 mK has been discussed. Optimum value of the SiN and Ti film thicknesses, leg width and sensor film TCR has been determined.

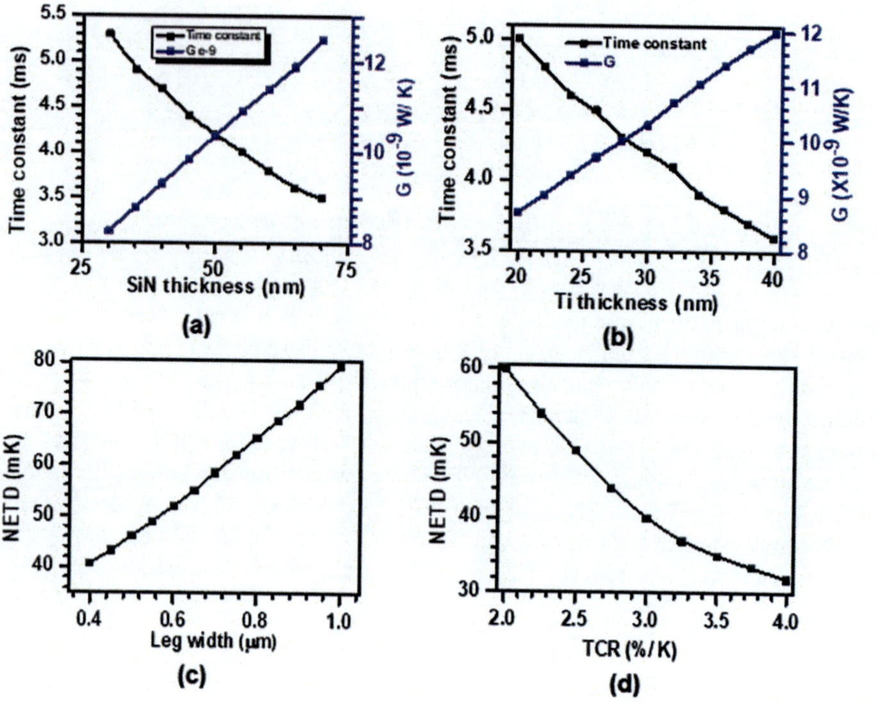

Fig. 3. Performance analysis results of TiO_x based MEMS bolometer pixel: Effect of variation of (a) SiN thickness; (b) Ti layer thickness; (c) Leg width; and (d) TCR.

Table 1. Optimized device performance parameters for the TiO_x based bolometer pixel

Sr. No.	Device performance parameters	Values
1	Thermal conductance (G)	2.08×10^{-08} (W/K)
2	Thermal Capacitance (C)	9.03×10^{-11} (J/K)
3	Time constant	4.12 (ms)
4	Temperature rise due to incoming heat flux	2.677 (K)
5	Figure-of-merit (TCR/G)	13.2×10^{7}
6	Voltage responsivity	1.62×10^{6} (V/W)
7	NETD	37.3 (mK)

Acknowledgments. The authors sincerely thank Dr. Seema Vinayak, Director SSPL for her continuous support and kind permission to publish this work.

References

1. Vollmer, M., Mollmann, K.P.: Infrared Thermal Imaging, Fundamentals, Research and Applications, 2nd edn. Wiley VCH, Verlag GmbH & Co, KGaA (2018)
2. http://i3system.com/uncooled-detector
3. Lin, A.L.: Bolometer having an amorphous titanium-oxide layer with high resistance stability. U.S. Patent (2008) 7442933
4. Reddy, Y.A.K., Kang, I.K., Shin, Y.B., Lee, H.C.: Bolometric properties of reactively sputtered TiO_{2-x} films for thermal infrared image sensor. J. Appl. Phys. D **48**, 355104 (2015)
5. Shinn, Y.B., Reddy, Y.A.K., Kang, I.K., Lee, H.C.: Nb doping effect on TiO_{2-x} films for bolometer applications. J. Phys. Chem. Solids **91**, 128–135 (2016)
6. Yadav, I., Jain, S., Lamba, S.S., Tomar, M., Gupta, S., Gupta, V., Jain, K.K., Dutta, S., Chatterjee, R.: Growth and electrical properties of TiO_x films and its effect on microbolometer design. J. Mat. Sci.: Mater. Electron. **31**, 6671–6678 (2020)
7. Yadav, I., et al.: Compositional, electrical and thermal properties of nonstoichiometric titanium oxide thin films for MEMS bolometer applications. Mat. Sci. Semicond. Proc. **148**, 106779 (2022)
8. Niklaus, F., Decharat, A., Jansson, C., Stemme, G.: Performance model for uncooled infrared bolometer arrays and performance predictions of bolometers operating at atmospheric pressure. Infrared Phys. Tech. **51**, 168–177 (2008)

FEM Analysis of Split Electrode IDTs Designed Lithium Tantalate-Polyaniline SAW Gas Sensor

Dhananjaya Panda and Koteswara Rao Peta(✉)

Department of Electronic Science, University of Delhi South Campus, Benito Juarez Road, New Delhi 110012, India

krao@south.du.ac.in

Abstract. The finite element method (FEM) underpinning COMSOL Multiphysics 6.0 has been used to model the sensing of various polar volatile organic compounds (VOCs) by the split electrode interdigital transducers (SEIDT) surface acoustic wave (SAW) sensor. On a $LiTaO_3$ base, split-electrode interdigital transducers were employed with a sensing layer of polyaniline (PANI). Acetone VOC gas was detected using a SAW sensor at RT with a concentration of 100 ppm. After the gas specimen interacted with the detecting polymer, a change in the density of the sensing surface was seen, and with it, a change in the Rayleigh wave's Eigen frequency. For a number of studies, we measured admittance, displacement, quality factor, and electric potential versus frequency in addition to plotting the deformed shape at the Eigen (resonance) frequency. The S-parameter and quality factor studies also demonstrate the benefits of using split electrode IDTs in SAW sensors for lowering reflection loss.

Keywords: VOCs · FEM · COMSOL · Split electrode IDT · Polyaniline (PANI)

1 Introduction

Gases in the atmosphere are essential for both living and nonliving organisms. Due to the growth in pollutant gases in the atmosphere, it is crucial for environmental and diagnostic reasons to detect dangerous gases at the micro level. Several gases may be detected by a gas sensor in our digitalized and technological age [1, 2]. The mass sensitivity, lesser environmental disturbance, lower power consumption, and mobility of the acoustic wave-based gas sensor make it very appealing.

Based on how the waves move, this acoustic wave sensor is divided into two parts: bulk acoustic wave (BAW) [3] and surface acoustic wave (SAW) sensor [4]. The SAW sensor device is made up of a substrate made of a piezoelectric material, an interdigital transducer (IDT) made of a conductive material, and a sensitive sensing layer that reacts to a specific gas. IDT is used to lay out the SAW device by putting the sensing layer on the pattern of the piezoelectric substrate. SAW based on a two-port Rayleigh resonator with a polymer-coated sensing layer has high sensitivity, higher stability, good response time, and low noise for gas detection [5, 6]. Due to its higher sensitivity and faster response time, SAW has been used for a wide range of applications, such as analysis of

A. K. Pandey et al. (Eds.): MAMM 2022, 126, pp. 250–265, 2023.
https://doi.org/10.1007/978-3-031-20353-4_20

gas and bio analyte in the lower levels of ppm, electronic nose in industrial installations, monitoring and protection of environmental issues, and medical applications [7–11]. The SAW sensor works because the response signal changes depending on how much the analyte weighs. When a specific analyte interacts with a surface acoustic wave (SAW), it gives off a specific frequency, which is picked up by the output circuitry.

Transducers, one of the three main parts of SAW sensors, have a big effect on how waves travel and how they interact with each other. Interdigitated transducers (IDTs) with specific shapes and sizes cause a certain wave to appear on the surface. The unique design of IDTs has solved some of the most important problems with SAW sensors, such as insertion loss, getting a high center frequency, etc. Different IDT parameters, like the width of the electrodes, the distance between the electrode fingers, the number of electrode fingers, and the size of the aperture, change the electrical impedance and frequency bandwidth, which affects how the wave spreads on the substrate. Compared to various IDTs structures, split electrodes have better control on insertion loss, eliminate reflections, self-resonance, and access to higher frequencies at the third harmonic [4, 12, 13]. The quality factor (Q Factor) is the parameter to measure the energy loss in the SAW device. The Q factor is the energy ratio between stored energy to the supplied energy. To get a high a stability of the SAW devices, the energy losses should be very less which is only possible by a high Q Factor. To achieve the high Q factor, a paired electrode has been developed in SAW device and named as split electrode interdigital transducers (SEIDT). The SEIDT configuration is also called meander line because of generation of third harmonic wave. These third harmonic wave have stronger response then the original wave.

Computer-aided design (CAD) using the finite element method (FEM) is a useful tool for developing action plans for research. COMSOL Multiphysics, a commercial finite element analysis product, provides a more robust modeling environment for developing SAW sensors [14, 15]. Zheng et al. [16] investigated sensing and how sensor layer spacing influenced gas detection. The SAW sensor was designed using layers of varying thicknesses on a $LiNbO_3$ substrate. The suggested SAW has a 100 nm, 200 nm, and 300 nm thick Si_3N_4 layer under a 100 nm to 1000 nm thick ZnO layer and a layered structure of ZnO, SiO_2, and $LiNbO_3$. The thickness of the sensing layer is directly proportional to the wave frequency and relies on the velocity of the surface waves, according to the study given here. Jakubik and colleagues [17] created a bilayer conductive material for SAW gas sensors. His study suggests that having many layers with distinct properties might increase sensitivity. Those researchers created various SAW structures for use in sensors. A lithium tantalate ($LiTaO_3$) piezoelectric substrate is topped with a polyaniline (PANI) sensing layer in this example ($C_{48}H_{38}N_8$).

2 Model Design

COMSOL is used for the design and modeling of SAW structures using a variety of physics. Lithium tantalate is used in this configuration as the piezoelectric substrate material for detecting gases. In order for surface acoustic waves to travel effectively, this $LiTaO_3$ material needs to have a high wave velocity. Since polyaniline (PANI) is very gas-sensitive, it is often utilized as a sensing layer [18–20]. In this paper, the device dimensions and the 3D and 2D structures of SAW devices are shown in detail for a better understanding of the devices. In the 3D layout of the SAW sensor, a thin layer of PANI sensing materials (with dimensions of 1 μm in width, 0.6 μm in depth, and 0.05 μm in height) is drawn beneath a wider layer of substrate (with dimensions of 4 μm in width, 1 μm in depth, and 0.5 μm in height). Aluminum electrode splits on substrates with dimensions of 0.02 μm in width, 0.5 μm in depth, and 0.045 μm in height. Twelve sets of IDTs (24 electrodes) are drawn in this 3D SAW sensor design, with each pair of electrodes considered to be one IDT of the same dimension. Figure 1a depicts the 3D SAW sensor described above, which uses split electrodes (double electrodes). The grating structure of the planned SEIDT has an electrode interval of $\lambda/8$, electrode width of $\lambda/8$, and center-to-center distance of $\lambda/4$ (Fig. 1b). Only with the correct electrode dimensions and spacing will the meander line be formed ($\lambda/8 + \lambda/8 + \lambda/8 = 3\lambda/8$), which is what generates the third harmonic wave and what ultimately causes the reflection loss to be lowered Reflection loss in a SAW device is shown as a function of the scattering parameter (S parameter), more specifically the input reflection coefficient (S_{11}).

Figure 1b depicts the IDTs with the correct dimensions and spacing. Six sets of electrodes pair serve as input IDTs, while the remaining six serves as output IDTs. Due to the symmetry of each IDT, only two sets of IDTs will function as a complete sensor in the 2D design of the SAW gas sensor. Figure 2 depicts the schematic geometry of the 2D version of the SAW gas sensor with sensing layer, with the correct leveling and indication.

3 Simulation Methodology

The COMSOL Multiphysics program, version 6.0, was utilized in the course of this study to carry out the design of the SAW gas sensor. On the basis of partial differential equations (PDEs), this piece of software has enabled every stage of the design and development process, including modeling, physics specifications, and graph plotting. Through the use of a variety of tools and the application of physical principles, this software is used to create both 3D and 2D SAW sensor designs. This section contains an explanation of a variety of design factors, mathematical formulae, and physics calculations that were done.

Before beginning to create the design, the initial step of SAW design is to integrate physics to the simulation. As the material of interest is piezoelectric and the analysis is based on frequency response, the piezoelectricity domain should be included. In structural mechanics physics, the piezoelectricity domain is accessible in electromagnetic-structure interaction. In this SAW gas sensor analysis, both piezoelectric behavior and gas detection occurred concurrently. In the input port of the SAW IDTs, the direct piezoelectric effect will generate a wave on the surface of the substrate, which will travel to the end (output) port of the SAW IDTs, where the inverse piezoelectric behavior will convert the wave into electric with a change in frequency, causing the signal processor to display an indication. During the study of gas sensing, it is essential to evaluate many characteristics for gas analysis, such as displacement curve, potential increase, and Eigen frequencies. The periodic unit cell of the 2D geometry SAW device is shown in Fig. 2. The 2D geometry is shown to illustrate the COMSOL design parameters. The SAW

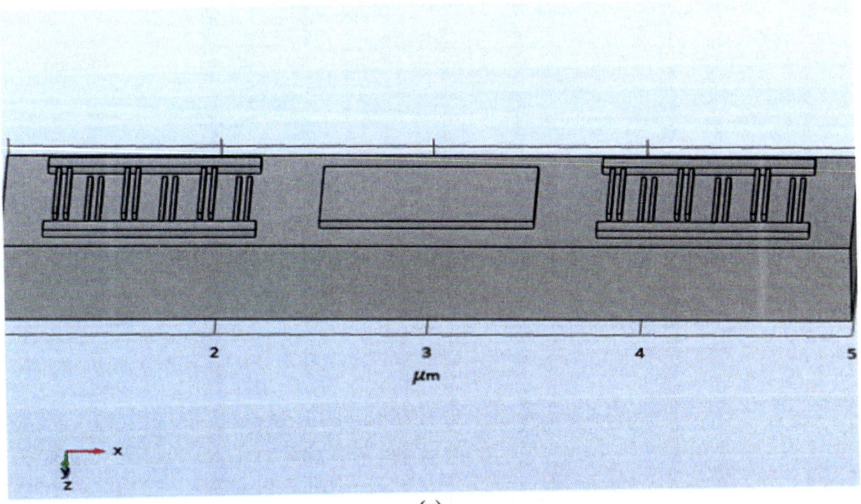

(a)

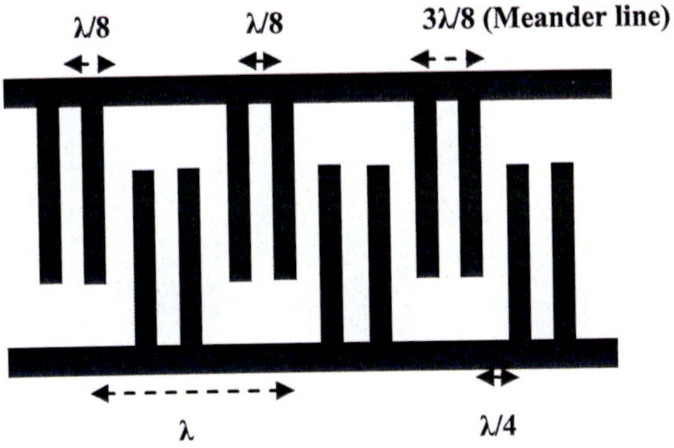

(b)

Fig. 1. Schematic of (a) SAW gas sensor design & (b) split electrodes with dimensions.

device's parameters are shown in Table 1. These parameters are the first input to the program before providing the SAW device's dimensions and meshing. The determined eigenfrequency represents the device's resonant frequency.

In order to comprehend the variation in frequency shift caused by both acetone gas and mass loading, it is essential to have a fundamental understanding of the mathematics behind the phenomenon. Some of the key points include wave propagation, gas concentration, the effect caused by mass loading, frequency shift, gas density, and gas ppm level. The equation that describes how waves move over the surface of the substrate is called the propagation equation (1).

$$T = C_E \times S - e^T \times E \tag{1}$$

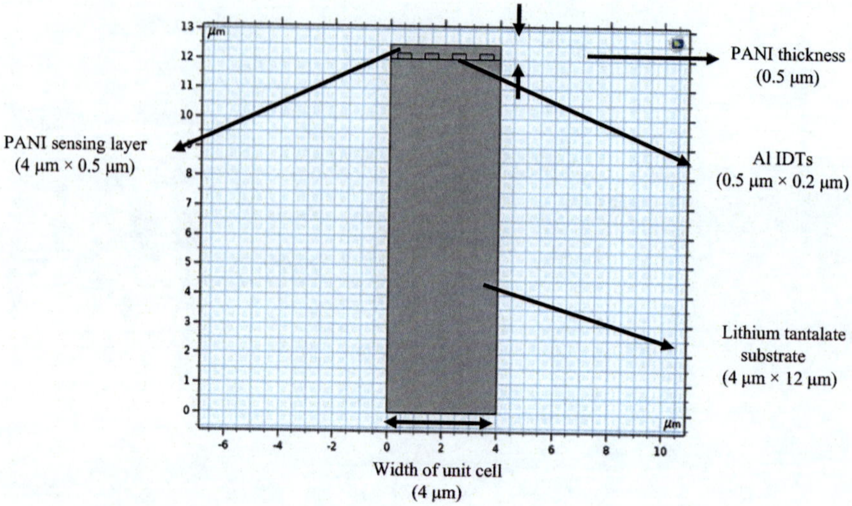

Fig. 2. Schematic diagram of 2D SAW gas sensor design.

Table 1. SAW Model parameters.

Variable	Values	Description
T	25 °C	Atmosphere temperature
P	1 atm	Atmosphere pressure
C_0	100	Acetone concentration (ppm)
C acetone air	$1e^{-6}*C_0*P/(R\text{-}Constant*T)$	Acetone concentration in air
M Acetone	58.08 g/mol	M Molar mass of gas
Rho-PANI	1.329 g/cm3	Density of PANI
E-PANI	4.42 GPa	Young's modulus of PANI
nu-PANI	0.3	Poisson's ratio of PANI
vR	3488 m/s	Rayleigh wave velocity
t-PANI	0.5 μm	PANI thickness
Width	4 μm	Width of unit cell
f_0	vR/width	Estimated SAW frequency

where, T = The stress matrix, C_E = Elasticity matrix, e^T = Piezoelectric matrix and E = electric field intensity.

Mass loading impact on a SAW sensor causes a change in the sensor's center frequency, among other effects. The molar mass of the gas has a significant impact on the resonance frequency shift of the piezoelectric materials. Ace-tone has a mol mass of 58.68 g per mole. To begin designing the sensor, it is necessary to consider the following parameters.

$$C = \left(Co \times 10^{-6} \times P\right) \times RT^{-1} \tag{2}$$

$$\rho_{gas} \text{ in PANI} = K \times M \times C \tag{3}$$

$$\rho_{total} = \rho_{PANI} + \rho_{gas/PANI} \tag{4}$$

where, P = Atmospheric pressure (760 mmHg), R = gas constant (8.31 $Jmol^{-1}K^{-1}$), T = Temperature of air (298.15 K), K = The air/PANI partition coefficient for gas, Co = Gas concentration (in ppm), M = Molar mass of gases, and C = Concentration of gas in air. ρPANI = Density of PANI film (In absence of acetone to substrate) and ρ_{gas}/PANI = Density of PANI film (in presence of acetone gas).

The density of acetone gas in the presence of the sensing layer may be found by using Eq. (3) [21] (PANI). The density of the sensing layer shifts as a result of changes brought about by the adsorption of gases on the substrate. Equation may be used to determine the resonance frequency equation (5).

$$\nu = \lambda \times f_o \tag{5}$$

where, λ = Wavelength of wave, f_o = Resonance frequency, and ν = 3.488 ms^{-1} is the wave velocity in the piezoelectric substrate ($LiTaO_3$).

Various forms of wave propagation were discovered during the study of gas sensing. The wave is generated in a 2D SAW with a split electrode arrangement, where the first two electrodes are fed at zero potential and the second two are linked to a floating potential. Waves with a central frequency are caused by this alternating potential between two sets of electrodes. To ensure accurate alignment in the SAW sensor setup, mesh generation is performed once the structure has been specified with input parameters and physics. The 2D geometry of the SAW sensor is shown below in Fig. 3 with a fine mesh. After the mesh mapping was complete, the SAW sensor ran a series of computations to analyze its electrical and mechanical components.

4 Results and Discussions

Within a SAW sensor, the gas that is being monitored by the acoustic wave will travel along the piezoelectric material. The change in frequency of the resonance that occurs in response to the mass load of the sensing gas. During the course of this investigation, a shift in frequency caused by gas was discovered close to the Rayleigh wave frequency (the central frequency), as well as six eigen frequency discoveries close to the central frequency. Changing the gas, its concentration level, and other input parameters associated with the study are examples of changes that can occur during a parametric study. However, a researcher can easily change the parameters by studying the change in frequency in accordance with their particular area of interest, which is sensing gas. In the work being done on this research project, acetone gas is considered a primary detecting gas in comparison to other gases. This gas sensing performance is carried out at room temperature.

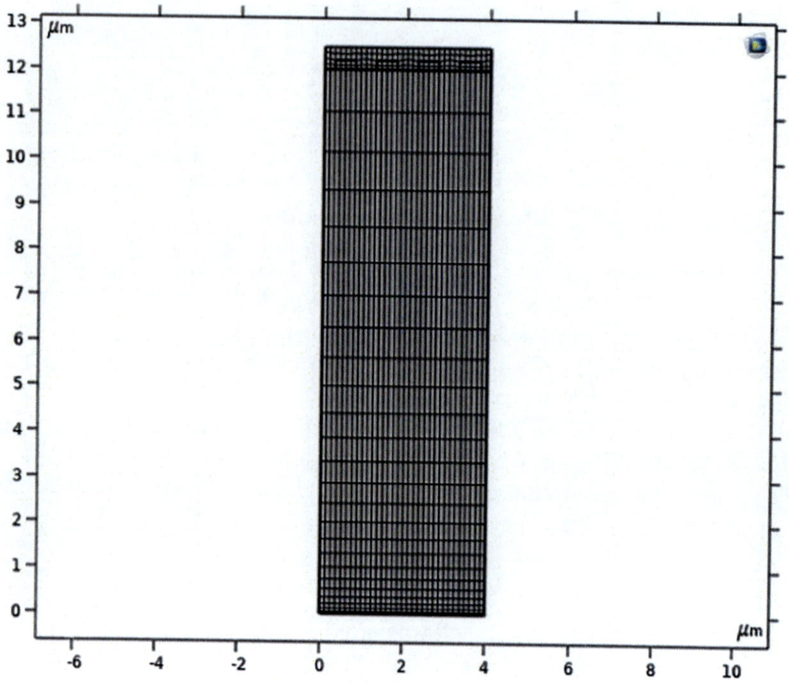

Fig. 3. Meshing to the 2D SAW geometry.

(a) Resonance and antiresonance plot.

The resonance and antiresonance frequencies of the SAW device are shown in the below Fig. 4. The real resonance frequency of the device was discovered to be 872 MHz while it was not exposed to gas; however, the resonance frequency that was computed after the device was exposed to gas was 825.67 MHz. It has been proved beyond a reasonable doubt that the acetone gas is responsible for the shift in resonance frequency, and the amount of the shift in resonance frequency of this SAW sensor is 46.33 MHz. Calculations of the SAW sensor's resonance frequency are shown in Table 2 with and without the presence of gas exposure.

Table 2. Frequency shift for acetone gas.

Gas	Resonance frequency (absence of acetone gas) (MHz)	Resonance frequency (presence of acetone gas) (MHz)	Frequency shift (MHz)
Acetone	872	825.67	46.33

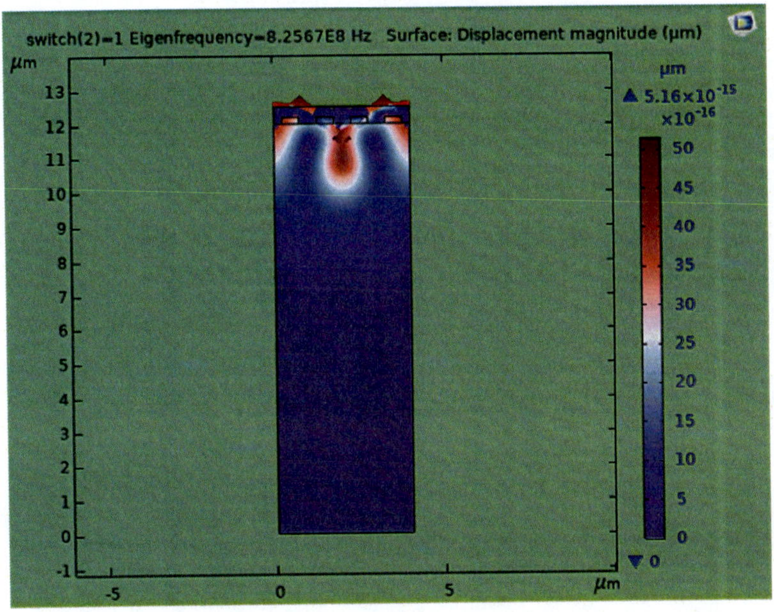

(a)

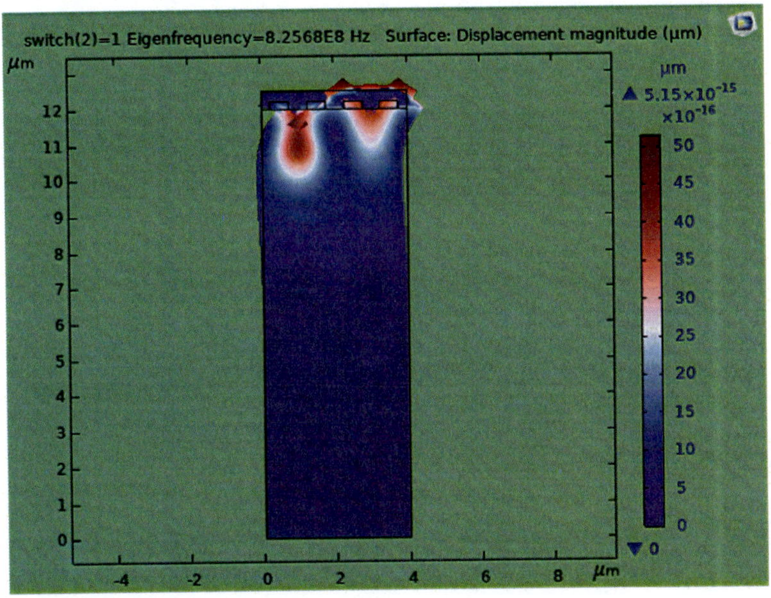

(b)

Fig. 4. Image of total displacement of (a) resonant frequency at 825.67 MHz and (b) anti-resonant frequency at 825.68 MHz of SAW senor.

(b) Electric potential distribution plot.

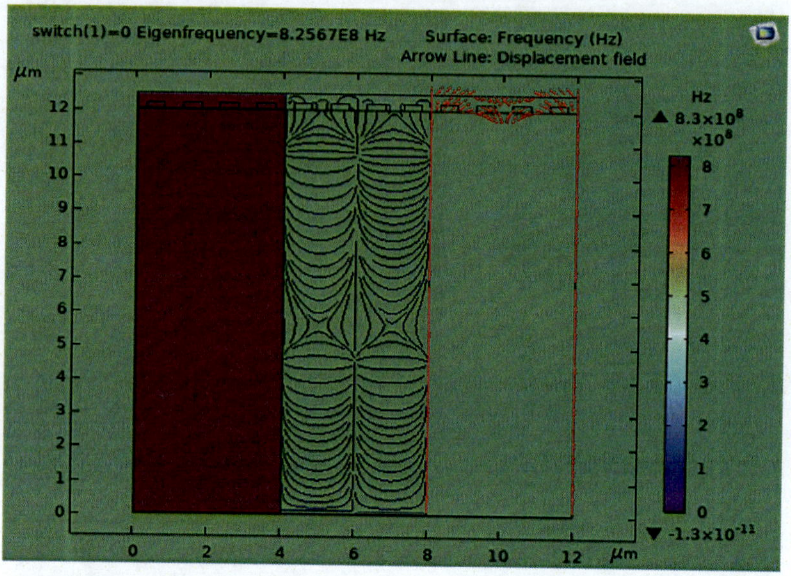

(a)

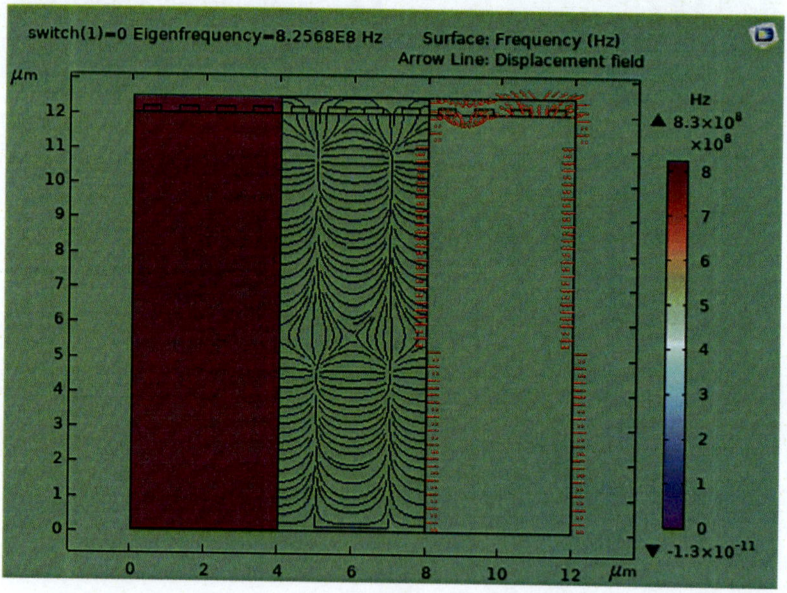

(b)

Fig. 5. Image of electric potential of (a) resonant frequency at 825.67 MHz and (b) anti-resonant frequency at 825.68 MHz of SAW sensor.

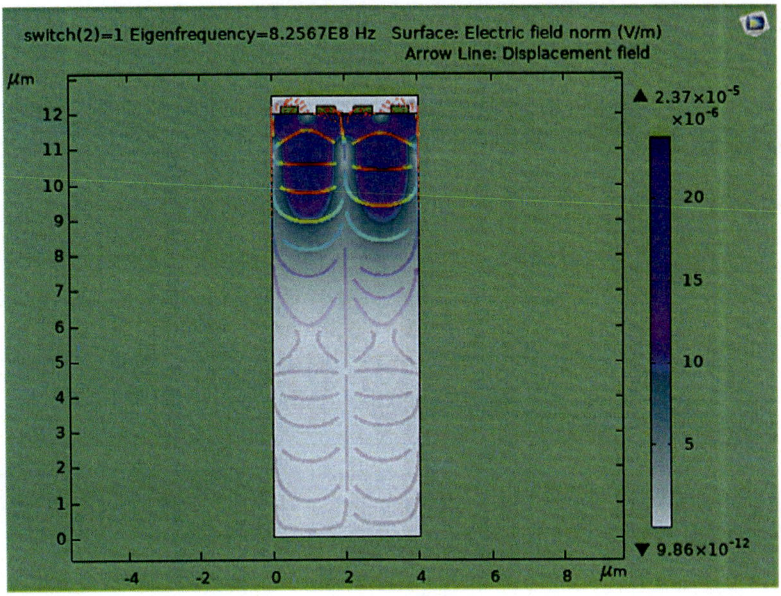

(a)

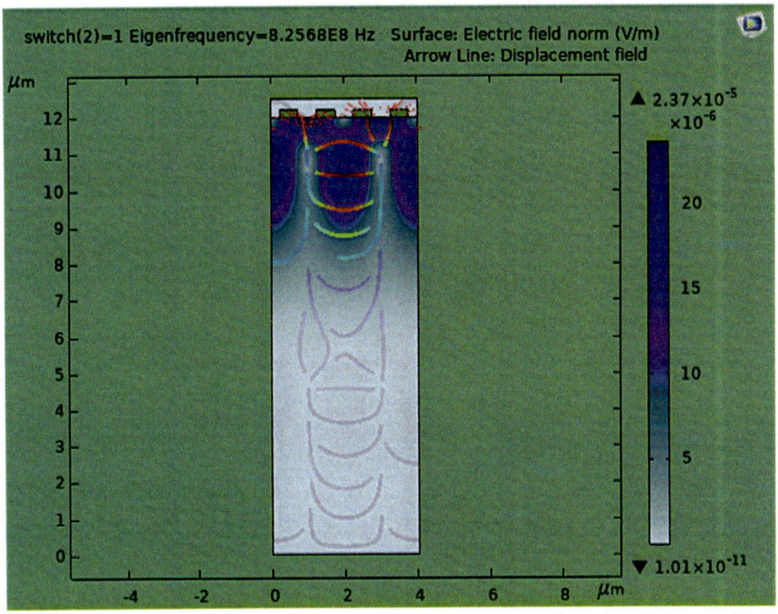

(b)

Fig. 6. Image of electric field normal at (a) resonant frequency at 825.67 MHz and (b) anti-resonant frequency at 825.68 MHz of SAW sensor.

(c) Electric field normal plot.

Figure 4 explains how far waves move on the surface at the frequency of resonance and the frequency of anti-resonance. The total displacement shows how mechanical waves travel through the substrate. Figure 5 shows the electric potential of a wave that is moving. Figure 5 shows how positive potential lines are spread out for resonant and anti-resonant frequencies. In the resonance frequency case, the electric potential is spread out properly. In the anti-resonance frequency case, the electric potential is also spread out along the side walls of the substrate. In this split electrode structure, two fingers are first connected to two floating potentials, and the other two are connected to the ground. Figure 6 shows the electric field normal of the propagating wave, which explains the tangential components of the surface for the flow of surface current.

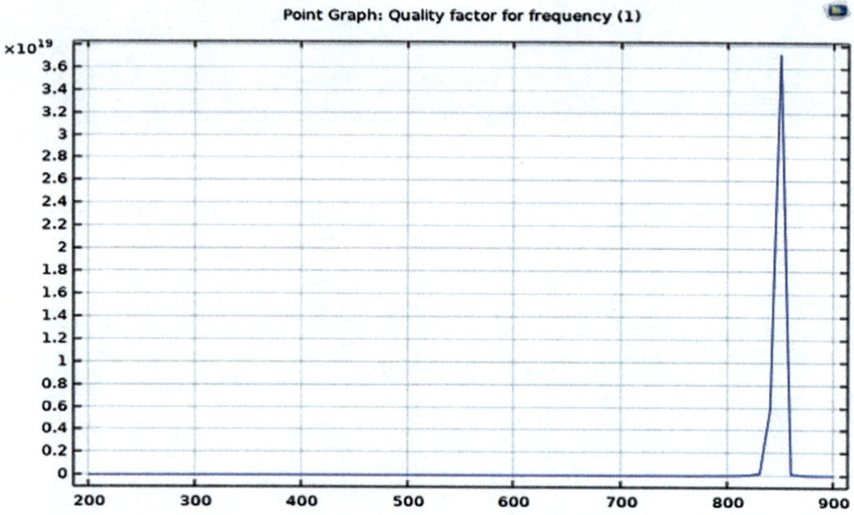

Fig. 7. Q Factor vs frequency curve for designed SAW device.

(d) Quality factor (Q factor) plot.

The Q Factor (Quality Factor) at resonant frequencies is seen in Fig. 7. As can be seen, the achieved Q factor for this surface wave resonant frequency is greater than 3.6×10^{19}. As can be seen in Fig. 7, a high Q factor results in less input-side reflection and thus reduced energy loss. Electric energy of the wave at the resonance frequency is seen in Fig. 8. Clearly seen in Fig. 8, the SAW wave has undergone a transformation in electric energy at its resonance frequency.

(e) Electric energy plot at resonance frequency.

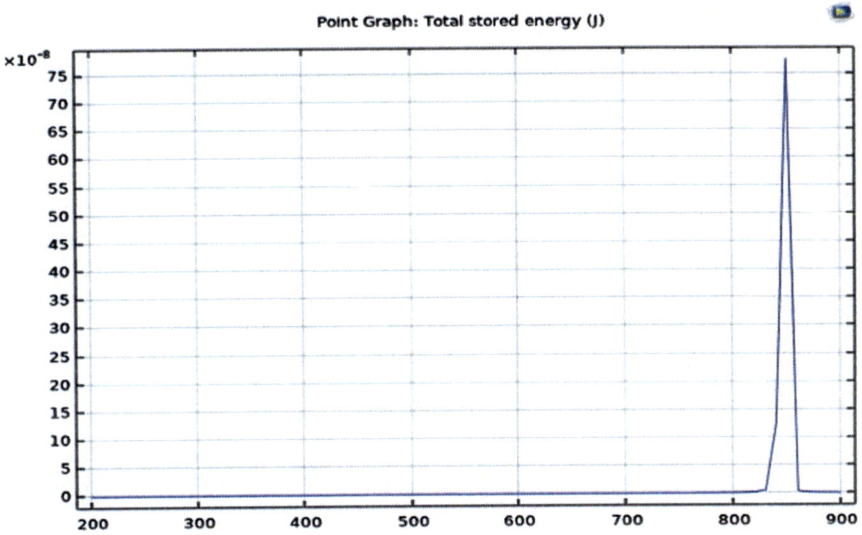

Fig. 8. Total electric energy curve at resonant frequency.

(f) Scattering parameter (S_{11} parameter) plot.

S-parameter analysis provides insight into the reflection coefficients, stability, power gains, etc. of any high-frequency circuit. To avoid problems with source-end reflection, a split-electrode IDTs configuration is provided here. Input reflection coefficient, expressed as a number between −11 dB and 0, is explained by the S_{11} parameter. At the resonance frequency, as shown in Fig. 9, the highest S-parameter is more than −11 dB, indicating that there is no reflection loss at the source. This SEIDT-based SAW gives improved control over the amount of reflection loss in comparison to bi-directional IDT-based SAW sensors [22]. The S_{11} value of the proposed SEIDT SAW sensor is lower than the S_{11} value (−8.856 dB) of the bi-directional IDT (two port) SAW sensor [22], which means that the bi-directional IDT (two port) SAW sensor has a higher S_{11} value. As a result of this study, it has been shown that SAW with SEIDT has a reduced S_{11}, which is essential for achieving a low reflection loss. In the experimental research work performed by Nguyen et al., the designed IDTs were double-split fingers with a finger length of 3.6 mm, a total IDT width of 3.2 mm, and space between IDTs of 4 μm [23] on the $LiTaO_3$ substrate. In millimeters, the finger length and total IDT width were measured, yielding an S_{11} value of −60 dB. The foremost reason of S_{11} is −60 dB because IDT was placed in a 4-(2-hydroxyethyl)-1-piperazineethanesulfonic acid (HEPES) buffer solution. When compared to the SEIDT SAW sensor, the high reflection coefficient is caused by the way the SAW wave moves through the HEPES buffer solution. In Table 3, we can see the differences between the proposed SEIDT SAW designs and the different IDT-based SAW structures.

Table 3. Comparison of different parameters of various SAW sensors.

Parameters	[22]	[23]	Present paper
Substrate	GaN/Si	36° Y-cut $LiTaO_3$	$LiTaO_3$
IDT structure	Bi-directional electrodes	Double split IDT	Split electrode IDT
S_{11} parameter (dB)	−8.856	−60	−11
Analyte	NA	NA	Acetone
Medium	Air	4-(2-hydroxyethyl)-1-piperazineethanesulfonic acid (Buffer solution)	Air
Resonance frequency (RF)	5.5 GHz	130 MHz	872 MHz 825.67 MHz (exposing gas)
Dimensions	(a) IDT range in "nm" (b) IDT Length = 100 μm, width = 200 nm	(a) IDT range in "mm" (b) IDT Length = 3.6 mm, width = 3.2 mm	(a) IDT range in "μm" (b) IDT Length = 0.5 μm, width = 0.2 μm
Remark	(a) More reflection at the output end because of simple IDTs (b) Due to the "nano meter" range IDT size, SAW device has high RF	(a) Very minimal reflection loss occurs as a result of the wave prorogate in the buffer medium (b) IDTs dimension are in "milli meter" range lead to very low RF	(a) The SEIDT topology significantly reduces reflection loss in comparison to [22] (b) Achieved higher reflection loss compared to [23] because of the air medium and higher RF

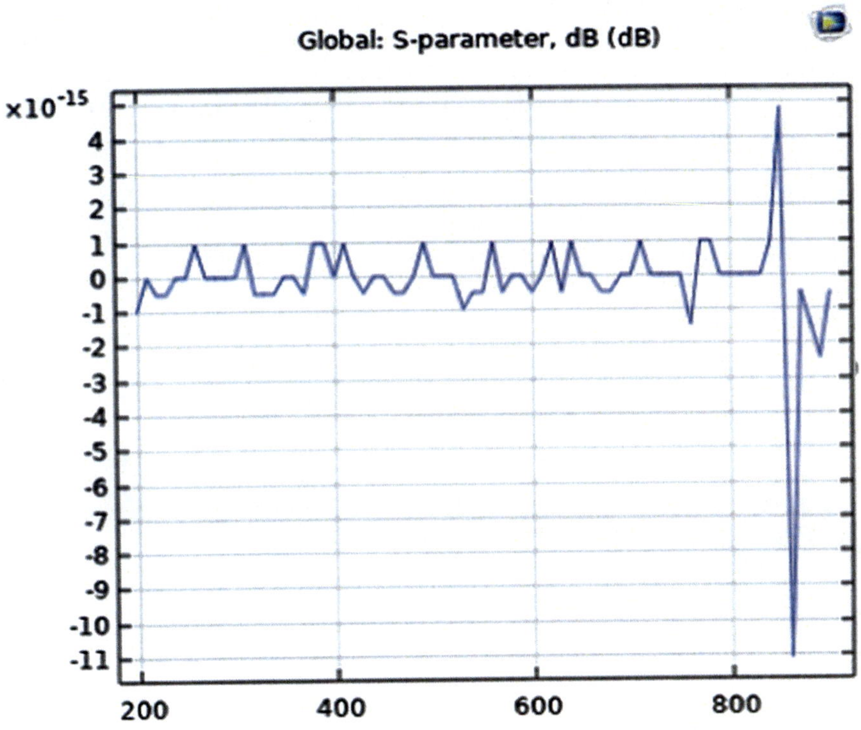

Fig. 9. S-parameter in dB at resonant frequency of SAW sensor.

5 Conclusions

In this study, COMSOL Multiphysics simulation software was used to plan the SAW gas sensor. Due to the symmetry of IDTs, the design of the SAW sensor is shown in both 3D and 2D, whereas all computations are done in the 2D version of the SAW sensor. Frequency-dependent analyses of SAW sensors, including electric potential distribution, total displacement, electric field norm, electric energy curve, and s-parameter, have been shown. The primary goal of this study is to develop a split electrode that may be used to lessen the reflection loss at the origin. The Q factors at resonance frequency, predicted to be 3.6×10^{19} for this research, are crucial to reducing reflection loss. Reduced energy dissipation and reflection result from a high Q factor. To further demonstrate the reduced energy loss or reflection loss, another research, such as S parameters, has been calculated. There is a sweet spot for the S_{11} parameter at -13 dB. According to the findings, the S_{11} parameter has a value of -11 dB, which is better than the minimum value of $-$ 13 dB required to achieve the lowest reflection loss on the transmitted signal at the input. The SAW sensor's resonant and anti-resonant frequencies to the acetone gas were 825.67 MHz and 825.68 MHz. In addition, this SAW layout may be put to use in practical gas sensing applications. The future of SAW gas sensors lies on the development of new IDT structures.

This effort will concentrate on constructing this through lithography methods in order to allow future SAW manufacture, and a vector network analyzer (VNA) will be utilized to compute the scattering parameters. This will be done in order to enable future SAW production.

Acknowledgements. The authors are pleased to acknowledge the Indian Science Technology and Engineering facilities Map (I-STEM) for providing the simulation platform facility and the University of Delhi, funded by the Institution of Eminence (IoE) under the faculty research programme (Ref.No./IoE/2021/12/FRP).

References

1. Kumar, M., Kumar, D., Gupta, A.K.: Fe-doped TiO_2 thin films for CO gas sensing. J. Electron. Mater. **44**(1), 152–157 (2015)
2. Kumar, M., Gupta, A.K., Kumar, D.: Mg-doped TiO_2 thin films deposited by low cost technique for CO gas monitoring. Ceram. Int. **42**(1), 405–410 (2016)
3. Liu, Y., Cai, Y., Zhang, Y., Tovstopyat, A., Liu, S., Sun, C.: Materials, design, and characteristics of bulk acoustic wave resonator: a review. Micromachines **11**(7), 630 (2020)
4. Mandal, D., Banerjee, S.: Surface acoustic wave (SAW) sensors: physics, materials, and applications. Sensors **22**(3), 820 (2022)
5. Barie, N., Rapp, M., Ache, H.J.: UV crosslinked polysiloxanes as new coating materials for SAW devices with high long-term stability. Sens. Actuators, B Chem. **46**(2), 97–103 (1998)
6. Tasaltin, C., Ebeoglu, M.A., Ozturk, Z.Z.: Acoustoelectric effect on the responses of saw sensors coated with electrospun ZnO nanostructured thin film. Sensors **12**(9), 12006–12015 (2012)
7. Ionescu, V.: Design and analysis of a Rayleigh saw resonator for gas detecting applications. Rom. J. Phys. **60**(3–4), 502–511 (2015)
8. Avramov, I., Radeva, E., Lazarov, Y., Grakov, T., Vergov, L.: Sensitivity enhancement in plasma polymer films for surface acoustic wave based sensor applications. Coatings **11**(10), 1193 (2021)
9. Esmeryan, K.D., Avramov, I.D., Radeva, E.I.: Temperature frequency characteristics of hexamethyldisiloxane (HMDSO) polymer coated Rayleigh surface acoustic wave (SAW) resonators for gas-phase sensor applications. Micromachines **3**(2), 413–426 (2012)
10. Parker, T.E., Schulz, M.B.: SiO_2 film overlays for temperature−stable surface acoustic wave devices. Appl. Phys. Lett. **26**(3), 75–77 (1975)
11. Eggins, B.R.: Chemical Sensors and Biosensors, vol. 2. Wiley (2002)
12. Banerjee, S., Leckey, C.A.: Computational Nondestructive Evaluation Handbook: Ultrasound Modeling Techniques. CRC Press, Boca Raton, FL, USA (2020)
13. Shreve, W.R.: U.S. Patent No. 4,081,769. U.S. Patent and Trademark Office, Washington, DC (1978)
14. Ramakrishnan, N., Namdeo, A.K., Nemade, H.B., Palathinkal, R.P.: Simplified model for FEM simulation of SAW delay line sensor. Procedia Eng. **41**, 1022–1027 (2012)
15. Atashbar, M.Z., Bazuin, B.J., Simpeh, M., Krishnamurthy, S.: 3D FE simulation of H_2 SAW gas sensor. Sens. Actuators, B Chem. **111**, 213–218 (2005)
16. Zheng, P., Greve, D.W., & Oppenheim, I.J.: Multiphysics simulation of the effect of sensing and spacer layers on SAW velocity. In: COMSOL Conference 2009, Boston (2009)
17. Jakubik, W.: Theory of saw gas sensor based on bi-layer conductivity changes. Procedia Eng. **47**, 1287–1290 (2012)

18. Nicolas-Debarnot, D., Poncin-Epaillard, F.: Polyaniline as a new sensitive layer for gas sensors. Anal. Chim. Acta **475**(1–2), 1–15 (2003)
19. Fratoddi, I., Venditti, I., Cametti, C., Russo, M.V.: Chemiresistive polyaniline-based gas sensors: a mini review. Sens. Actuators, B Chem. **220**, 534–548 (2015)
20. Sengupta, P.P., Barik, S., Adhikari, B.: Polyaniline as a gas-sensor material. Mater. Manuf. Processes **21**(3), 263–270 (2006)
21. Ho, C.K., Lindgren, E.R., Rawlinson, K.S., McGrath, L.K., Wright, J.L.: Development of a surface acoustic wave sensor for in-situ monitoring of volatile organic compounds. Sensors **3**(7), 236–247 (2003)
22. Stefanescu, A., Muller, A., Konstantinidis, G., Buiculescu, V., Dinescu, A., Stavrinidis, A., Cismaru, A.: SAW GaN/Si based resonators: modeling and experimental validation. In: CAS international semiconductor conference, vol. 1, pp. 193–196. IEEE (2012)
23. Nguyen, V.H., Kaulen, C., Simon, U., Schnakenberg, U.: Single interdigital transducer approach for gravimetrical SAW sensor applications in liquid environments. Sensors **17**(12), 2931 (2017)

Thermal Study of Thin-Film Heater for PCR Reaction-Based Applications

Andleeb Zahra[1](✉), Giampiero de Cesare[2], Domenico Caputo[2], and Zia Abbas[1]

[1] CVEST, International Institute of Information Technology (IIIT), Hyderabad, India
andleeb.zahra@research.iiit.ac.in
[2] DIET, Sapienza University of Rome, Rome, Italy

Abstract. This paper presents the design, fabrication, and characterization of a thin-film Cr-Al-Cr (300 A°-1500 A°-300 A°) metal stack heater specially designed for chemical reactions which occur at a uniform temperature in the lab on a chip platform such as polymerase chain reaction (PCR) applications. The heater has been designed using COMSOL Multiphysics. The simulated design has been fabricated using lithography and patterning on a 5 × 5 cm^2 glass substrate. For the validation of the proposed design thermal study of the fabricated heater has also been done using a FLIR IR camera. A very good agreement between the thermal image of modeled and fabricated heater has been achieved. This demonstrates the suitability of the proposed heater for PCR reaction applications.

Keywords: Lab-on-chip · Thin film heater on the glass · Temperature distribution · COMSOL Multiphysics · Electro-thermal simulations

1 Introduction

Lab-on-chip (LoC) technology is receiving a lot of attention in different possible applications of biology and genomics. Among the different functionalities of the LoC Polymerase Chain Reaction (PCR) procedure for the molecular amplification of DNA has received a lot of attention. PCR is essential to all genetic analysis applications due to its double-strand DNA amplification capability.

The reaction progress in repeated thermal cycles with three different temperature steps. In the first step called denatures, at temperatures between 94 and 96 °C, denatures the DNA template strands; in the second step called annealing, typically at a temperature in the range of annealing 45–60 °C, the primers hybridize to their complementary sequences on the parent strand; during the third temperature step called extension, usually at 70–72 °C, the DNA polymerase forms new daughter strands extending the primer sequences by adding individual dNTPs from solution. Therefore, repetition of the sequence generates 2^n daughter strands, where n is the number of cycles [1, 2]. However, temperature uniformity is a critical issue in on-chip PCR applications. Therefore, the integration of a fully integrated LoC system for PCR is still a challenging task [3].

Different heater geometries have been studied by Hsieh et al. [4], by comparing their temperature response. Selva et al. [5, 6] showed also that it is possible to optimize heater

A. K. Pandey et al. (Eds.): MAMM 2022, 126, pp. 266–274, 2023.
https://doi.org/10.1007/978-3-031-20353-4_21

shape to generate different temperature profiles by making a finite element study of the thermal response of the heater. In this paper, we will discuss the thermal study of the Cr-Al-Cr metal stack thin film heater dedicated to PCR reaction applications, stacking layers means we are using different metals to fabricate a thin film heater. In our heater design first, we deposited a thin Chromium (Cr) layer, then an Aluminium (Al) thick layer, and then another thin Cr layer. The typical thicknesses of the Cr-Al-Cr layers are 300 A°/1500 A°/300 A°. The same heater can also be used for another lab on a chip application in which uniform temperature is required.

The reason to consider Cr-Al-Cr stack layers are as follows-

- The middle Al layer is used because of its higher conductivity than another available alloy of titanium in tungsten (TiW). Also, the cost is low in comparison to gold (Au) [7].
- The Upper thin Cr layer over Al is because of its relatively large atomic size, which will not generally diffuse in many substrate materials when heated up by means such as friction and heating by irradiation and chemical recombination energy release [8].
- However, the lower thin Cr layer is because of adhesion of Al with glass is not good the bottom layer of Cr is to make a good adhesion on the glass substrate. So, this stack of Cr-Al-Cr gives better stability to the microheater in many applications [9].

In this paper, a Cr-Al-Cr (300 A°/1500 A°/300 A°) thin film heater on a 5*5 cm^2 glass substrate was specially designed and fabricated to obtain a uniform temperature for LoC applications.

The heater has been designed using finite element simulations COMSOL Multiphysics, coupling the electrostatic and the heat transfer problems and obtaining the temperature profiles due to a potential difference applied across the resistor. The structure showing a more uniform temperature distribution has been fabricated using photolithography and patterning. To validate the modeled results, an infrared thermo-camera (FLIR IR) has been used to study the thermal distribution of the fabricated heater. To the best of our knowledge, this is the first time a thermal study of Cr-Al-Cr (300 A°/1500 A°/300 A°) thin-film heater for PCR reaction application has been carried out.

The organization of the paper is as follows. In Sect. 2 the thin film heater design and electro-thermal simulation are described. Section 3 shows the fabrication of the optimize thin-film heater. Section 4 thermal study of the fabricated heater and Section 5 discusses the main conclusions.

2 Thin Film Heater Design

A heater, in general, based its operation on the Joule effect: if we apply a potential difference V across a circuit and the amount of current flowing is I. The amount of electrical power supply transformed into heat. In the case of a resistor for which Ohm's law is applicable:

$$P = V \cdot I = V^2 R = I^2 \cdot R \tag{1}$$

where P is the electric power. V is the potential difference, I is the electric current, and R is the resistance.

The resistance value R can be expressed as a function of the type of metal used, the film thickness, and the geometry, according to the equation:

$$R = \rho L/S \tag{2}$$

where ρ is the electrical resistivity, L is the length of the resistive path and S is the cross-section area perpendicular to the direction of the current. For the flow of current in a resistor, electric energy is transformed into heat and the dependence on the resistivity usually is linearized with the thermal expansion:

$$R = R_0[1 + \alpha(\Delta T)] \tag{3}$$

where R_0 is the value of the resistance at a given temperature T_0 and α is the coefficient of thermal resistance that depends on the type of material. The formula can also be expressed in terms of resistivity, which in the case of a metal increases with temperature as given below-

$$\rho = \rho_0[1 + \alpha(T - T_0)] \tag{4}$$

where ρ is the resistivity of the metal at a temperature T_0.

In this research work, our goal is to achieve a novel Cr/Al/Cr metal stack heater fabricated on a glass substrate, which can produce uniform heating inside the reaction chamber required to perform on-chip PCR. The design parameters are area, the thickness of the metal layer, metals used, and the substrate.

The heater geometry has been optimized, to get a uniform temperature distribution in the specific area of the heater called the active area in our proposed design its value is 8700 μm × 8500 μm. Therefore, after optimization of heater geometry for uniform temperature distribution in this area, if we will put the reaction chamber in this particular area, the temperature inside the reaction chamber will be uniform. The optimization has been carried out using the Multiphysics finite element simulation program COMSOL Multiphysics, which couples the electrostatic and heat transfer.

Different heaters' geometries have been designed for obtaining uniform temperature profiles due to a potential difference applied across the heater resistor.

2.1 Serpentine Geometry of Heater

The first geometry tested was the Cr/Al/Cr (300 A°/1500 A°/300 A°) serpentine geometry [10] of the heater with vertical arm coils as shown in Fig. 1(a). Each coil is connected to the other as shown in Fig. 1 (a) and finally attached to larger metal areas, called pads, whose task is to facilitate contact with the heater and on whose lower edges the potential difference is applied.

The initial choice of the vertical arm coil is based on fundamental reasons:

- its geometric symmetry makes construction and calculation of the resistance easier.
- the symmetry of the generated temperature field across the active area.

The voltage of 5 V is applied across the heater pads and the temperature distribution has been measured see Fig. 1(b). We draw a line (ab) in the middle of the active area i.e., 4250 μm from the top of the heater & its length covers all figures of the heater. The temperature distribution across this line has been measured as shown in Fig. 1(c). Therefore, from the temperature distribution line graph it's clear that the temperature is not uniform along this line. To make a uniform temperature distribution, the serpentine geometry of the heater has to be optimized.

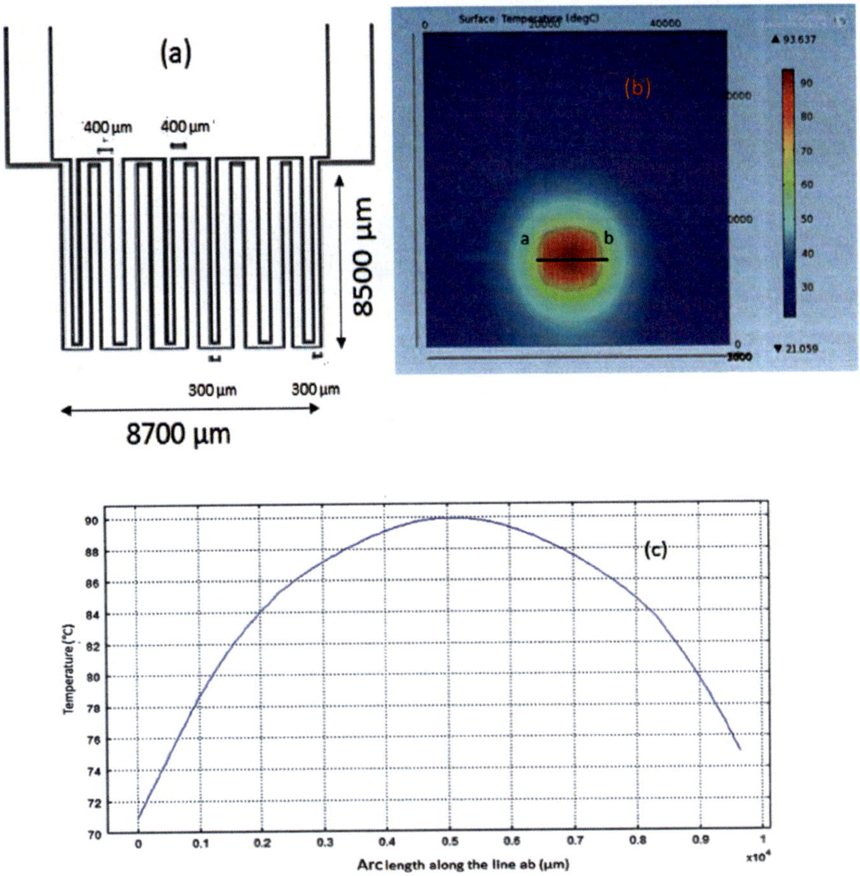

Fig. 1. (a). Serpentine geometry of heater, (b). Temperature distribution along the active area (8700 μm × 8500 μm) of the heater and (c) corresponding line graph along line 'ab' (middle line along the active area)

2.2 Structure Optimization of the Serpentine Heater

The heater is Cr/Al/Cr (300 A°/1500 A°/300 A°) metal stack layers as described above. To make a uniform temperature distribution, the geometry of the serpentine heater has been optimized varying both the width and the spacing of the fingers all the modified dimensions are shown in Fig. 2. We have seen an enhanced uniformity in the modified serpentine geometry of the heater as shown in Fig. 3. We found that the temperature uniformity is better than 2% (only less than 2% variation in the temperature uniformity) over the entire active area. Therefore, the temperature distribution of modified serpentine geometry makes it suitable to maintain the almost uniform temperature inside the reaction chamber during PCR reaction.

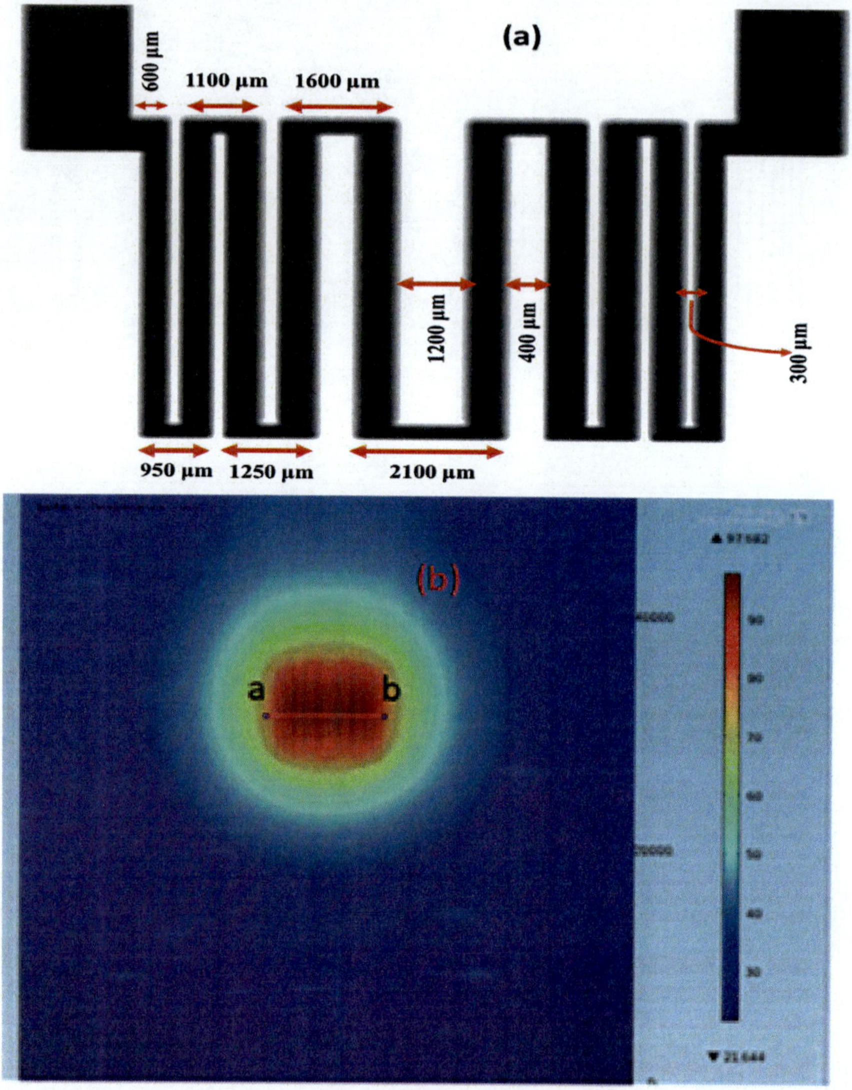

Fig. 2. (a). The modified serpentine heater geometry with optimized dimension produces almost uniform temperature in the heater's active area (8700 µm × 8500 µm). (b). Temperature distribution of the modified heater.

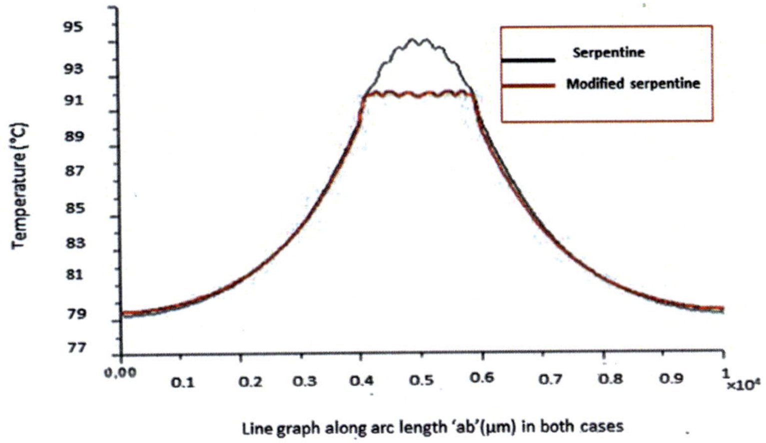

Fig. 3. Comparison of temperature distribution profile along arc 'ab' in serpentine and modified serpentine structure.

3 Fabrication of Thin Film Heater

The fabrication of the heater has been performed with the following technological steps:

- Cleaning of the glass substrate of size 5*5 cm^2 by standard cleaning procedure using piranha solution.
- Deposition of Cr/Al/Cr (300 A°/1500 A°/300 A°) metal layers on glass substrate by the thermal evaporation system.
- Patterning of the heater geometry (as shown in Fig. 2) by using the photolithography process of positive photoresist AZ 1518. The following steps have been followed-

Spin coating: photoresist is deposited over the glass substrate coated with Cr-Al-Cr in two steps; 5 s at 500 rpm and 30 s at 3000 rpm the thickness of photoresist achieved over the substrate was almost 1.5 μm [11].

Soft bake: we have put the substrate on a hot plate for 1 min at 100 °C to evaporate the residual photoresist.

Exposure: Then the structure has been exposed to UV rays for 18 s.

Development: the substrate is immersed in a developing solution, consisting of AZ351B developer and deionized water in the proportion 1: 4, during this phase the portion of the photoresist that is exposed to light becomes soluble to the photoresist developer.

Hard bake: to make the final stable structure, hard baking has been carried out on a hot plate at 115 °C for 2 min.

After lithography, the next step is the removal of three metal layers from areas affected by UV rays; this allows to obtain of the layer of Cr/Al/Cr only in the desired geometry. This process involves selective attacks on the three metals by the acidic solutions.

For the removal of the chromium layer, the following composition has been used-

- 30 g $Ce(NH_4)_2(NO_3)_6$ (ammonium nitrate and cerium)
- 9 ml CH_3COOH (acetic acid)
- 200 ml deionized water

The speed of attack is about 5 Å/s.

For the removal of the aluminum layer, the composition is the following-

- 80 ml H_3PO_4 (85% phosphoric acid)
- 5 ml HNO_3 (nitric acid)
- 10 ml deionized water

The speed of the attack is about 10 Å/s.

The process times for the removal of three metal layers are:

- 1 min in the acidic solution for the removal of Cr.
- 3 min in the acidic solution for the removal of Al.
- 1 min in the acidic solution for the removal of Cr.

The next step involves the removal of the residual photoresist, with a solution of AZ 100 REMOVER for 5 min, then placing the whole on a hot plate for 5 min at 30 °C.

The fabricated heater and its microscopic image are shown in Fig. 4.

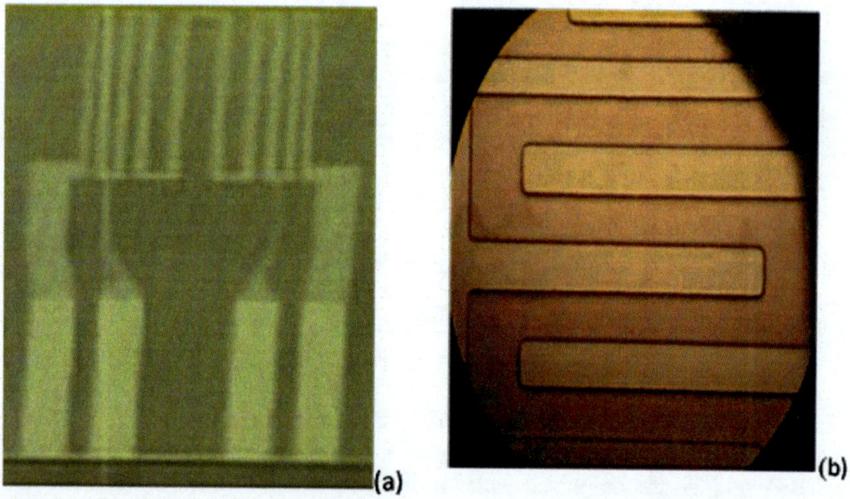

Fig. 4. (a). Modified serpentine design of microheater in Cr/Al/Cr (300 A°/1500A°/300A°) layers (b) Microscopic image of the fabricated heater.

4 Thermal Study of Thin Film Heater

The temperature distribution of the fabricated heater has been seen under the FLIR IR camera. When a voltage of 5-V is applied through the power supply on the heater pads the maximum temperature in the active area (8700 μm × 8500 μm) of the heater is almost 96 °C as shown in Fig. 5. The response time of the heater to reach this temperature is almost 5 s. The thermal image of the modified heater design through an IR camera has good matches the thermal image of the simulated design as shown in Fig. 2(b). This shows the feasibility of the proposed heater for PCR reaction applications in the Lab on Chip (LoC) system.

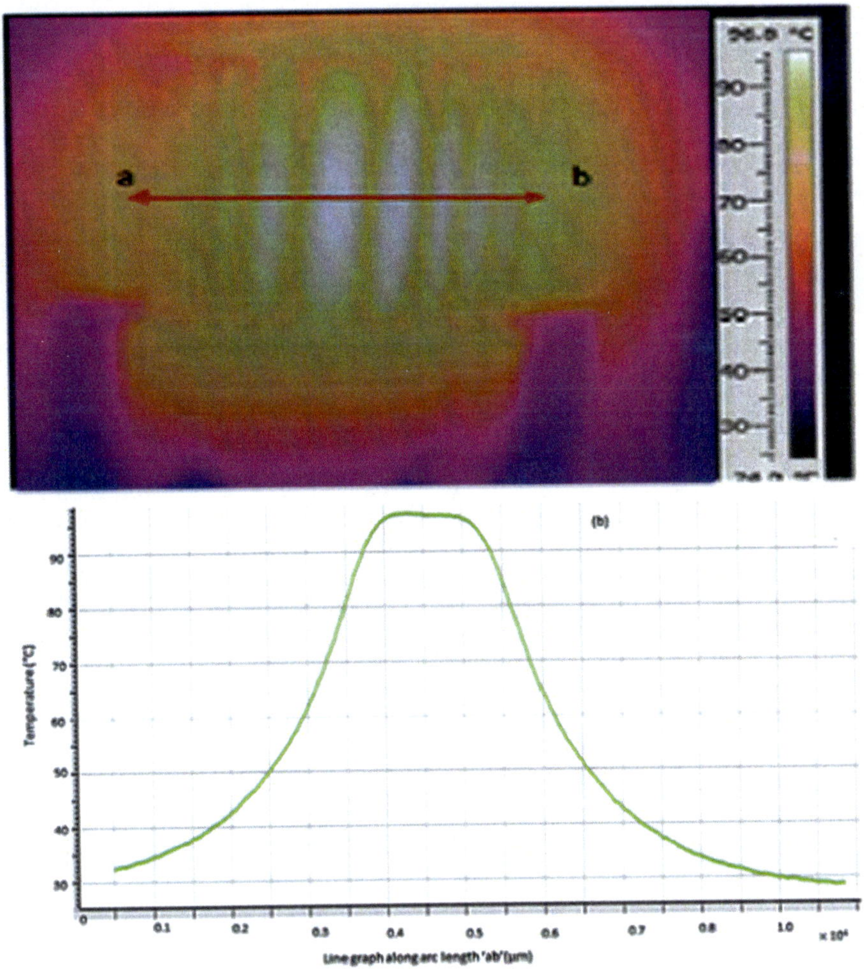

Fig. 5. (a). Thermal image of the fabricated heater using the FLIR IR camera. (b) and the corresponding line graph along line 'ab'.

5 Conclusions

The thin film Cr-Al-Cr (300 A°/1500 A°/300 A°) metal stack heater on a 5 × 5 cm^2 glass substrate for PCR reaction applications has been designed using COMSOL Multiphysics. The modified serpentine structure of the heater shows the uniform temperature over the active area of the heater (8700 μm × 8500 μm). The heater has been fabricated using lithography & patterning of positive photoresist AZ1518. A thermal image of the fabricated heater has been seen using a FLIR IR camera which shows a good agreement with the simulated design. Therefore, by patterning the substrate with an optimized resistor, it is possible to generate a uniform temperature distribution on a selected area with great accuracy and with a short response time.

References

1. Niu, Z.Q., Chen, W.Y., Shao, S.Y., Jia, X.Y., Zhang, W.P.: DNA amplification on a PDMS–glass hybrid microchip. J. Micromech. Microeng. **16**(2), 425–433 (2006)
2. Zahra, Z., Caputo, D., Nascetti, A., Petrucci, G., Lovecchio, N., Scipinotti, R., de Cesare, G.: Thermally actuated microfluidic system for lab on chip applications. In: Proceedings of the 18th AISEM Annual Conference, pp. 1–4 (2015)
3. Zahra, A., Scipinotti, R., Caputo, D., Nascetti, A., de Cesare, G.: Design and fabrication of microfluidics system integrated with temperature actuated microvalve. Sens. Actuators A: Phys. **236**, 206–213 (2015)
4. Hsieh, T.-M., Luo, C.-H., Wang, J.-H., Lin, J.-L., Lien, K.-Y., Lee, G.-B.: Enhancement of thermal uniformity for a micro thermal cycler and its application for polymerase chain reaction. Microfluid. Nanofluid. **6**, 797–809 (2009)
5. Selva, B., Marchalot, J., Jullien, M.-C.: An optimized resistor pattern for temperature gradient control in microfluidics. J. Micromech. Microeng. **19**, 065002:1–065002:10 (2009)
6. Selva, B., Mary, P., Jullien, M.-C.: Integration of a uniform and rapid heating source into microfluidic systems. Microfluid. Nanofluid. **8**, 755–765 (2010)
7. https://www.engineeringtoolbox.com/thermal-conductivity-d_429.html
8. Vogt, J.-B., Serre, I.P.: A review of the surface modifications for corrosion mitigation of steels in lead and LBE. Coatings **11**, 53 (2021)
9. Onyiriuka, E.C., Kinney, L.D., Binkowski, N.J.: Adhesion and delamination of tantalum and chromium films on glass. J. Adhesion Sci. Technol. **11**(7), 929–940 (1997)
10. Wu, Y., Du, X., Li, Y., et al.: Optimization of temperature uniformity of a serpentine thin film heater by a two-dimensional approach. Microsyst. Technol. **25**, 69–82 (2019). https://doi.org/10.1007/s00542-018-3932
11. https://www.microchemicals.com/products/photoresists/az_1518.html

Etching Characteristics of Si{110} in NaOH Based Solution

Soumya Purohit, Veerla Swarnalatha(✉), and Prem Pal

MEMS and Micro/Nano Systems Laboratory, Department of Physics, Indian Institute of Technology Hyderabad, Hyderabad, India
swarnalatha.veerla@phy.iith.ac.in

Abstract. In the present work, we have studied the etching characteristics of Si{110} in 10M NaOH solution without and with addition of NH_2OH. The etch rate of silicon and thermal oxide, and the undercutting rate at convex corners, which are important parameters to be known in the fabrication of MEMS structures using silicon wet bulk micromachining, have been studied in modified NaOH solution. The etch rate of silicon and the undercutting at convex corners increase significantly with the addition of NH_2OH, while the etch rate of silicon dioxide reduces considerably with the addition of NH_2OH etchant.

Keywords: NH_2OH · NaOH · Etch rate · Selectivity · Undercutting

1 Introduction

Wet etching is a prominent technique for the fabrication of microstructures and sub divided into two categories namely isotropic etching and anisotropic etching. Anisotropic etching is an important step to fabricate the MEMS components for various applications [1]. Potassium hydroxide (KOH) and tetramethyl ammonium hydroxide (TMAH) are widely employed in the fabrication of microstructures via wet anisotropic etching [1–10]. Recently, NH_2OH-added TMAH and KOH are explored for applications in wet bulk micromachining for the formation of MEMS structures [2, 3, 11–15]. Although the addition of NH_2OH-added KOH/TMAH provides very high etch rate, the etch rate is reduced with the age of the etchant [14, 15]. NaOH, a low-cost etchant, has been explored for the fabrication of microstructures and surface texturing for silicon solar cell [17–19].

The silicon wafers with {100}, {110}, and {111} surface orientations are mainly produced by wafer manufacturers. Other orientations are produced based on the specific requirements. In wet anisotropic etching, sidewalls of the stable etched profile are formed by slowest {111} planes. The exposure of the number of {111} planes and their angle with wafer surface during etching depend on the orientation of wafer surface. The presence of vertical {111} planes make {110} silicon wafers very useful for some specific applications such as fabrication of microstructures with vertical and smooth sidewalls [20–27].

A. K. Pandey et al. (Eds.): MAMM 2022, 126, pp. 275–280, 2023.
https://doi.org/10.1007/978-3-031-20353-4_22

In the present work, we have investigated the etching characteristics of Si{110} in 10M NaOH without and with the addition of NH_2OH. The main objective of this work is to improve the etch rate and undercutting at convex corners which are inevitable for high productivity.

2 Experimental Details

Cz-grown one side polished 4-inch diameter p-type {110} oriented silicon wafers of resistivity from 1 to 10 Ωcm are used. One-micron thick oxide layer is deposited using thermal oxidation process. Oxidized wafers are patterned using lithography followed by oxide etching in buffered hydrofluoric (BHF) acid. Thereafter, wafers are cleaned thoroughly in deionized (DI) water. Now the photoresist is removed using acetone followed by cleaning in running DI water. The wafers are then diced into small chips using a dicing saw. The diced samples are cleaned in piranha bath (H_2SO_4:H_2O_2::1:1) followed by DI water rinse. After rinsing with DI water, samples are dipped into 1% HF to remove oxide layer grown during piranha cleaning followed by rinsing in DI water. Thereafter, etching experiments are carried out in 10M NaOH without and with addition of 12% NH_2OH using constant temperature bath. A reflux condenser with a narrow opening is used to avoid the concentration variation of the etchant during etching. Etch depth and undercutting at convex corners are measured using 3D laser scanning microscope (OLYMPUS OLS4000), an optical microscope (OLYMPU MM6C-PC), and Spectroscopic Ellipsometry (J. A. Woolman Co. Inc).

3 Results

Etch rate is an important parameter in MEMS fabrication. It is calculated by measuring the etch depth for different times of etching. Figure 1 presents the etch rate of Si{110} in pure and modified NaOH solutions. It can easily be noticed from Figure 1 that the etch rate is significantly increased by the addition of NH_2OH to NaOH. High etch rate reduces etching time and therefore very useful for industrial production.

Silicon dioxide (SiO_2) is widely used as mask and/or structural material in MEMS fabrication. Therefore, etch selectivity between Si and SiO_2 is another important factor in the fabrication of cavities/grooves using oxide layer as mask and freestanding structures (e.g., cantilevers) using oxide layer as structural material. The etch selectivity is defined as the ratio of the etch rate of the Si to the etch rate of SiO_2. In this work, SiO_2 is used as mask material. Figure 2 presents the etch rate of SiO_2 and its selectivity with Si{110}. It can obviously be noticed from the figure that the etch rate of SiO_2 is decreased when NH_2OH is added to 10M NaOH. At the same time there is two-fold increment in the etch selectivity, which is a desirable etching characteristic of an etchant. In the case of NH_2OH-added TMAH solution, the etch selectivity is 2.2 times more than the etch selectivity in pure TMAH solution [3, 11]. However, in NH_2OH-added NaOH solution, the etch selectivity is 2.4 times more than the etch selectivity in pure NaOH solution. Therefore, the etch selectivity of NH_2OH added NaOH solution is comparable with that of the TMAH-NH_2OH solution.

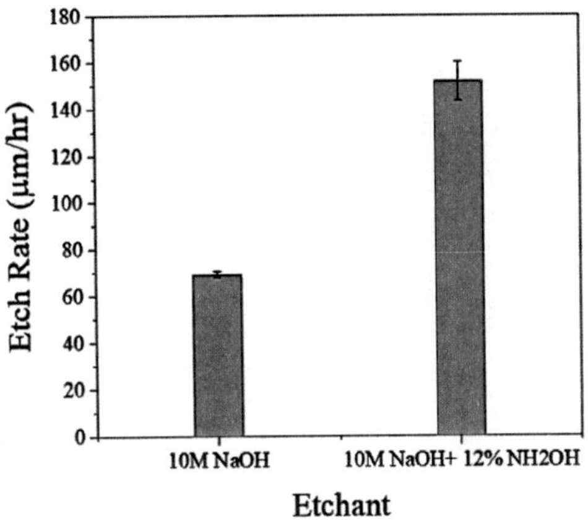

Fig. 1. Etch rate of Si{110} in pure and 12%NH_2OH-added 10M NaOH solution at 70 ± 1 °C.

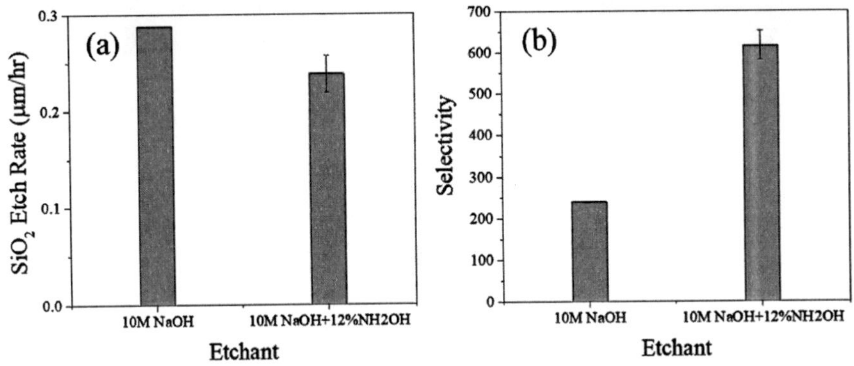

Fig. 2. (a) Etch rate of SiO_2 and (b) Etch selectivity between Si{110} and SiO_2 in pure and 12% NH_2OH-added 10M NaOH solution at 70 ± 1 °C.

In the formation of suspended structures, undercutting plays an important role in removing underneath material to release the structures from the substrate [28–31]. In the mask pattern, the intersection of crystallographic directions (i.e., <110> and <112> directions on Si{110} surface) on silicon wafer surface forms either a concave corner (<180° i.e. corners turning inside) or a convex corners (>180° i.e. corners turning outside). Although both types of corners are bound by <110> and <112> directions, their etching characteristics are exactly opposite to each other. The concave corner remains intact during the etching, while convex corners are immediately attacked by the etchant leading to undercut. Figure 3 presents the schematic view of undercutting at convex corners on Si{110} surface. The intersection of <112> directions on Si{110} surface forms a rhombus pattern that comprises acute and obtuse corners as presented in Fig. 3. On etching in anisotropic etchant, vertical {111} planes are exposed at the mask edges of this rhombus pattern. The undercutting rate at acute and obtuse corners are presented in Fig. 4. It is clear from the results that the undercutting rate increases considerably on the addition of NH_2OH. In the case of NH_2OH-added TMAH and KOH solutions, the undercutting is increased [28–31]. In TMAH-based solution, the undercutting rate is increased by a factor of more than three when NH_2OH is added, whereas in NH_2OH-added KOH solution, it is around four times than that in pure KOH [11, 13]. In the case of NH_2OH-added NaOH, the undercutting rate is 3.6 times more than that in pure NaOH as shown in Fig. 4. Hence, it can be concluded that the undercutting rate is strongly improved by the addition of NH_2OH.

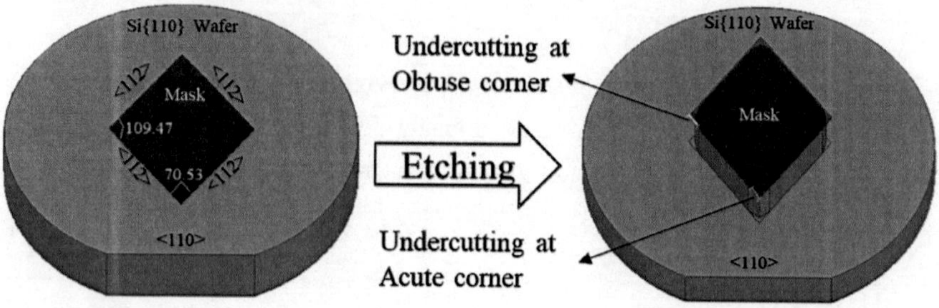

Fig. 3. Schematic view of undercutting at acute and obtuse corners on Si{110}.

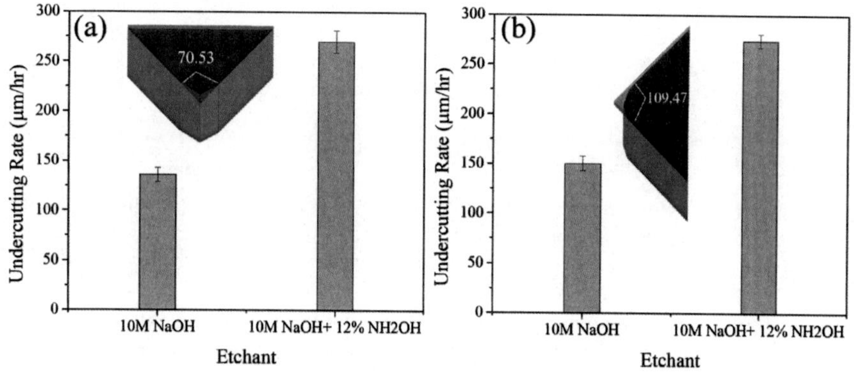

Fig. 4. Undercutting rate on Si{110} at (a) acute and (b) obtuse corners in pure and 12%NH_2OH-added 10M NaOH solution at 70 ± 1 °C.

4 Conclusions

Etching characteristics of Si{110} in 10M NaOH doped with 12% NH_2OH as a reducing agent have been investigated for the fabrication of microstructures for MEMS applications. The etchant is characterized by evaluating the etch rate of Si and SiO_2, etch selectivity between Si and SiO_2, and the undercutting at convex corners. Etching characteristics of NaOH are improved significantly when NH_2OH is incorporated. Hence this study is very important for applications in MEMS fabrication via wet anisotropic etching. Ultimately, it can be concluded that the addition of NH_2OH in alkaline solution favourably influences the etching characteristics.

Acknowledgements. This work was supported by research grant from the Council of Scientific and Industrial Research (CSIR, Ref: 22(0824)/19/EMR-II,0527/NS), We would greatly acknowledge CSIR for the support.

References

1. Pal, P., Sato, K.: Silicon Wet Bulk Micromachining for MEMS. Pan Stanford Publishing, Singapore (2017)
2. Narasimha Rao, A.V., Swarnalatha, V., Ashok, A., Singh, S.S., Pal, P.: Effect of NH_2OH on etching characteristics of Si{100} in KOH solution. ECS J Solid-State Sci Technol **6**(9), 609–614 (2017)
3. Swarnalatha, V., Narasimha Rao, A.V., Ashok, A., Singh, S.S., Pal, P.: Modified TMAH based etchant for improved etching characteristics on Si{100} wafer. J. Micromech. Microeng. **27**(8), 085003 (2017)
4. Zubel, I.: Anisotropic etching of Si. J. Micromech. Microeng. **29**(9), 093002 (2019)
5. Yang, C.R., Chen, P.Y., Chiou, Y.C., Lee, R.T.: Effects of mechanical agitation and surfactant additive on silicon anisotropic etching in alkaline KOH solution. Sens. Actuators A, Phys. **119**(1), 263–270 (2005)
6. Pal, P., Sato, K., Gosalvez, M.A., Shikida, M.: Study of rounded concave and sharp edge convex corners undercutting in CMOS compatible anisotropic etchants. J. Micromech. Microeng. **17**(11), 2299–2307 (2007)
7. Pal, P., et al.: Surfactant adsorption on single crystal silicon surfaces in TMAH solution: orientation-dependent adsorption detected by in-situ infra-red spectroscopy. J. Microelectromech. Syst. **18**, 1345–1356 (2009)
8. Pal, P., Sato, K.: Various shapes of silicon freestanding microfluidic channels and microstructures in one step lithography. J. Micromech. Microeng **19**(5), 05003 (2009)
9. Shikida, M., Sato, K., Tokoro, K., Uchikawa, D.: Differences in anisotropic etching properties of KOH and TMAH solutions. Sens. Actuators A **80**(2), 179–188 (2000)
10. Xu, Y.W., Michael, A., Kwok, C.Y.: Formation of ultra-smooth 45° micromirror on (100) silicon with low concentration TMAH and surfactant: techniques for enlarging the truly 45° portion. Sens. Actuators A **166**, 164–171 (2011)
11. Swarnalatha, V., Narasimha Rao, A.V., Pal, P.: Effective improvement in the etching characteristics of Si{110} in low concentration TMAH solution. Micro Nano Lett. **13**(8), 1085–1089 (2018)
12. Swarnalatha, V., et al.: Systematic study of the etching characteristics of Si{111} in modified TMAH. Micro Nano Lett. **15**(1), 52–57 (2020)
13. Narasimha Rao, A.V., Swarnalatha, V., Pal, P.: Etching characteristics of Si{110} in 20 wt% KOH with addition of hydroxylamine for the fabrication of bulk micromachined MEMS. Micro Nano Syst. Lett. **5**(23), 1–9 (2017)
14. Narasimha Rao, A.V., et al.: Aging effects of KOH+NH_2OH solution on the etching characteristics of silicon. ECS J. Solid State Sci. Technol. **8**(11), 685–692 (2019)
15. Pal, P., Swarnalatha, V., Narasimha Rao, A.V., Pandey, A.K., Tanaka, H., Sato, K.: High speed silicon wet anisotropic etching for applications in bulk micromachining: a review. Micro Nano Syst. Lett. **9**(4), 1–59 (2021)

16. Vazsonyi, E., Adam, M., Ducso, C., Vizvary, Z., Toth, A.L., Barsony, I.: Three-dimensional force sensor by novel alkaline etching technique. Sens. Actuators A **620**(6), 123–124 (2005)
17. Vazsonyi, E., Vertesy, Z., Toth, A.L.: Anisotropic etching of silicon in a two-component alkaline solution. J. Micromech. Microeng. **13**, 165–169 (2003)
18. Gangopadhyay, U., Dhungel, S.K., Kim, K.: Novel low cost chemical texturing for very large area industrial multi-crystalline silicon solar cells. Semicond. Sci. Technol. **20**, 938–946 (2005)
19. Akhter, P., Baig, A., Mufti, A.: Dissolution of Si (100) layers in NaOH aqueous solutions. J. Phys. D: App. Phys. **22**(12) (1989)
20. Holke, A., Henderson, H.T.: Ultra-deep anisotropic etching of (110) silicon. J. Micromech. Microeng. **9**, 51–57 (1999)
21. Tolmachev, V.A., et al.: One-dimensional photonic crystal obtained using vertical anisotropic etching of silicon. Semiconductors **36**, 932–935 (2002)
22. Zubel, I., Kramkowska, M.: Possibilities of extension of 3D shapes by bulk micromachining of different Si (hkl) substrates. J. Micromech. Microeng. **15**, 485–493 (2009)
23. Lee, D., Yu, K., Krishnamoorthy, U., Solgaard, O.: Vertical mirror fabrication combining KOH etch and DRIE of (110) silicon. J. Microelectromech. Syst. **18**, 217–227 (2009)
24. Pal, P., Gosalvez, M.A., Sato, K., Hida, H., Xing, Y.: Anisotropic etching on Si {110}: experiment and simulation for the formation of microstructures with convex corners. J. Micromech. Microeng. **24**(12), 125001–125012 (2014)
25. Ahn, M., Heilmann, R.K., Schattenburg, M.L.: Fabrication of ultrahigh aspect ratio Freestanding gratings on silicon-on-insulator wafers. J. Vacuum Sci. Technol. B. **25**(6), 2593–2597 (2007)
26. Kendall, D.L.: Vertical etching of silicon at very high aspect ratios. Ann. Rev. Mat. Sci **9**(1), 373–403 (1979)
27. Kim, S.H., Lee, S.H., Lim, H.T., Kim, Y.K., Lee, S.K.: Anisotropic bulk etching of (110) silicon with high aspect ratio. IEEJ Trans Sens. Micromach. **118**, 32–36 (1997)
28. Pal, P., Sato, K.: A comprehensive review on convex and concave corners in silicon bulk micromachining based on anisotropic wet chemical etching. Micro Nano Syst. Lett. **3**(1), 1–42 (2015). https://doi.org/10.1186/s40486-015-0012-4
29. Pal, P., Sato, K., Chandra, S.: Fabrication techniques of convex corners in (100)-silicon wafer using bulk micromachining: a review. J. Micromech. Microeng. **17**(10), 111–133 (2007)
30. Pal, P., Haldar, S., Singh S.S., Ashok, A., Xing, Y., Sato, K.: A detailed investigation and explanation to the appearance of different undercut profiles in KOH and TMAH. J. Micromech. Microeng. **24**(9), 095026-45 (2014)
31. Pal, P., Singh, S.S.: A new model for the etching characteristics of corners formed by Si{111} planes on Si{110} wafer surface. Engineering **5**(11A), 1–8 (2013)

Effect of IPA on Micromachining Characteristics of Silicon in KOH-Based Solution

Avvaru Venkata Narasimha Rao[1(✉)] and Prem Pal[2]

[1] Department of Physics, Rajiv Gandhi University of Knowledge Technologies Ongole, Ongole, Andhra Pradesh, India
avnarasimha111@gmail.com

[2] MEMS and Micro/Nano Systems Laboratory, Department of Physics, Indian Institute of Technology Hyderabad, Hyderabad, India

Abstract. Microstructures for microelectromechanical system (MEMS) applications are widely fabricated using silicon wet bulk micromachining technique. In this method, potassium hydroxide (KOH) is one of the most used anisotropic etchants. It provides high etch rate when it is modified by the addition of hydroxylamine (NH_2OH). Moreover, it shows high undercutting which is a desirable property for the fabrication of overhanging microstructures. In this paper, the effect of *isopropyl alcohol* (IPA) in NH_2OH-added KOH on the etching characteristics of Si{100} and Si{110}) is systematically studied. The results are compared with the etching characteristics of IPA-added pure KOH. The undercutting rate reduces drastically when IPA is added to NH_2OH + KOH which protects the convex corners of the structures during etching process. At the same time, the etch rate of {110} plane is suppressed considerable which is exploited to expose {110} plane at the mask edges aligned along <100> direction on {100} surface that makes 45° angle with {100} surface and act as a micromirror.

Keywords: Wet bulk micromachining · MEMS · High speed anisotropic etching · 45° micromirrors · KOH · NH_2OH · IPA

1 Introduction

Wet anisotropic etching is a simple and cheap fabrication process, which is extensively used in silicon micromachining for the fabrication of microstructures for microelectromechanical systems (MEMS) [1–8]. Potassium hydroxide (KOH) and tetramethylammonium hydroxide (TMAH) are most used alkaline solutions in silicon wet anisotropic etching [1]. KOH exhibits higher etch rate and improved etch selectivity between Si{100}/Si{110} and Si{111} in comparison to TMAH. Etch rate is a very important etching parameter that affects the industrial production. To improve the etch rate, various methods such as microwave irradiation, ultrasonic agitation, etching temperature at the boiling point of etchants, etc. are investigated [9]. To obtain smooth etched surface morphology, effects of alcohols and surfactants are investigated [10–15]. Recently, we reported the etching characteristics of silicon in various concentrations

A. K. Pandey et al. (Eds.): MAMM 2022, 126, pp. 281–289, 2023.
https://doi.org/10.1007/978-3-031-20353-4_23

of NH_2OH-added 20 wt% KOH [4, 16–19]. The addition of NH_2OH in 20 wt% KOH considerably increases the etch rate and the undercutting at convex corners. Moreover, it improves etch selectivity between Si and SiO_2.

The addition of IPA in KOH suppresses the etch rate of Si{110} planes. Therefore {110} planes expose at <100> mask edge on Si{100} surface that makes an angle of 45° with wafer surface. As 15% NH_2OH-added 20 wt% KOH provides very high etch rate and undercutting, the study of the effect of IPA in this etchant composition is required. In this paper, the effect of IPA on the etching characteristics of Si{100} and Si{110} in 20 wt% KOH and 15% NH_2OH + 20 wt% KOH is studied.

2 Experimental Details

4-inch Cz-grown {100} and {110} oriented p-type silicon wafers with resistivity of 5–10 Ω cm are used. A thermally grown silicon dioxide of 1 μm thickness is employed as etch mask for the selective etching of silicon. The first step of the experimental process is to pattern the oxide layer by photolithography. Silicon dioxide is etched out in buffered hydrofluoric acid (BHF) followed by cleaning in running DI water. Thereafter, the wafer is diced into small chips (2 × 2 cm^2). Diced chips are cleaned in a piranha bath (H_2O_2:H_2SO_4::1:1), then thoroughly rinsed in running DI water. Prior to dipping these chips in an etchant, oxide layer grown in piranha bath is removed in 1% hydrofluoric acid (HF) followed by cleaning in DI water. To prepare 20 wt% KOH solution, KOH pellets (99.99%, Alfa Aesar) are dissolved in DI water. Aqueous NH_2OH solution is used to prepare 15% NH_2OH + 20 wt% KOH. To study the effect of IPA on the etching characteristics of Si{100} and Si{110}, different concentration IPA is added to KOH and NH_2OH + KOH solutions. All experiments are performed at a fixed temperature of 75 ± 1 °C. A reflux condenser is used to avoid any change in etchant concentration during the etching process. Etch depth, etched surface roughness, and undercutting at convex corners are measured using a 3D laser scanning microscope (Olympus, OLS4000). Scanning electron microscope (SEM) is used to inspect various kinds of microstructures.

3 Results and Discussion

Etching characteristics of Si{100} and Si{110} (i.e., etch rates, etched surface morphologies, undercutting at convex corners) are investigated in KOH and NH_2OH + KOH without and with addition of different concentration IPA. The results obtained in different etchant compositions are compared. Etching characteristics are methodically presented in the following subsections.

3.1 Etch Rate

The study of etch rate is very important in the fabrication of MEMS structures such as grooves and cavities. Figure 1 shows the etch rates of {100} and {110} planes in 20 wt% KOH and 15% NH_2OH-added 20 wt% KOH without and with the addition of different concentration IPA. It can easily be noticed that the etch rates of both Si{100} and Si{110} drastically decreases as the concentration of IPA increases and attain almost minimum value if the concentration of added IPA increases more than 6%. In the case of Si{110}, the decrease in etch rate significantly higher than that of Si{100}. This property of etching is very useful to fabricate 45° mirror on Si{100} surface by exposing {110} planes at <100> mask edges [11].

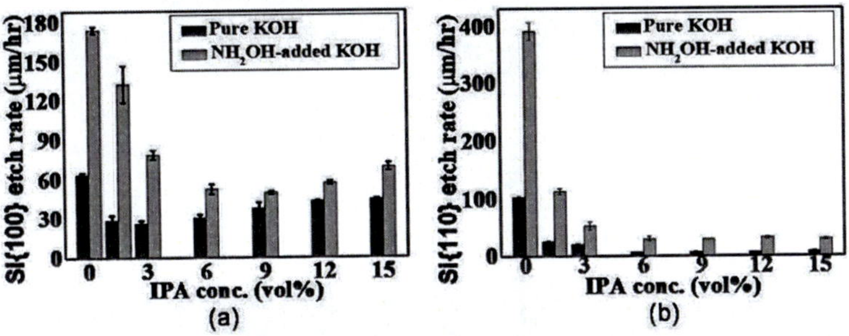

Fig. 1. Etch rate of (a) Si{100}, (b) Si{110} in 20 wt% KOH and 15% NH_2OH-added 20 wt% KOH solutions without and with different concentration IPA at 75 ± 1 °C.

In IPA-added KOH solution, IPA molecules form a layer on the silicon surface, which protects the silicon surface from the reactants [12, 13, 20]. Therefore, the etch rate of silicon reduces on the addition of IPA. Recently, we reported the etching mechanism of silicon in NH_2OH-added KOH [16–19]. In NH_2OH-added KOH, H_2O is the main reactive molecule, while NH_2O^- and OH^- work as catalysts. When IPA is added to $NH_2OH + KOH$, the etch rate of silicon is suppressed as shown in Fig. 1. It may be due to the adsorption of IPA molecules on the silicon surface that reduces the diffusion of reactants and catalysts to silicon surface. This is schematically presented in Fig. 2.

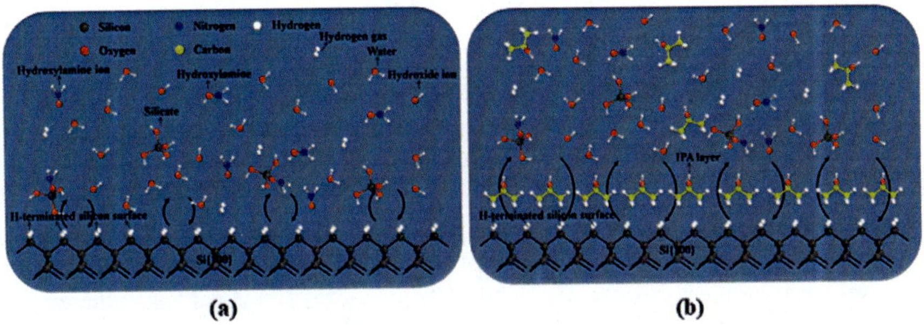

Fig. 2. Schematic diagrams of Si{100} surface (a) without and (b) with isopropyl alcohol (IPA) layer in IPA + NH_2OH + KOH solution during etching process.

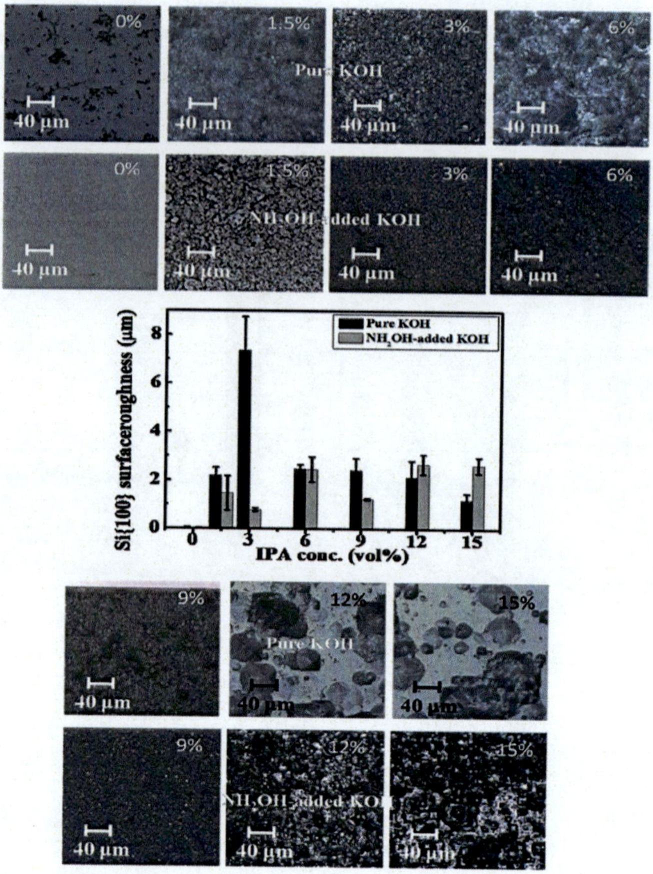

Fig. 3. Surface roughness and corresponding 3D laser scanning microscope images of Si{100} etched in 20 wt% KOH and 15% NH_2OH-added 20 wt% KOH solutions without and with the addition of different concentration IPA at 75 ± 1 °C (etching time: 2 h).

3.2 Etched Surface Roughness and Morphology

Surface roughness is one of the important concerns, especially when it is used for optical and solar cell applications [21, 22]. Figures 3 and 4 show the surface roughness of Si{100} and Si{110}, respectively, planes in 20 wt% KOH and 15% NH_2OH-added 20 wt% KOH without and with the addition of different concentration IPA. In both cases, surface roughness is improved when 6% (or more) IPA is added in the etchant. The main cause of the formation of hillocks on the surface is the generation of hydrogen bubbles on the surface being etched and/or the stiction of impurities and byproducts on the surface [1]. The addition of IPA might reduce the surface tension of the etchant, or the IPA layer does not allow the formation of bubbles on the surface [1, 20].

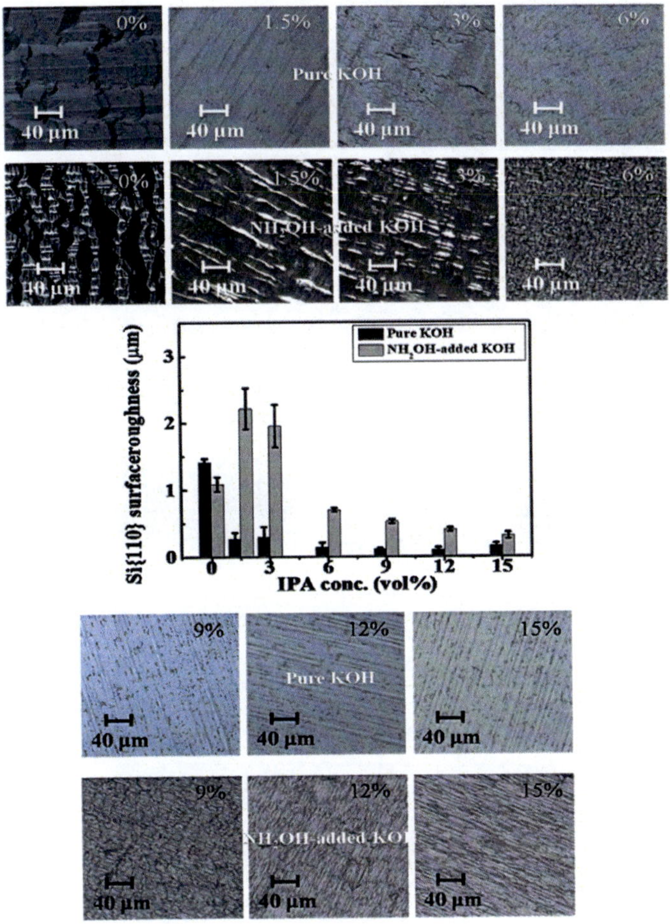

Fig. 4. Surface roughness and corresponding 3D laser scanning microscope images of Si{110} etched in 20 wt% KOH and 15% NH_2OH-added 20 wt% KOH solutions without and various concentration IPA at 75 ± 1 °C (etching time: 2 h).

3.3 Undercutting

Undercut rate is an important etching characteristic that is advantageously used to release the suspended structures [23–25]. On the other hand, it is an unwanted effect when mesa structures are fabricated with protected convex corners. Figure 5 shows the undercut rate and corresponding 3D laser scanning microscope images of the convex corners on Si{100} etched in pure and NH_2OH-added KOH solution without and with addition of different concentration IPA. It can easily be noticed in Fig. 5 that the undercut rate suppressed significantly when IPA is added in pure and NH_2OH-added KOH. It is well known that the undercut rate reduces due to the reduction of etch rate of high index planes that are exposed at the convex corner during the wet anisotropic etching process

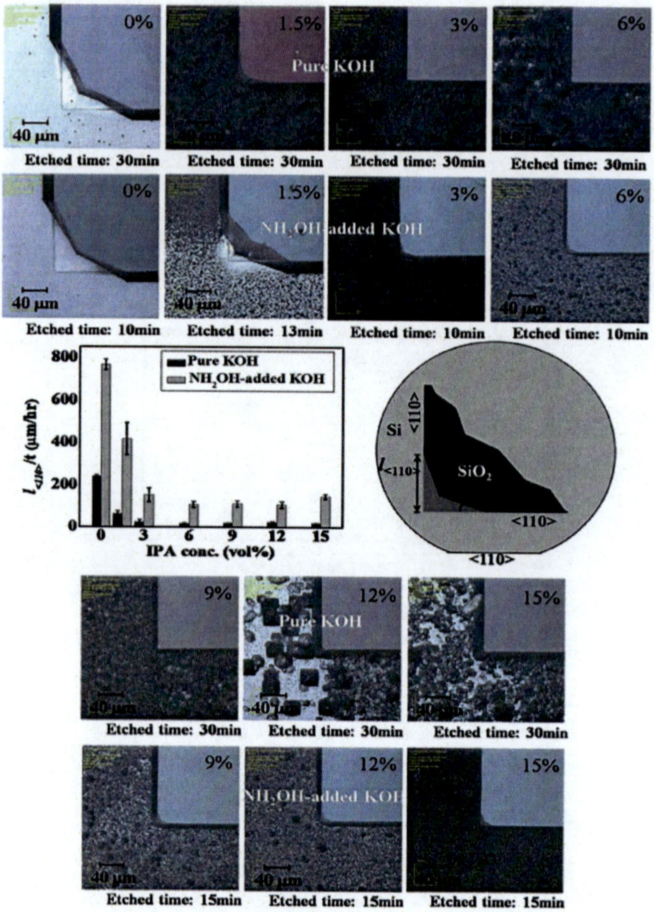

Fig. 5. Undercutting rate and corresponding 3D laser scanning microscope images of convex corner on Si{100} after etching in 20 wt% KOH and 15% NH_2OH-added 20 wt% KOH solutions without and with addition of IPA at 75 ± 1 °C.

[1, 20]. In this case, IPA layer forms on the high index planes appearing at convex corners during etching process that protect the convex corner from the main reactive elements.

Mesa and cavity structures are successfully fabricated on Si{100} in IPA-added 20 wt% KOH and 15% NH_2OH + 20 wt% KOH solutions as presented in Fig. 6. It can easily be observed that the fabrication of mesa/cavity with same height/depth requires less etching time in IPA + NH_2OH + KOH in comparison to that in IPA + KOH solution. Thus, the fabrication of mesa and cavity structures with a higher etch rate, is very useful for the fabrication microstructures with 45° sidewalls that act as micromirror [11].

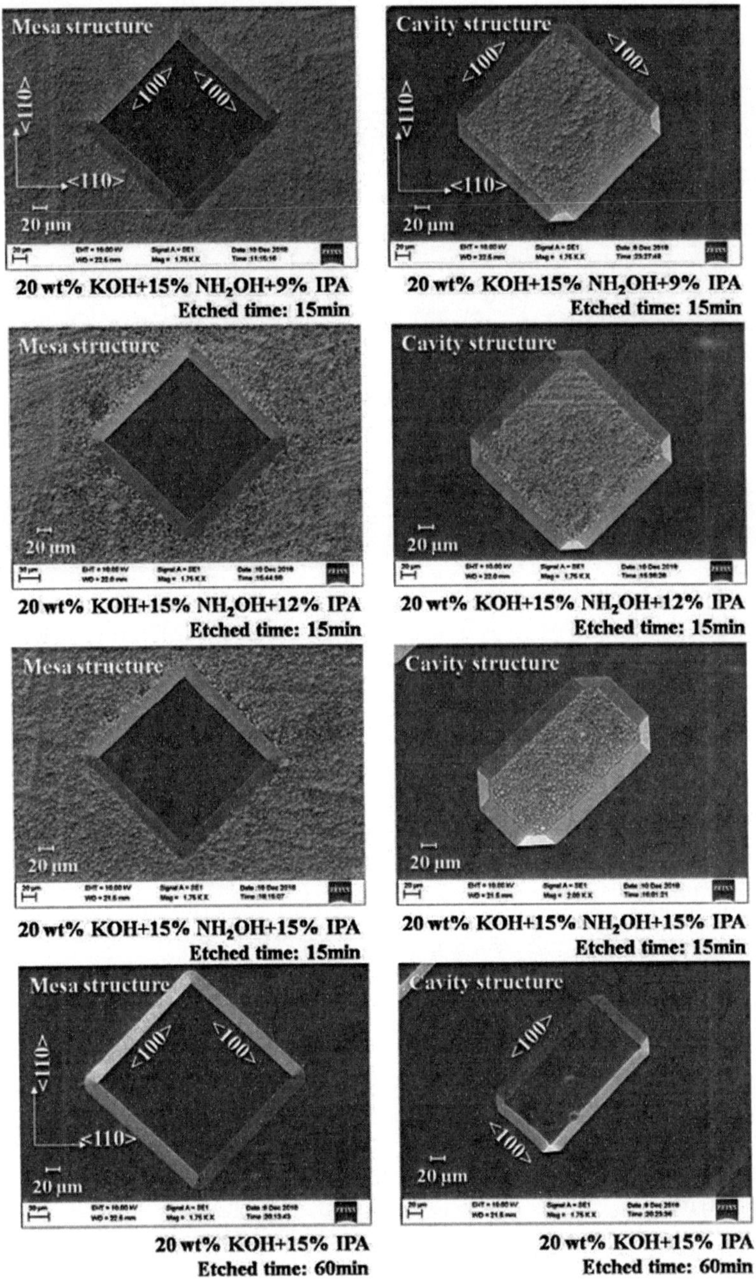

Fig. 6. SEM images of mesa and cavity structures with <100> edges on Si{100} formed in 20 wt% KOH and 15% NH_2OH-added 20 wt% KOH solutions with the addition of various concentrations IPA at 75 ± 1 °C.

4 Conclusions

The effect of IPA on the etching characteristics of silicon (i.e., etch rate, surface roughness, and undercutting) in 20% KOH and 15% NH_2OH + 20 wt% KOH solutions is methodologically investigated. The addition of IPA suppresses the etch rate and improves the etched surface morphology. The etch rate of Si{110} is more drastically reduced in comparison to that of Si{100}. This property is exploited to form 45° mirror on Si{100} surface. Moreover, the undercutting rate in both KOH and NH_2OH + KOH is influenced dramatically on the addition of IPA, which is desirable property to form well shaped mesa structures on Si{100}. Surface morphology in IPA + NH_2OH + KOH follows the same trend as IPA + KOH. Finally, it can be concluded that the addition of IPA affects the etching characteristics of both KOH and NH_2OH + KOH significantly and the selection of the etchant depend upon the requirement of undercutting, etch rate and surface morphology.

References

1. Pal, P., Sato, K.: Silicon Wet Bulk Micromachining for MEMS. Pan Stanford Publishing, Singapore (2017)
2. Koo, K.I., Chung, H., Yu, Y., Seo, J., Park, J., Lim, J.M., Paik, S.J., Park, S., Choi, H.M., Jeong, M.J., Kim, G.S.: Fabrication of pyramid shaped three-dimensional 8 × 8 electrodes for artificial retina. Sens. Actuators A Phys. **130**, 609–615 (2006)
3. Narasimha Rao, A.V., Swarnalatha, V., Pandey, A.K., Singh, S.S., Pal, P.: Determination of precise crystallographic directions on Si{111} wafers using self-aligning pre-etched pattern. Micro Nano Syst. Lett. **6**(1), 4 (2018)
4. Menon, P.K., Narasimha Rao, A.V., Murthy, A.L., Pandey, A.K., Pal, P.: High speed etching of silicon in KOH+NH_2OH solution at lower temperatures for the fabrication of through holes in silicon wafer. Micro Nano Lett., 1–6 (2020)
5. Yang, H., et al.: A novel method to fabricate single crystal nano beams with (111)-oriented Si micromachining. Microsyst. Technol. **14**(8), 1185–1191 (2008)
6. Pal, P., Sato, K.: Various shapes of silicon freestanding microfluidic channels and microstructures in one step lithography. J. Micromech. Microeng. **19**(5), 055003 (2009)
7. Rola, K.P., Zubel, I.: Impact of alcohol additives concentration on etch rate and surface morphology of (100) and (110) Si substrates etched in KOH solutions. Microsyst. Technol. **19**(4), 635–643 (2013)
8. Seidel, H., Csepregi, L., Heuberger, A., Baumgärtel, H.: Anisotropic etching of crystalline silicon in alkaline solutions I. Orientation dependence and behavior of passivation layers. J. Electrochem. Soc. **137**(11), 3612–3626 (1990)
9. Pal, P., Swarnalatha, V., Narasimha Rao, A.V., Pandey, A.K., Tanaka, H., Sato, K.: High speed silicon wet anisotropic etching for applications in bulk micromachining: a review. Micro Nano Syst. Lett. **9**(1), 1–59 (2021)
10. Pal, P., Sato, K., Gosalvez, M.A., Kimura, Y., Ishibashi, K., Niwano, M., Hida, H., Tang, B., Itoh, S.: Surfactant adsorption on single crystal silicon surfaces in TMAH solution: orientation-dependent adsorption detected by in-situ infra-red spectroscopy. J.Microelectromech. Syst. **18**, 1345–1356 (2009)
11. Xu, Y.W., Michael, A., Kwok, C.Y.: Formation of ultra-smooth 45° micromirror on (100) silicon with low concentration TMAH and surfactant: techniques for enlarging the truly 45° portion. Sens. Actuators A Phys. **166**(1), 164–171 (2011)

12. Zubel, I., Kramkowska, M.: The effect of alcohol additives on etching characteristics in KOH solutions. Sens. Actuators A Phys. **101**(3), 255–261 (2002)
13. Zubel, I., Rola, K., Kramkowska, M.: The effect of isopropyl alcohol concentration on the etching process of Si-substrates in KOH solutions. Sens. Actuators A Phys. **171**, 436–445 (2011)
14. Pal, P., Sato, K., Gosalvez, M.A., Shikida, M.: Study of rounded concave and sharp edge convex corners undercutting in CMOS compatible anisotropic etchants. J. Micromech. Microeng. **17**(11), 2299–2307 (2007)
15. Narasimha Rao, A.V., Swarnalatha, V., Pal, P.: Effect of surfactant and alcohol additives on etching characteristics in aqueous potassium hydroxide solutions. ECS Trans. **77**(11), 1761–1769 (2017)
16. Narasimha Rao, A.V., Swarnalatha, V., Ashok, A., Singh, S.S., Pal, P.: Effect of NH_2OH on etching characteristics of Si{100} in KOH solution. ECS J. Solid State Sci. Technol. **6**(9), P609–P614 (2017)
17. Narasimha Rao, A.V., Pal, P., Pandey, A.K., Swarnalatha, V., Menon, P, K., Tanaka, H., Sato, K.: Aging effects of KOH + NH_2OH solution on the etching characteristics of silicon. ECS J. Solid State Sci. Technol. **8**(11), P685–P692 (2019)
18. Narasimha Rao, A.V., Swarnalatha, V., Pal, P.: Etching characteristics of Si {110} in 20 wt% KOH with addition of hydroxylamine for the fabrication of bulk micromachined MEMS. Micro Nano Syst. Lett. **5**(1), 1–9 (2017)
19. Swarnalatha, V., Vismaya, K.T., Narasimha Rao, A.V., Pal, P., Pandey, A.K., Tanaka, H., Sato, K.: Etching mechanism behind the high speed etching of silicon in NH_2OH-added alkaline solutions. IEEJ Trans. Sens. Micromach. **140**(1), 24–30 (2020)
20. Zubel, I.: Anisotropic etching of Si. J. Micromech. Microeng. **29**(9), 93002 (2019)
21. Gupta, A., Pal, P., Sharma, C.S.: Surface texturing of Silicon {100} in an extremely low concentration TMAH for minimized reflectivity. ECS J. Solid State Sci. Tech. **8**(10), 622–628 (2019)
22. Rai-Choudhury, P.: MEMS and MOEMS Technology and Applications. SPIE Press (2000)
23. Pal, P., Sato, K.: A comprehensive review on convex and concave corners in silicon bulk micromachining based on anisotropic wet chemical etching. Micro Nano Syst. Lett. **3**(1), 1–42 (2015). https://doi.org/10.1186/s40486-015-0012-4
24. Pal, P., Sato, K., Chandra, S.: Fabrication techniques of convex corners in (100)-silicon wafer using bulk micromachining: a review. J. Micromech. Microeng. **17**(10), R111–R133 (2007)
25. Pal, P., Sato, K., Shikida, M., Gosalvez, M.A.: Study of corner compensating structures and fabrication of various shapes of MEMS structures in pure and surfactant added TMAH. Sens. Actuators A **154**(2), 192–203 (2009)

Deep Grooves in Borofloat Glass by Wet Bulk Micromachining

Adarsh Nigam(✉), Swarnalatha Veerla, and Prem Pal

MEMS and Micro/Nano Systems Laboratory, Department of Physics, Indian Institute of Technology Hyderabad, Hyderabad, Telangana, India
adarsh.nigam@phy.iith.ac.in

Abstract. In this work, deep grooves of more than 300 μm depth are fabricated in a high-quality Borofloat glass wafer using wet bulk micromachining in buffered HF (BHF) solution. The Cr/Au/photoresist layers are used as the etching mask. These masking layers showed very good chemical resistance to etching in BHF solution. The etchant provides smooth etched surface morphology with an excellent etch rate of 11.1 μm/min. The proposed process is very useful for the formation of deep grooves with smooth vertical sidewalls and uniform bottom surfaces.

Keywords: Buffered HF · Wet chemical etching · Borofloat glass · Masking layers · Etch-rate

1 Introduction

In the microfabrication process, the glass is the second most widely used material after silicon owing to various interesting properties such as high chemical resistance, high heat resistance, high electrical isolation, low optical absorption, etc. Micromachining is an approach that is widely used for the development of microstructures in glass wafers. Borofloat glass is broadly utilized for the formation of various microstructures for microelectromechanical systems (MEMS) applications such as Bio-MEMS and optical-MEMS because of its superior physical, chemical, and mechanical properties [1, 2]. The optical properties of Borofloat glass-like high transparency and less absorption makes it suitable for the fabrication of lenses and other optical applications where low optical loss is essential. Also, the excellent mechanical properties of Borofloat glass-like high thermal expansion coefficient and good electrical insulation properties make it an ideal candidate for the fabrication of different MEMS sensors and actuators [3].

There are various approaches like mechanical, plasma, and chemical utilized for the development of deep structures like cavities and through-hole in glass wafers. Mechanical approaches comprise traditional and ultrasonic drilling, electrochemical discharge, and powder blasting. These processes provide very rough surfaces [4]. Plasma is also used for glass etching, but it provides a very slow etch rate of the order of 10 nm/min [5]. Besides, chemical processes such as dry and wet chemical etching are widely used. Among them, dry etching has the advantages of providing smooth surfaces and high uniformity, but it is expensive and exhibits a very low etch rate. Wet bulk micromachining

A. K. Pandey et al. (Eds.): MAMM 2022, 126, pp. 290–295, 2023.
https://doi.org/10.1007/978-3-031-20353-4_24

is a cheap and easy approach to form microstructures in glass and silicon wafers [6–14]. It is utilized in several applications for glass micromachining such as wafer thinning, through-hole development, microchannels fabrication for different MEMS applications [1, 15]. Generally, hydrofluoric acid (HF) is most commonly used for glass etching [5]. However, the aggressive behavior of concentrated HF can severely damage the associated masking layer. In addition, it provides rough surface morphology and creates surface nonuniformities. To perform selective etching of glass, various kinds of masking layers such as Cr/Au, multiple metal stack Cr/Au/Cr/Au, amorphous silicon, polysilicon, and silicon carbide are employed [6, 15–18].

In this work, we have developed a process for the formation of deep grooves in a Borofloat glass wafer using wet bulk micromachining in Buffered HF (BHF) solution.

2 Experimental Details

In this study, a 500 μm thick, 4-inch diameter Borofloat glass (HK9L) wafer is used. Firstly, Cr/Au masking layers with 30 nm and 200 nm thickness, respectively, are deposited on both sides of the wafer using the DC magnetron sputtering process. Here, the deposition of metals is performed at room temperature. The thickness of deposited films is observed by an in-situ thickness monitor.

After the deposition of Cr/Au masking layers, pattering of the masking layer is performed using photolithography. In this process, the hexamethyldisilazane (HMDS) is spin-coated on Cr/Au at 3000 rpm for 30 s followed by spin coating of the positive photoresist AZ1512HS at 3000 rpm for 30 s. Here, the HMDS works as a primer to enhance the adhesion of the photoresist layer. The photoresist-coated wafer is pre-backed in an oven at 90 °C for 30 min. Thereafter, the wafer is exposed under UV illumination in a mask aligner after aligning it with an appropriately designed mask pattern. Then the wafer is developed and post-backed for 1 h at 120 °C. After this step, Cr and Au layers are etched using wet chemicals. In this process, first, the top Au layer is etched in potassium iodide (KI) solution. Subsequently, the Cr layer is etched in the Cr etchant prepared by mixing 1.645 g ceric ammonium nitrate ($Ce(NH_4)_2(NO_3)_6$) with 0.43 ml nitric acid (HNO_3) in 100 ml DI water. After the selective etching of Cr and Au layers, the wafer is thoroughly cleaned in DI water. A Teflon container, as shown in Fig. 1, is used for glass etching. To fabricate deep grooves in the Borofloat glass wafer, the wet bulk micromachining is performed in buffered HF solution (BHF), which is prepared by mixing HF and ammonium fluoride (NH_4F) in a ratio of 1:7. After preparation of BHF, it is shifted in Teflon container for subsequent etching process. In this work, BHF is chosen for glass etching because it makes very low damage to the photoresist layer, therefore the masking layer durability in the etching process is sufficiently improved. In this process, the backside of the wafer is protected by Cr/Au and photoresist layers. Process steps used in this work are schematically illustrated in Fig. 1.

3 Results and Discussion

Etch rate is outlined as the vertical distance etched per unit time and is calculated by measuring etch depth for a particular etch time. In the semiconductor industries, a high

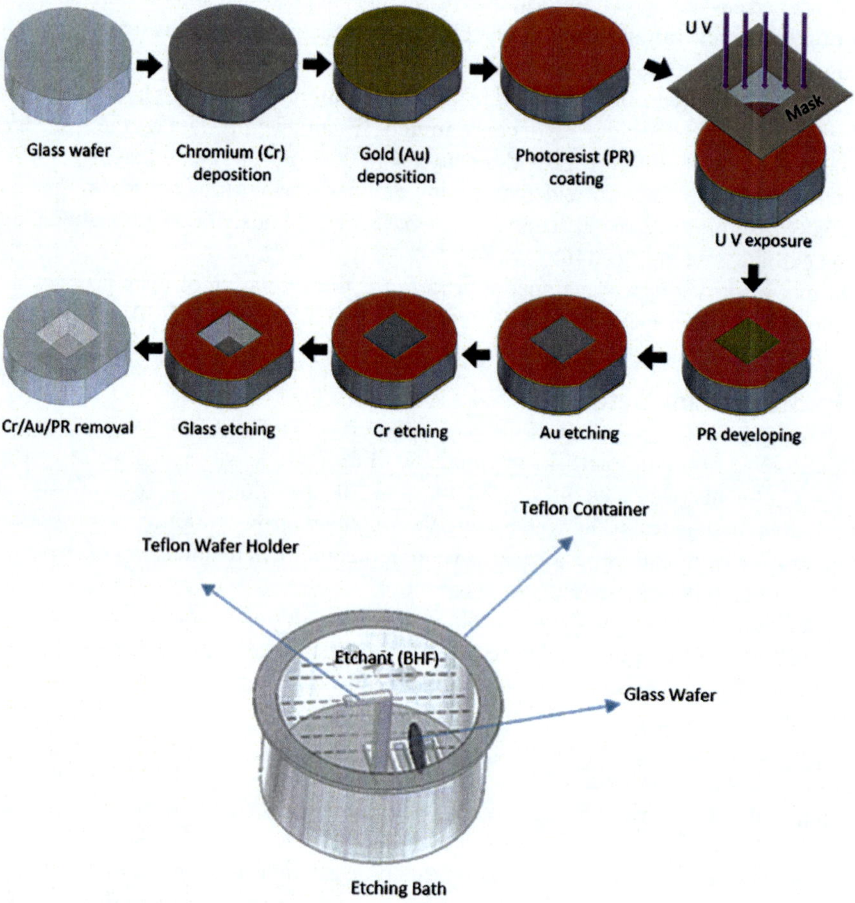

Fig. 1. Schematic representation of process steps used to form deep grooves in Borofloat glass wafer using wet bulk micromachining in BHF.

etch rate is required to enhance productivity and to reduce the overall cost of the product simultaneously. Figure 2 shows the 3-D optical image of an etched sample measured using a 3D laser scanning microscope (OLYMPUS OLS4000). It is evident from Fig. 2 that the sidewalls of the groove are very smooth. Also, there are no notching defects observed on the edges of the pattern. Moreover, it can be noticed from Fig. 2 that the mask layer (Cr/Au/photoresist) is steadily intact with the wafer surface.

Furthermore, the depth and edge profile of the etched groove is also measured using the 3D laser scanning microscope, and it is depicted in Fig. 2. Here red line signifies the depth profile of the developed groove, yellow line provides the depth and scan length measurement, whereas the blue line is the marker line which covers the scan area. It can be easily seen that the square groove possesses sharp edges with minimal lateral etching. The etch depth of the sample after 30 min of etching is measured to be 332.792 μm, which gives the etch rate 11.1 μm/min. This etch rate is two times higher than the previously reported results for soda-lime glass [19]. Moreover, the etched depth profile in Fig. 2 indicates that the bottom surface of the groove is uniform. Furthermore, the average etched surface roughness (R_a) measured using a 3D laser microscope is 0.26 μm, which is useful for the formation of various microstructures with the smooth surface finish for applications in MEMS.

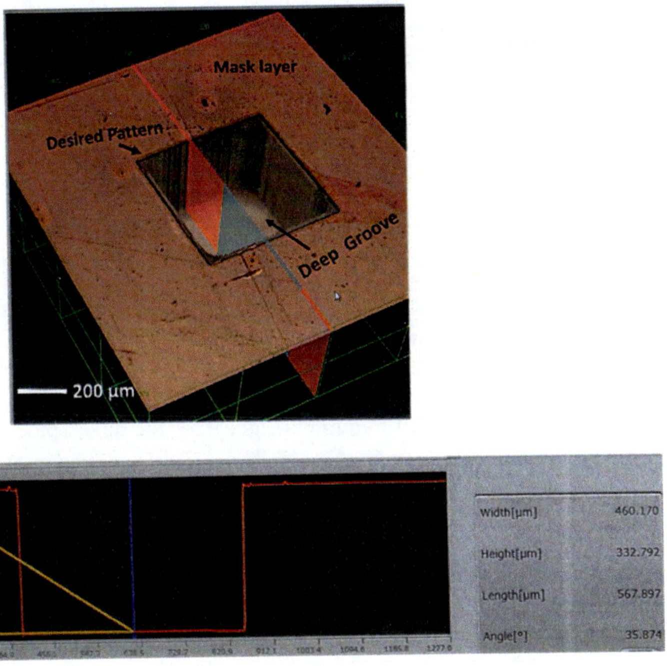

Fig. 2. Optical photograph and etch depth profile of a deep groove formed in Borofloat glass wafer using wet bulk micromachining in BHF solution (Etch time: 30 min, etch depth: 332.792 μm).

After performing glass etching, the total width of the window of the square groove is calculated and undercutting is calculated, which was measured to be around 3μm. Further analysis is performed after removing the photoresist layer and metal mask layer to observe the under etching and the results are shown in Fig. 3. The total under etching after removing of all masking layers is measured to be ~30 μm. The present study reveals that, the BHF provides excellent etching with minimal lateral etching for Borofloat Glass.

4 Conclusions

In this work, a simple method has been developed to fabricate deep grooves with a higher aspect ratio and smooth surface finish in the Borofloat glass wafer. The thin films of Cr/Au are deposited by DC sputtering and used as the main masking layer. A positive photoresist, which is used for the patterning of Cr/Au using photolithography, is worked as an additional mask layer over Cr/Au thin film. These layers served as a very good

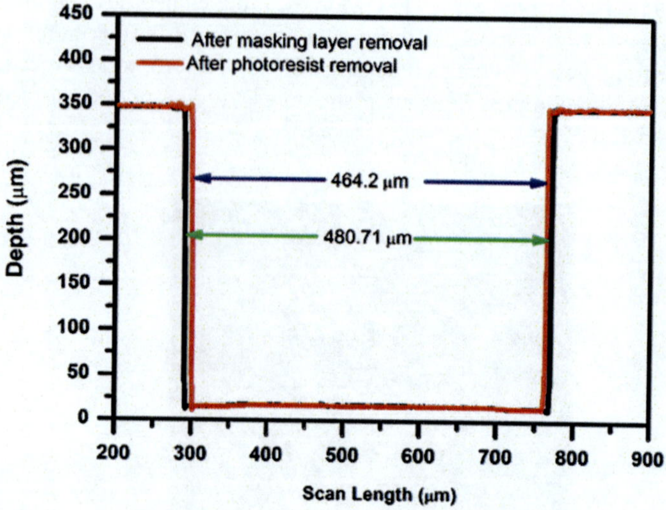

Fig. 3. Etched profile after removing photoresist and metal masking layers.

masking layer for the selective etching of glass for the development of deep grooves in Borofloat glass using BHF as an etchant. Thus, the developed approach paves a simple and efficient way to produce deep grooves with a higher aspect ratio in a low-cost and rapid manner.

References

1. Tiwari, S.K., Bhat, S., Mahato, K.K.: Design and fabrication of low-cost microfluidic channel for biomedical application. Sci. Rep. **10**(1), 1–14 (2020)
2. Moridi, M., Tanner, S., Wyrsch, N., Farine, P. A., Rohr, S.: An amorphous silicon photodiode array for glass-based optical MEMS application. In: IEEE Sensors, 25–28 Oct 2009, pp. 1604–1608 (2009)
3. Ding, G., Ma, B., Yan, Y., Yuan, W., Deng, J., Luo, J.: Through glass vias by wet-etching process in 49% HF solution using an AZ4620 enhanced Cr/Au mask. In: 2021 IEEE 16th International Conference on Nano/Micro Engineered and Molecular Systems (NEMS), 25–29 April 2021, pp. 872–875 (2021)
4. Van Toan, N., Toda, M., Ono, T.: An investigation of processes for glass micromachining. Micromachines **7**(3), 51 (2016)
5. Iliescu, C., Jing, J., Tay, F.E., Miao, J., Sun, T.: Characterization of masking layers for deep wet etching of glass in an improved HF/HCl solution. Surf. Coat. Technol. **198**(1–3), 314–318 (2005)
6. Bu, M., Melvin, T., Ensell, G.J., Wilkinson, J.S., Evans, A.G.: A new masking technology for deep glass etching and its microfluidic application. Sens. Actuators A Phys. **115**(2–3), 476–482 (2004)
7. Ashok, A., Pal, P.: Silicon micromachining in 25 wt% TMAH without and with surfactant concentrations ranging from ppb to ppm. Microsyst. Technol. **23**(1), 47–54 (2017)
8. Elwenspoek, M., Jansen, H.V.: Silicon Micromachining, vol. 7. Cambridge University Press (2004)

9. Lee, H.W., Bien, D., Badaruddin, S.A.M., Teh, A.S.: Silver (Ag) as a novel masking material in glass etching for microfluidics applications. Microsyst. Technol. **19**(2), 253–259 (2013)
10. Lee, H.W., Bien, D.C., Badaruddin, S.A.M., Teh, A.S.: Thin film Ag masking for deep glass micromachining. Electrochem. Solid-State Lett. **13**(11), H399 (2010)
11. Nagarah, J.M., Wagenaar, D.A.: Ultradeep fused silica glass etching with an HF-resistant photosensitive resist for optical imaging applications. J. Micromech. Microeng. **22**(3), 035011 (2012)
12. Pal, P., Sato, K.: Complex three-dimensional structures in Si 100 using wet bulk micromachining. J. Micromech. Microeng. **19**(10), 105008 (2009)
13. Pal, P., Sato, K.: Silicon Wet Bulk Micromachining for MEMS. Jenny Stanford Publishing (2017)
14. Tay, F.E., Iliescu, C., Jing, J., Miao, J.: Defect-free wet etching through pyrex glass using Cr/Au mask. Microsyst. Technol. **12**(10), 935–939 (2006)
15. Iliescu, C., Chen, B., Miao, J.: On the wet etching of Pyrex glass. Sens. Actuators A Phys. **143**(1), 154–161 (2008)
16. Iliescu, C., Chen, B., Miao, J.: In Deep wet etching-through 1mm pyrex glass wafer for microfluidic applications. In: 2007 IEEE 20th International Conference on Micro Electro Mechanical Systems (MEMS), pp. 393–396. IEEE (2007)
17. Grétillat, M.A., Paoletti, F., Thiébaud, P., Roth, S., Koudelka-Hep, M., de Rooij, N.F.: A new fabrication method for borosilicate glass capillary tubes with lateral inlets and outlets. Sens. Actuators A Phys. **60**(1), 219–222 (1997)
18. Corman, T., Enoksson, P., Stemme, G.: Deep wet etching of borosilicate glass using an anodically bonded silicon substrate as mask. J. Micromech. Microeng. **8**(2), 84 (1998)
19. Lin, C.-H., Chen, K.-W., Li, T.-Y.: Rapid soda-lime glass etching process for producing microfluidic channels with higher aspect ratio. Microsyst. Technol. **20**(10–11), 1905–1911 (2013). https://doi.org/10.1007/s00542-013-1980-z

Micro-piezo Actuator for Cell Lysis

Sushama Agarwalla and Suhanya Duraiswamy(✉)

Department of Chemical Engineering, Indian Institute of Technology Hyderabad, Hyderabad, India
suhanya@che.iith.ac.in

Abstract. In this study, we demonstrate a piezoelectric actuator for the lysis of bacteria using Travelling Surface Acoustic Waves (TSAWs) produced by gold IDTs (Interdigitated Transducers) fabricated on lithium niobate substrate. We are able to achieve ~99% lysis using IDTs of width 50 μm each, at 19.9 MHz. The developed piezoelectric actuator is a chemical free technology and can be used to lyse any type of bacteria as well as eukaryotic cells.

Keywords: Cell lysis · Piezoelectric actuators · TSAW

1 Introduction

Cell lysis is a process that involves destroying or breaking down the outer boundary of a cell to release its intercellular contents, such as DNA, RNA, and proteins. Lysis is an important tool in the field of molecular diagnostics of pathogens and in the development of various laboratory equipment such as point of care tests and protein purification kits. In addition, it is a critical step in several diagnostic and therapeutic efforts, since many disease biomarkers are typically found within the cell membrane [1]. The presence of a multi-layered cell wall in microorganisms requires advanced processes to lyse them and some of the currently used processes along with their disadvantages are as follows: (1) Thermal lysis which is not suitable for all organisms and can damage certain heat sensitive proteins and nucleic acids. (2) Electroporation, known for its versatility and low DNA requirements though cell damage and nonspecific transport of molecules are some of the disadvantages associated with it. (3) Mechanical lysis, the most preferred technique, which typically involves shear forces to physically damage the cell membrane. (4) Chemical lysis, the most effective strategy, though chemicals might interfere with downstream processing. Although several such techniques can lyse bacteria, a single method is not widely used for all types of bacteria like gram negative (*E. coli, Pseudomonas*) and gram positive (*Staphylococcus, Bacillus*) [2, 3].

In this study, we have developed a piezoelectric actuator for lysis using TSAWs (Travelling Surface Acoustic Waves) that are produced by gold IDTs (Interdigitated Transducers) on lithium niobate substrate. A piezoelectric actuator is a type of device that converts electrical energy into mechanical stress and the lithium niobate substrate is the piezoelectric material used in our study [4, 5]. We believe that this technique can be used for the lysis of not only bacteria but also other microorganisms and eukaryotic cells, without compromising the quality of the released cytoplasmic content.

A. K. Pandey et al. (Eds.): MAMM 2022, 126, pp. 296–300, 2023.
https://doi.org/10.1007/978-3-031-20353-4_25

2 Methodology

2.1 Sample Preparation

We use *E. coli* (DH 5 alpha) as a model for our study. We prepared an overnight bacterial culture (broth) and measured its optical density using UV-Vis spectrophotometer (UV1780, Shimadzu, Japan) at 600 nm. An absorbance 1 at OD_{600} was considered to be 10^9 CFU/ml where CFUs are colony forming units, to calculate the concentration of the culture.

2.2 IDT Fabrication

Fabrication schematic is depicted in Fig. 1. The lithium niobate wafer (128° Y-X cut, University Wafer, Inc. South Boston, USA) was first washed with IPA (Isopropyl alcohol) and blow dried, followed by sputter (AJA International, Inc. USA) deposition of 20 nm titanium and 80 nm gold layers. Then positive photoresist (S1813, Shipley, USA) was spin coated at 3000 rpm and baked for 1 min at 120 °C. 405 nm laser source was used to pattern IDT structure on the substrate. After the patterning, the positive resist was removed using MF319 developer (Dow, USA) while the gold and titanium coating were etched off to get the final IDT structures.

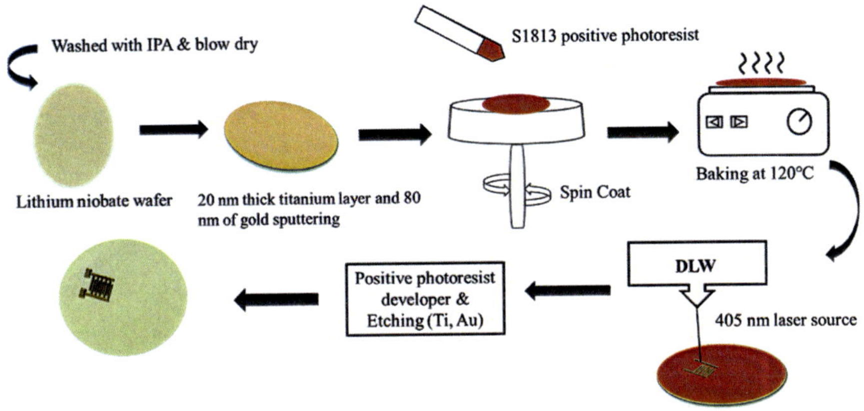

Fig. 1. Schematic of process flow for IDT fabrication

2.3 PDMS Microchannel Fabrication

The microchannel was designed in Auto CAD and printed using SLA (Stereolithography) 3D printer (Project 6000 HD, USA). The 3D printed wafer was then rinsed with IPA and further cured in UV chamber for 30 min to get the master patterns. Poly (dimethyl siloxane) (PDMS) (Sylgard 185, Dow, USA) replica was then moulded from these master patterns. The procedure in brief: PDMS base and cross-linker were mixed in 10:1 ratio and degassed for 15 to 20 min. They were then poured onto the master patterns and cured in a hot air oven at 70 °C for 3 h. The PDMS layer was then peeled off from the master patterns, inlet and outlets holes punched and bonded to the lithium niobate substrate (having the fabricated IDT patterns) using a plasma cleaner (PDC-VCG-2, Harrick Plasma, USA) as shown in the schematic in Fig. 2.

2.4 Device Setup and Operation

The individual fingers of the IDTs were interconnected using bus bars as shown in the inset in Fig. 2. We used RF signal generator (DSG815, Rigol, China) in modulation mode to send pulsed, high-frequency signal to the IDT. Samples of required concentration (10^9 CFU/ml) were pumped through the PDMS channels using at various flowrates (from 1 to 20 μl/min) using a syringe pumps (Fusion 101, Chemyx Inc., USA). They were then collected and plated on LB agar (M1151, HiMedia, India) to determine the cell viability by colony counting (using a pen and a click-counter).

3 Results and Discussion

We fabricated a hybrid device with IDTs patterned to a piezoelectric substrate and a PDMS microchannel bonded onto the same substrate as shown in Fig. 2. The IDTs were fabricated onto a 128° Y-X cut lithium niobate substrate, which were excited to generate the surface acoustic waves. Bacteria were exposed to TSAW that can cause them to lyse, while they pass through the channel. The IDT patterns fabricated on Lithium Niobate substrate have parallel finger pairs of different finger width and gap and can operate at different frequencies. We calculated the SAW wavelength generated from different structures of IDT using Eq. 1;

$$\lambda_s = 2(w_i + g_i) \tag{1}$$

where, λ_s is SAW wavelength and *wi* and *gi* are the width of the finger and gap between the adjacent fingers of IDT respectively. We fabricated IDTs of 50 μm finger spacing and varied the frequency of the signal to the IDTs. The device can produce SAW wavelength of 200 μm (using Eq. 1). The operating frequency can be calculated by using Eq. 2

$$\mathrm{f_O} = v_s/\lambda_s \tag{2}$$

where, *fo* is operating frequency and v_s is the acoustic velocity of the substrate (3980 m/s for the current substrate).

The device can be used for an applied operating frequency of 19.9 MHz calculated using Eq. 2 for SAW wavelength of 200 μm [6, 7]. We varied the flowrate of the sample through the microchannel at constant applied frequency to analyze the effect on lysis. The experimental results are summarized in Fig. 3.

We also performed a control with 'TSAW off' at all flowrates and analyzed the cell viability. We see that majority of the cells are dead (cell lysis > 97%) after being subjected to TSAW, (indicated by black bars in Fig. 3a) for all flow rates tested. Cell lysis is ~100% at flowrates <5 μl/min (Fig. 3a) since the bacteria are exposed to TSAW

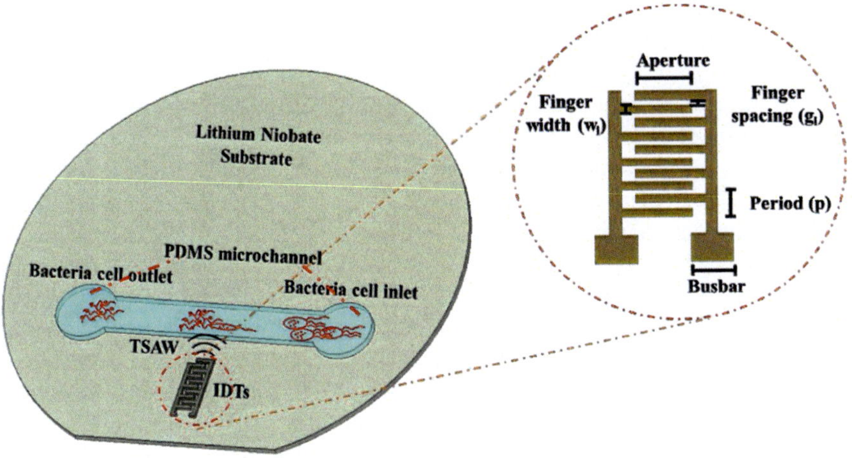

Fig. 2. Schematic of piezoelectric actuator for microbe lysis

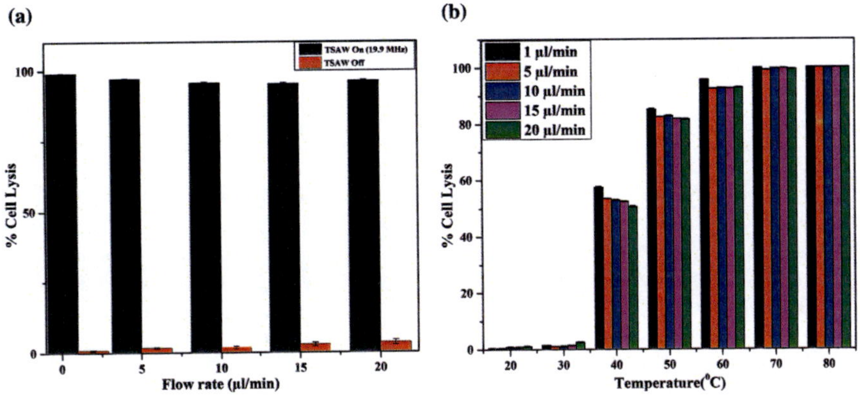

Fig. 3. Bar chart showing the lysis of bacteria (in %) plotted against (**a**) different flowrate at two different conditions TSAW on (black) and off (red) at 19.9 MHz frequency to generate 200 µm wavelength of TSAW (**b**) different temperature, without TSAW.

much longer (~13.7 s) at these flowrate. As the flowrate increases, to 20 µl/min, we observe a slight reduction in viability from ~0 to ~4% due to reduction in residence time (3.4 s).

We also observed temperature increase with TSAW application and performed an experiment to check the heating effect on lysis in the microchannel, as shown in Fig. 3b. We placed the hybrid device on a hotplate, pumped the bacteria through the microchannel, at varying flow rates, with 'TSAW off' at all conditions, and observed the cell lysis pattern. Cell lysis is low at 20 to 30 °C (~1–4%) while it increases from ~4% to 100% with increase in temperature from 30 °C to 80 °C. As observed previously, cell lysis was higher at lower flowrate and higher temperature (for example, ~53.5% at 5 µl/min and 40 °C as compared to ~50.6% at 20 µl/min and 40 °C). This again is due to the increased residence time of bacteria in the microchannel thus, their increased exposure time to high temperature, affecting their lysis.

Experiments are in progress to determine the mechanism of cell lysis. The possible reasons for lysis of cells treated with SAW is the energy transfer to the cells that are located at the pressure nodes of the SAW waves as well as the temperature. Several factors such as the frequency, the power, and the volume of the cell suspension, flowrate, distance between the channel and the IDTs, will affect the efficiency of the lysis process. The other major challenge with the cell lysis component of the device is determining the right dimensions for the IDTs. Although the exact mechanism of the interaction between the device and the bacteria is not known, it is believed that the resulting cell lysis can be performed efficiently. We are working on the numerical simulation and will be fabricating IDTs based on the simulation result and will come up with the strategies of lysis of bacteria with varying IDT designs and input operating frequencies. The current study is a proof of principle for this particular work. In conclusion, this work shows in principle that diverse bacterial population can be lysed using SAW generated by IDTs fabricated onto a piezoelectric substrate and this can be a generic lysis strategy.

Acknowledgements. The authors acknowledge PMRF for their financial support.

References

1. Shehadul Islam, M., Aryasomayajula, A., Selvaganapathy, P.R.: A review on macroscale and microscale cell lysis methods. Micromachines **8**(3), 83 (2017)
2. Wang, S., et al.: Electrochemical cell lysis of gram-positive and gram-negativebacteria: DNA extraction from environmental water samples. Electrochim. Acta **338**, 135864 (2020)
3. Mahalanabis, M., et al.: Cell lysis and DNA extraction of gram-positive and gram- negative bacteria from whole blood in a disposable microfluidic chip. Lab Chip **9**(19), 2811–2817 (2009)
4. Lu, H., et al.: Rapid additive-free bacteria lysis using traveling surface acoustic waves in microfluidic channels. Lab Chip **19**(24), 4064–4070 (2019)
5. Ha, B.H., et al.: Acoustothermal heating of polydimethylsiloxane microfluidic system. Sci. Rep. **5**(1), 1–8 (2015)
6. Packard, M.M., et al.: Performance evaluation of fast microfluidic thermal lysis of bacteria for diagnostic sample preparation. Diagnostics **3**(1), 105–116 (2013)
7. Kondoh, J., et al.: Development of temperature-control system for liquiddroplet using surface acoustic wave devices. Sens. Actuators, A **149**(2), 292–297 (2009)

Wax-Printed Microfluidic Paper Analytical Device for Viscosity-Based Biosensing in a 3D Printed Image Analysis Platform

Mrunali D. Wagh[1,2], S. B. Puneeth[3], Subhendu Kumar Sahoo[2], and Sanket Goel[2(✉)]

[1] MEMS, Microfluidics and Nanoelectronics Laboratory, Birla Institute of Technology and Science Pilani, Hyderabad 500078, India

[2] Department of Electrical and Electronics Engineering, Birla Institute of Technology and Science Pilani, Hyderabad 500078, India
sgoel@hyderabad.bits-pilani.ac.in

[3] Departments of Electronics and Communication Engineering, Presidency University, Bangalore 560064, India

Abstract. In recent years, several techniques for detecting relative viscosity in the microfluidic environment have been described as having their advantages and limitations. In this context, the present method of microfluidic viscometer provides a very feasible and cost-effective method, in terms of device fabrication detection and analysis. Herein, the hydrophobic patterns of paraffin wax on cellulose paper were created using an inexpensive laminator. In the fabricated microscale paper-based analytical device (μPAD), two points in a microchannel, with pre-defined length, were created and the average time taken for sample fluid to cover this length was leveraged to compute viscosity. These two points in the microchannel were color-coded regions of interest (ROIs) and video frames were captured during the fluid flow across the microchannel. An image processing algorithm with the greyscale alteration in the ROIs was used to compute the time taken by the fluid to cover the two points with known distances. A 3D printed platform, comprising a μPAD, a commercial onboard camera, and an LCD display, was established to implement the image processing mechanism and automate the overall workflow. This inexpensive, user-friendly and adaptable wax lamination technology is an important alternative to existing approaches, and it opens up a world of possibilities for researchers working in resource-strapped labs. Such microfluidic-based viscometers have shown to have great potential for regular monitoring the customized point-of-care (PoC) devices in a regulated and turnkey manner.

Keywords: Microscale paper-based analytical devices (μPAD) · Microfluidic viscometer · Point of care (PoC)

1 Introduction

Numerous tabletop viscometers are available on the market with different models. Automation of such devices is carried out for increasing the accuracy and precision

A. K. Pandey et al. (Eds.): MAMM 2022, 126, pp. 301–309, 2023.
https://doi.org/10.1007/978-3-031-20353-4_26

of the system which reduces the error probability. As Ostwald viscometer [1], viscometers based on the scattering light [2], also image processing based viscometer [3], capillary-based micro viscometer [4] has some drawback of large sample volume requirement, bulkiness, and least accuracy. Currently, to harness an appropriate viscometer for diverse point-of-care (PoC) applications, a suitable fabrication method with microfluidic technology is needed.

Microfluidic technology has prospered to provide accurate, cost-effective, rapid, and user-friendly solutions even in biomedical and biochemical field domains. PoC devices were established by integrating such microfluidic devices with other components to provide sample-to-answer while using a very small volume (μL-pL range). The microfluidic viscometer was employed for the viscosity estimation based on the distance (width) covered by the fluid and reference and that was tested for sensing the blends of fuel adulteration [5] and milk adulteration [6] and also for biosensing application [7].

Conventionally, several fabrication techniques, such as Hot embossing [8], photolithography [9], direct UV laser writing [10], and CO_2 laser printing technology [11], were used to fabricate such microfluidic devices. These devices have apparently been used to measure a range of physical properties, including viscosity, diffusion coefficient, and conductivity. More recently, microfluidic paper-based analytical devices (μPAD) have been developed to provide a clean-room-free, cost-effective, disposable, self-pumping, and compact rapid prototyping methodology that can be optimized for biochemical testing and diagnostics in PoC settings [11–13].

In our previous work, the 3D printed μPAD, with hydrophobic boundaries, was developed using polycaprolactone (PCL) filament for viscosity measurement by image processing approach [3, 14]. In this proposed work, the micro viscometer platform consists of wax-printed μPAD which is suitably aligned with the Raspberry Pi camera module and LCD display. This image processing-based approach is employed for the velocity measurement on the porous chromatograph platform have the advantages of the capillary effect which is helpful for pump-free operation [1]. After absorption of several biochemical reagents, various platforms for sensing employing electrochemistry were designed on paper microstrips, resulting in a decrease in complexity and a fast analysis time. The developed micro viscometer mini-platform was employed to calculate the viscosities of numerous biological samples like lysozymes and Bovine Serum Albumin (BSA).

2 Materials and Methods

2.1 Materials

Grade-I filter paper was procured from Whatman, India. The Para-film wax was procured from Unicorn Petroleum Pvt. Ltd., India. The chemicals for instance BSA, Lysosomes were purchased from Sigma Aldrich, India. Two circular provinces were made and that was using green and blue markers (Camlin, India). Ostwald viscometer was used to compute the viscosity values for benchmarking.

2.2 Fabrication of the Microfluidic Wax-based Paper-based Analytical Device (μ-PAD)

The design of a microchannel with triangular reservoir was created and printed on grade I filter paper based on our previous work [15]. In our earlier work, various microchannel characteristics, such as filter paper grade, microchannel width, hydrophobic boundary width, wax temperature, heating duration, and microchannel reservoir shape, have been optimized. Such optimized parameters were used to create the paper-based microchannel with two regions-of-interest (ROI) [1]. First, the microchannel was designed in AutoCAD software to print the desired design using CO_2 laser engraver with maximum 30 W power and 10.6 μm wavelength (VLS 3.30 from Universal Laser, AZ, USA). The design was transferred on the GSM paper through CO_2 laser engraver cut. The laser cut was carried out with optimized parameters speed and power 6.5 and 4.5% of maximum rating [16]. The design was created with a 3.98 mm channel width and a 4 mm diameter with two separate spots which were later filled with different colors (blue and green) and work as ROIs. A micro-reservoir of 7.27 mm in length was designed. The μPAD was printed on Grade-I filter paper with pore size of 40 μm. Consequently, the fabricated design was immersed into para-film molten wax for 10 s and was left to dry at the room temperature. The detailed step-wise method to fabricate μPAD is depicted in Fig. 1.

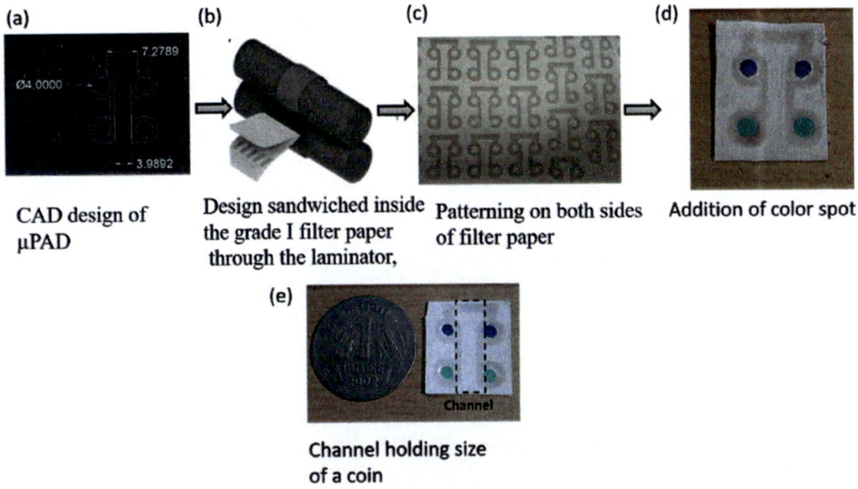

Fig. 1. Schematic representation of fabrication process of μPAD **a** CAD design of μPAD **b** Design sandwiched inside the grade I filter paper through the laminator, **c** patterning on both sides of filter paper **d** Final wax produced μPAD with two different color spot **e** Channel holding the size of a coin

Figure 1a design of the mold with the triangular channel shape has been designed in the AutoCAD software and the mold with that design was sandwiched inside the grade I paper through the laminator for the further printing of the channel boundary as shown the Fig. 1b. The channel with the wax boundary has been fabricated, and the triangular microchannel was used for the bio-fluid to flow through the system as in the captured image of Fig. 1c. A single what-man (4 cm × 3 cm) filter paper can accommodate 15 such microchannels. As presented in Fig. 1d the two-color spots were added, and the Region of interest (ROI) is developed with the detection of the color spot by the raspberry pie system. Bio-fluid (BSA and Lysozyme) was poured into the channel and the change in the paper brightness was monitored for the time measurement with the help of the ROI. Figure 1e shows the final ready-printed μPAD with the addition of the color spot and holding size of a coin.

3 Experimentation

3.1 Integrated Image Processing Platform with Raspberry Pie Module and μPAD

Figure 2a shows the photograph of an integrated image processing platform with a raspberry pie module and μPAD. The grade I filter paper (pore size 11 μm) was used as the microchannel and a wax boundary was created to encapsulate blue and green color spots. When sample fluid was flowing in the microchannel, the brightness of the paper changed, and the region of interest (ROI) is developed which is converted into a grayscale image as shown in Fig. 2b, c. Once ROI gets developed, images get converted into grayscale images. Using optimized distance and alignment between the camera and micro-pads, the image of the microchannel gets captured clearly. For the first 15 s, a reference conforming to the dry filter paper was measured, and an average of the mean brightness value of ROI is calculated. When the sample liquid moves toward the contact with the microchannel, the brightness of the paper changes. To validate the miniaturized viscosity measurement platform, two biological fluidics, bovine serum albumin (BSA) and Lysosomes were tested.

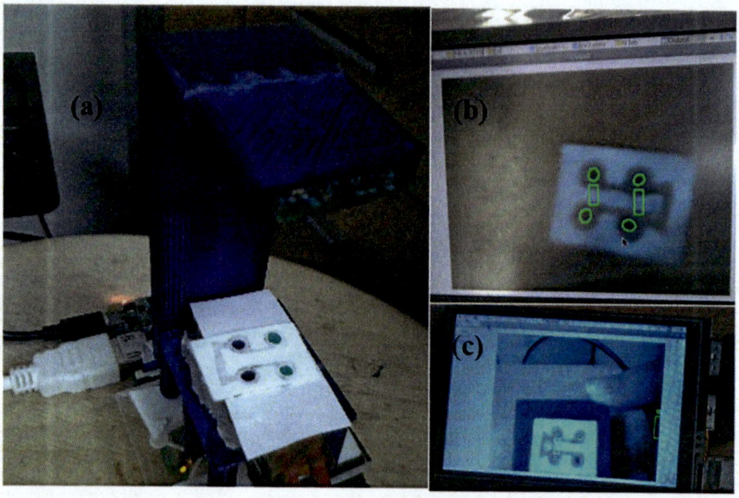

Fig. 2. Photograph of integrated image processing platform with **a** raspberry pie module and μPAD **b** developed ROI μPAD **c** captured μPAD image on the LCD display

Because the BSA and lysozymes were folded before denaturation, the change in viscosity level, which gets unfolded when heated at 80 °C, was used to confirm the denaturation of the protein. Denaturation of the biofluids causes the change in the viscosity at different concentration ranges. The measured viscosity values were benchmarked with the values measured using a benchtop Ostwald viscometer. These experiments were conducted at a room temperature of 25 °C.

3.2 Estimation of Viscosity

To prove the competency of the fabricated µPADs along with the optimized parameters, viscosity values were evaluated. Water was used as a reference fluid to measure the relative viscosity, and the time required by water to cover a fixed length in the microchannel was estimated from three trials and it was found to be 0.89 cP. The benchtop (Ostwald) viscometer was used for comparison with the viscosity values. And, the error of the presented µPADs was showing nearly equal value as stated in the Ostwald viscometer [17] and relative data is shown in Tables 1 and 2.

Table 1. Relative viscosity of BSA for various concentrations for n = 3 trial

BSA before denaturation								
Concentration mg/ml	Time taken (s)			Average time (s)	Standard deviation	Viscosity (cP)	Ostwald viscometer (cP)	Error (%)
5	21.52	21.03	21.6	21.38	0.25	1.98	2.2	−9.70
10	22.6	22.35	23	22.65	0.26	2.10	2.53	−16.8
15	24.75	24.3	24.65	24.56	0.19	2.28	2.89	−21.0
BSA after denaturation								
Concentration mg/ml	Time taken (s)			Average time (s)	Standard deviation	Viscosity (cP)	Ostwald Viscometer (cP)	Error (%)
5	12.37	12.89	12.63	12.63	0.21	1.17	1.18	−0.56
10	12.05	12.12	11.36	11.84	0.34	1.10	1.3	−15.3
15	15.6	15.36	15.45	15.47	0.09	1.43	1.45	−0.88

4 Result and Discussion

The display unit and spotted µPAD were located below the Raspberry Pi camera and integrated into a single 3D printed platform. The pattern was recognized on the µPAD, Sample volume used for the testing viscosity was 20 µL.

Table 2. Relative viscosity of Lysozymes for various concentrations for n = 3 trial

Lysozymes before denaturation								
Concentration mg/ml	Time taken (s)			Average time (s)	Standard deviation	Viscosity (cP)	Ostwald viscometer (cP)	Error (%)
5	10.9	10.23	11.32	10.84	0.45	1.00	1.02	−1.26
10	11.4	11.32	11.36	11.36	0.03	1.05	1.06	−0.41
15	11.9	11.41	11.71	11.71	0.22	1.08	1.11	−1.99
Lysozymes after denaturation								
Concentration mg/ml	Time taken (s)			Average time (s)	Standard Deviation	Viscosity (cP)	Ostwald Viscometer (cP)	Error (%)
5	12.8	12.36	12.45	12.53	0.18	1.16	1.19	−2.12
10	13.7	13.56	13.47	13.6	0.12	1.26	1.28	−1.29
15	14.5	14.36	14.89	14.59	0.22	1.35	1.35	0.42

4.1 Denaturation of BSA

Different samples of various concentrations (5, 10 and 15 mg/ml) were prepared with before and after denaturation. The viscosity of the BSA sample was monitored before and after denaturation/folding. The solution was heated at 80 °C for 30 min. The solution was cooled to room temperature before testing in μPAD to determine its viscosity. The difference in contact angle and surface tension before and after denaturation was determined, which were found to be insignificant and was so discarded through the given Lucas–Washburn Eq. (1) governs the flow observed in the microchannel.

$$L^2 = \frac{\gamma . t . \cos\theta}{\eta} \tag{1}$$

where, L represent the length of the microchannel, γ represents surface tension, θ represents contact angle of the fluid with the filter paper and η as the viscosity of the fluid (Fig. 3).

4.2 Denaturation of Lysozymes

Various concentrations of lysozymes (5, 10 and 15 mg/ml) were prepared. The heating temperature was maintained at 80 °C for 30 min. Viscosity values of all the samples before and after denaturation were measured and standard deviation for the same was observed with an least error percentage [17]. The specimens were tested in μPAD for both states (before and after denaturation), and the average time and viscosity were calculated. With the control of the experiment, the viscosity value was measured and found to be 1.00 cP, 1.05 cP, 1.08 cP after denaturation and 1.16 cP, 1.26 cP, 1.35 cP before denaturation for 5 mg/ml, 10 mg/ml, 15 mg/ml respectively. As from the measured data, denaturation proved through the significant change in the viscosity (Figs. 3 and 4).

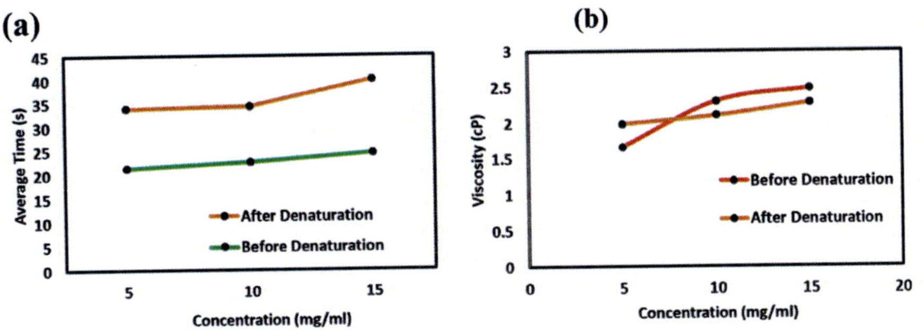

Fig. 3 (**a**) Average time by fluid to cover microchannel (**b**) Relative Viscosity of BSA before and after denaturation for different concentration of n = 3 trials

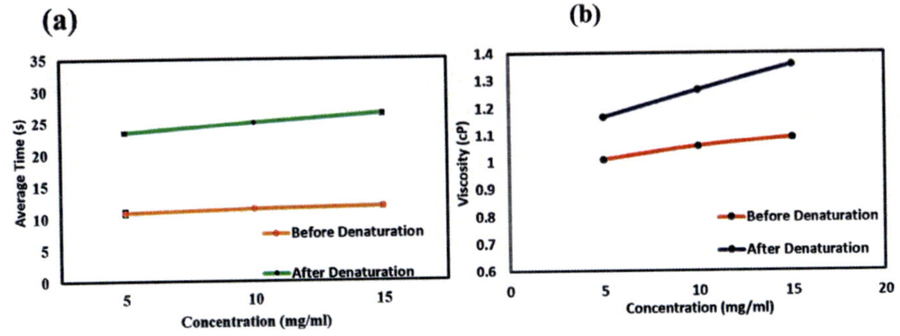

Fig. 4 (**a**) Average time taken by fluid to cover microchannel (**b**) Relative Viscosity of Lysozymes before and after denaturation for different concentrations of n = 3 trials

5 Conclusion

In this work, a wax and laminator-based fabrication method was proposed to develop microscale paper-based analytical devices (μPAD) utilizing wax mold and a laminator. Two regions of interest (RoI) with known distances were created and marked with different colors to measure the time of flight during the fluid flow using image processing analysis. A Raspberry Pi module and μPAD were integrated into the single module and used as the complete portable platform for onboard testing of the viscosity. The proposed scheme to fabricate μPAD has a several advantages, notably cost-effectiveness, ease of printing non-toxic wax patterns, and the capability to be quickly adapted for a variety of PoC testing and investigative applications. Herein, the fabrication of Microfluidic paper-based Analytical Devices (μPAD) utilizing a paraffin wax-coated molds and laminator has been demonstrated in this study. Furthermore, every μPAD will be fabricated for less than US $0.001. The versatility of the device allows it to work with not only biological samples but also fluids like oil and milk.

References

1. Rayaprolu, A., Srivastava, S.K., Anand, K., Bhati, L., Asthana, A., Rao, C.M.: Fabrication of cost-effective and efficient paper-based device for viscosity measurement. Anal. Chim. Acta **1044**, 86–92 (2018). https://doi.org/10.1016/j.aca.2018.05.036
2. Sankaran, V.P., Dubey, S., Sharma, A., Agarwal, A., Goel, S.: Stereolithographic 3D printed microfluidic viscometer for rapid detection of automobile fuel adulteration. Sens. Lett. **15**(7), 545–551 (2017). https://doi.org/10.1166/SL.2017.3848
3. Puneeth, S.B., Munigela, N., Puranam, S.A., Goel, S.: Automated mini-platform with 3-D printed paper microstrips for image processing-based viscosity measurement of biological samples. IEEE Trans. Electron Devices **67**(6), 2559–2565 (2020). https://doi.org/10.1109/TED.2020.2989727
4. Bamshad, A., Nikfarjam, A., Sabour, M.H.: Capillary-based micro-optofluidic viscometer. Meas. Sci. Technol. **29**(9), 95901 (2018). https://doi.org/10.1088/1361-6501/aace7d
5. Venkateswaran, P.S., Kashyap, D., Agarwal, A., Goel, S.: Computational analysis of a microfluidic viscometer and its application in the rapid and automated measurement of biodiesel blending under pressure driven flow. J. Comput. Theor. Nanosci. **12**(9), 2311–2317 (2015). https://doi.org/10.1166/JCTN.2015.4026
6. Venkateswaran, P.S., Sharma, A., Dubey, S., Agarwal, A., Goel, S.: Rapid and Automated Measurement of Milk Adulteration Using a 3D Printed Optofluidic Microviscometer (OMV). IEEE Sens. J. **16**(9), 3000–3007 (2016). https://doi.org/10.1109/JSEN.2016.2527921
7. Vallero, A., et al.: Memristive biosensors integration with microfluidic platform. IEEE Trans. Circuits Syst. I Regul. Pap. **63**(12), 2120–2127 (2016). https://doi.org/10.1109/TCSI.2016.2608959
8. Deshmukh, S.S., Kar, T., Som, S., Goswami, A.: Investigation of replication accuracy of embossed micro-channel through hot embossing using laser patterned copper mold. Mater. Today Proc., (2022). https://doi.org/10.1016/J.MATPR.2022.03.128
9. Acuautla, M., Bernardini, S., Gallais, L., Fiorido, T., Patout, L., Bendahan, M.: Ozone flexible sensors fabricated by photolithography and laser ablation processes based on ZnO nanoparticles. Sensors Actuators B Chem. **203**, 602–611 (2014). https://doi.org/10.1016/J.SNB.2014.07.010
10. Srikanth S. et al.: Optimization and characterization of direct UV laser writing system for microscale applications. J. Micromech. Microeng. **30**(9) (2020). https://doi.org/10.1088/1361-6439/AB92EA
11. Wagh, M.D., P.S.B., Goel, S., Sahoo, S.K.: Development of laser-induced graphene-based automated electro microfluidic viscometer for biochemical sensing applications. IEEE Trans. Electron Devices, pp 1–8 (2021). https://doi.org/10.1109/TED.2021.3107374
12. P.S.B., Kulkarni, M.B., Goel, S.G.: Microfluidic viscometers for biochemical and biomedical applications: a review. Eng. Res. Express (2021). https://doi.org/10.1088/2631-8695/abfd47
13. Noviana, E., et al.: Microfluidic paper-based analytical devices: from design to applications. Chem. Rev. **121**(19), 11835–11885 (2021). https://doi.org/10.1021/ACS.CHEMREV.0C01335/ASSET/IMAGES/MEDIUM/CR0C01335_0023.GIF
14. Shaker, M., Sundfor, E., Farine, G., Slater, C., Farine, P.A., Briand, D.: Design and optimization of a low power and fast response viscometer used for determination of the natural gas wobbe index. IEEE Sens. J. **19**(23), 10999–11006 (2019). https://doi.org/10.1109/JSEN.2019.2928479
15. Puneeth, S.B., Salve, M., Akshatha, R., Goel, S.: Realization of microfluidic paper-based analytical devices using a 3-D printer: characterization and optimization. IEEE Trans. Device Mater. Reliab. **19**(3), 529–536 (2019). https://doi.org/10.1109/TDMR.2019.2927448

16. Kothuru, A., Hanumanth Rao, C., Puneeth, S.B., Salve, M., Amreen, K., Goel, S.: Laser-induced flexible electronics (LIFE) for resistive, capacitive and electrochemical sensing applications. IEEE Sens. J. **20**(13), 7392–7399 (2020). https://doi.org/10.1109/JSEN.2020.2977694
17. Puneeth, S.B., Goel, S.: Handheld and 'Turnkey' 3D printed paper-microfluidic viscometer with on-board microcontroller for smartphone based biosensing applications. Anal. Chim. Acta **1153**, 338303 (2021). https://doi.org/10.1016/j.aca.2021.338303

Detection of Volatile Organic Compounds Using Solution Processed Organic Field-Effect Transistors

Suresh Kumar Garlapati[1,2,3](✉), Sheida Faraji[1,2], Daniel Tate[1], Aiman Rahmanudin[1], Palaniappan Valliappan[2], Alessandro Patti[2], Krishna Persaud[2], and Michael Turner[1](✉)

[1] Department of Chemistry, OMIC, University of Manchester, Manchester M13 9PL, UK
gsuresh@msme.iith.ac.in, michael.turner@manchester.ac.uk
[2] Department of Chemical Engineering, University of Manchester, Manchester M13 9PL, UK
[3] Department of Materials Science and Metallurgical Engineering, Indian Institute of Technology Hyderabad, Hyderabad 502285, India

Abstract. We report chemical sensors based on solution processed organic field-effect transistors (OFETs) for detecting volatile organic compounds (VOCs). The OFETs were prepared on flexible plastic substrates and operated at low voltages (≤3 V) with high performance. These OFETs can be used to detect a range of VOCs, including alcohols and ketones. The sensitivity of these devices depends on the polarity of the analytes, and more hydrophobic analytes with longer alkyl chains show higher sensitivities than those with shorter alkyl chains. Arrays of these devices are suitable for portable sensor applications, such as environmental monitoring, food quality monitoring, and medical diagnostics.

Keywords: OFETs · Gas sensors · VOCs · Alcohols · Ketones · Sensing mechanisms

1 Introduction

Volatile organic compounds (VOCs) are released from many sources, including industrial production, chemical handling, commercial operations, foods/food production, and even people [1, 2]. VOCs released in the environment can have different impacts; for example, toxic VOCs released from industries or nature can cause sickness and possibly death. In contrast, VOCs released from foods or food products indicate the freshness or spoilage state to the consumer; for instance, an increase in ethanol concentration indicates spoilage in certain fruits and vegetables [3, 4]. Humans also emit numerous VOCs through breath, skin, blood, urine, and faeces. These odors can indicate the metabolic state of individuals. VOCs found in exhaled breath include alcohols, ketones, aldehydes

The original version of this chapter was revised: The Author name has been corrected from "Alessanndro Patti" to "Alessandro Patti". The correction to this chapter is available at https://doi.org/10.1007/978-3-031-20353-4_33

A. K. Pandey et al. (Eds.): MAMM 2022, 126, pp. 310–322, 2023.
https://doi.org/10.1007/978-3-031-20353-4_27

[5, 6], and the concentrations of these compounds can vary depending on the health of the individual. Analysis of breath using sensors for these VOCs is attracting considerable current academic and commercial interest as it may replace invasive methods (e.g., blood tests) currently in use for certain health conditions that are time-consuming, costly, and often uncomfortable to users [7]. Therefore, developing improved VOC sensors will be important for many possible applications such as environmental monitoring, food quality monitoring, and medical diagnostics.

Gas sensors based on chemiresistors have been extensively studied over many decades and have been applied to many applications [8–12]. More recently, the use of field-effect transistors (FET) as gas sensors has been growing as they show several advantages over the use of chemiresistors. These include multiparametric outputs, higher sensitivity levels through the inherent amplification of the transistor operation and faster response/recovery times [13–17]. Although metal oxide sensor technology has been well established, lowering the process temperature and promoting selectivity in the detection of VOCs is still a challenge due to the non-selective nature of metal oxides [18]. In contrast, OFETs can be selective in the detection of different classes of VOCs, due to the interactions between the gaseous analyte and the molecular structure of the organic semiconductor (OSC). In addition, OFETs based sensors can be fabricated on flexible plastic substrates at low temperatures by inexpensive solution-based processing techniques over large areas. There is now an extensive literature on the application of OFETs as sensors to detect VOCs [19–25], and the response of OFETS to VOCs is thought to be due to an interaction between the OSC layer and the analyte molecule [26]. However, the adsorption and penetration of volatile molecules through the semiconductor layer is not well understood. This paper discusses the fabrication of OFETs using poly(3,6-di(2-thien-5-yl)-2,5-di(2-octyldodecyl)-pyrrolo[3,4-c]pyrrole-l, 4-dione)thieno[3,2-b] thiophene) (DPPT-TT) in a sensor platform for detecting alcohols and ketones, as these analytes are important VOCs in exhaled breath and are markers of food spoilage. It investigates the selectivity of the response of these sensors to analytes of increasing alkyl chain length and uses Molecular Dynamics (MD) simulations of a polymeric matrix exposing a DPPT-TT surface to a set of VOCs as a tool to understand how different analyte molecules may elicit different responses from the OFET devices.

2 Materials and Methods

Poly(vinylidene fluoride-trifluoroethylene-chlorofluoroethylene (PVDF-TrFE-CFE) terpolymer was purchased from Piezotech incorporation. The poly(methyl methacrylate) (PMMA, $M_w = 120$ kDa), benzophenone, and all required solvents were purchased from Merck(Sigma-Aldrich). Gold metallic wire for evaporation was purchased from Cooksongold precious metals limited company. The organic semiconductor (DPPT-TT) was prepared in-house using the procedure described elsewhere [27]. Bottom-gate bottom-contact OFETs were prepared on flexible polyethylene naphthalate (PEN) substrates that were bonded to glass substrates (24 × 24 mm^2) using a cool-off tape. The PEN substrates were detached from glass after fabrication. The substrates were thoroughly cleaned with isopropanol and dried at 80 °C for 5 min, followed by UV-ozone treatment for 5 min. Aluminium (≈50 nm thick) gate electrodes were deposited through a shadow mask by thermal evaporation. A 5 wt.% solution of PVDF-TrFE-CFE was prepared in dimethyl formamide (DMF). It was filtered through a 0.45μm PTFE filter and spin-coated on the Al gate electrodes. The spin-coating parameters were optimized (3000 rpm, 2 min) to obtain a thickness of 170 ± 5 nm, and the films were annealed at 110 °C for 2 h to evaporate the solvent and to crystallize the polymer. A 2 wt.% PMMA solution was prepared by dissolving the polymer with a photoinitiator (benzophenone) in anisole. It was filtered through a 0.45 μm PTFE filter and spin-coated at 3000 rpm for 1 min to obtain a thickness of 30 ± 5 nm. These films were dried at 80 °C for 10 min, treated with a UV lamp (wavelength of 254 nm) for 10 min, and annealed at 90 °C for 30 min. Gold source and drain electrodes (50 nm) were deposited using a shadow mask by thermal evaporation. The channel length and widths were fixed to 140 μm and 3600 μm, respectively. The gold electrodes were treated with a 2 mM 2,3,4,5,6-pentafluorothiophenol (PFBT) solution in isopropanol for 2 min and dried at 100 °C for 5 min. Finally, a 0.5 wt.% DPPT-TT solution in dichlorobenzene (DCB), was spin-coated over the device at 3000 rpm for 2 min. The films were annealed at 100 °C for 1 h under N_2.

Atomic force microscopy (AFM) and thickness measurements on films were performed using a non-contact mode Park AFM and a Dektak-XT profilometer, respectively. Capacitance measurements and the electrical characterization of OFETs were carried out using an Agilent LCR meter and Agilent network analyzer, respectively. Sensing experiments were carried out using an in-house sensing setup in which mass flow controllers were used to mix the vapor with synthetic air to the correct concentration. Contact angle measurements (sessile drop method) were performed using Attension Theta Lite optical tensiometer.

2.1 Model and Simulation Methodolgy

To assess the sensitivity of DPPT-TT towards ketones and alcohols at the molecular scale, Molecular Dynamics (MD) simulations were performed on a polymeric matrix exposing a DPPT-TT surface to a set of VOCs. More specifically, all-atom MD simulations were performed by employing the 2018.4 GROMACS package [28] with intra and intermolecular potentials of the GROMOS 54A7 force-field [29]. Initial configurations of 2770 PMMA monomers and 120 DPPT-TT dimers were separately created by using the PACKMOL package [30] and then compressed at 298 K and atmospheric pressure to calculate the equilibrium density, which is 1.1 g cm^{-3} for both DPPT-TT and PMMA. In particular, initial cubic simulation cells with periodic boundaries were constructed by the random distribution of polymer chains at relatively low density to avoid overlap, and the steepest descent method was employed to minimize the energy. The resulting 2.5 nm thick DPPT-TT layer was then arranged on top of a PMMA layer; also 2.5 nm thick, and the bi-layered polymeric matrix was allowed to equilibrate by further energy minimization. The leap-frog MD algorithm was used to integrate Newton's equations of motion, with a time step of 2 fs, and the stochastic velocity rescaling thermostat [31] to keep the temperature at 298 K. The cut-off for the computation of the electrostatic and dispersion interactions was set to 1.4 nm. The equilibrated polymer substrate was arranged at the bottom of a rectangular prism with square cross section and height approximately eight times larger than the base. In the resulting simulation box, 80 ketone (acetone, 2-octanone, butanone) or alcohol (methanol, butanol, octanol) molecules were distributed randomly, and their dynamics were studied over 10 ns. In order to estimate the binding free energy, ΔG, between these VOCs and DPPT-TT, the potential of mean force which harmonically restrains one molecule of each VOC at increasing distances from the DPPT-TT layer was calculated, by employing the umbrella sampling simulation technique [32].

3 Results and Discussion

The structure of the bottom-gate, bottom-contact OFET is shown schematically in Fig. 1. The devices operate at low voltage as the PVDF-TrFE-CFE layer is relatively thin and has a high-k dielectric constant of 50, leading to a high areal capacitance. The PVDF-TrFE-CFE layer is covered by a thin low-k PMMA layer to reduce the device hysteresis and to improve the surface wettability for the deposition of the OSC layer. The gold source-drain electrodes were treated with a self-assembled monolayer (PFBT) to tune the work function and facilitate better charge injection into the semiconductor.

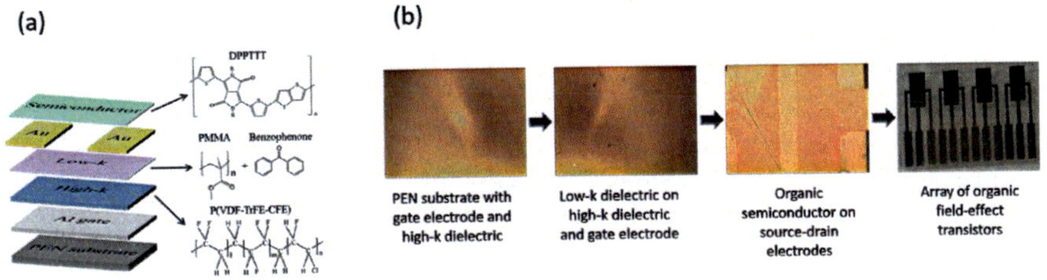

Fig. 1. (**a**) Schematic of the OFET device and the chemical structures of the individual components, (**b**) the optical microscopy images of high-k and low-k dielectrics, organic semiconductor and array of field-effect transistors.

The surface morphology of the polymer films was characterized by AFM and the images are collected in Fig. 2. The surface of the high-k terpolymer (Fig. 2a) is polycrystalline with a root mean square (RMS) roughness of 2 nm. Crystallization of the terpolymer leads to good ferroelectric relaxor behaviour with a high energy density[33]. The areal capacitance of the terpolymer films was as high as 290 ± 10 nF/cm^2 and this enables the OFET devices to operate at low operating voltages. After coating with a thin (30 nm) layer of PMMA, the AFM (Fig. 2b) shows a smoother surface topography with a reduced RMS roughness of 0.5 nm and uniform surface coverage. The capacitance of the complete dielectric stack was measured as 70 ± 5 nF/cm^2 and although this is much reduced from the original terpolymer, OFET devices using the bilayer operate at $\leq$ 3 V. After deposition of the DPPT-TT layer (Fig. 2c), the surface topography shows a homogeneous nano-grain morphology with a low RMS roughness (0.4 nm).

Electrical characterization of the resulting OFET devices was carried out, and the transfer and output curves are shown in Fig. 3. The transfer curve (Fig. 3a) indicated that the devices are p-type and normally-off with negative threshold voltages (V_T = –1.1 V), and the on/off ratio for source-drain current (I_{DS}) is 10^3. The transfer curve's hysteresis is minimal, which indicates a low trap density at the semiconductor-dielectric interface.

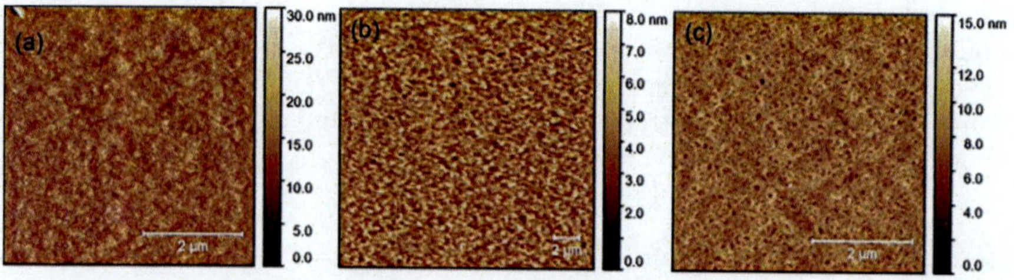

Fig. 2. AFM images of as-prepared films of (**a**) PVDF-TrFE-CFE, (**b**) PMMA and (**c**) DPPT-TT

The field-effect mobility (μ_{sat}) in the saturation region was calculated using the standard expression:

$$\mu_{sat} = \frac{2L}{WC}\left(\frac{\partial\sqrt{I_D}}{\partial V_G}\right)^2 \tag{1}$$

where L is channel length, W is width, C is capacitance, I_D is source-drain current and V_G is gate-source voltage[34]. The values for μ_{sat} extracted are in the range of 0.1–0.2 cm^2/Vs, which is comparable to values determined for similar low voltage devices reported previously [14, 35–37]. The linear relationship between V_G and $\sqrt{I_D}$ (green curve in Fig. 3a) indicates that the devices have negligible contact resistance and the mobility value is not overestimated[34]. The negligible contact resistance is due to modification of the gold source-drain electrodes with PFBT solution, which lead to the work function of the gold approaching that of the OSC. The output curves for these devices (Fig. 3b) show the expected linear and saturation regions, confirming an Ohmic contact between the semiconductor and gold electrodes. These are indications that the prepared devices have good charge injection and charge transport.

Sensing measurements were performed by exposing the OFETs to the vapors of different analytes (alcohols and ketones). Alcohols with increasing alkyl chain lengths were studied, namely methanol, ethanol, butanol, and octanol. The response of the OFET to these vapours was determined by calculating the percentage change in the measured source-drain current using the equation,

$$\%Response = \left(\frac{I_{D,initial} - I_{D,analyte}}{I_{D,initial}}\right) \times 100 \tag{2}$$

where $I_{D,initial}$ is the initial source-drain current and $I_{D,analyte}$ is the new source-drain current due to analyte exposure. The OFETs respond to all four alcohols (Fig. 4a-d), and the responses increase proportionately with concentration. Among these four alcohols, the OFETs displayed the highest sensitivity (limit of detection (LOD)≈1 ppm) towards octanol.

The sensitivities depend on the alkyl chain length with the steeper slopes observed for the plot of concentration versus response for the higher alcohols in Fig. 5a. This relationship between the alkyl chain and sensor sensitivity is consistent with the dependence of the polarity of the alcohol with the alkyl chain length, as plotted in Fig. 5b. The sensors responded rapidly to analyte exposure, but the recovery after exposure depended

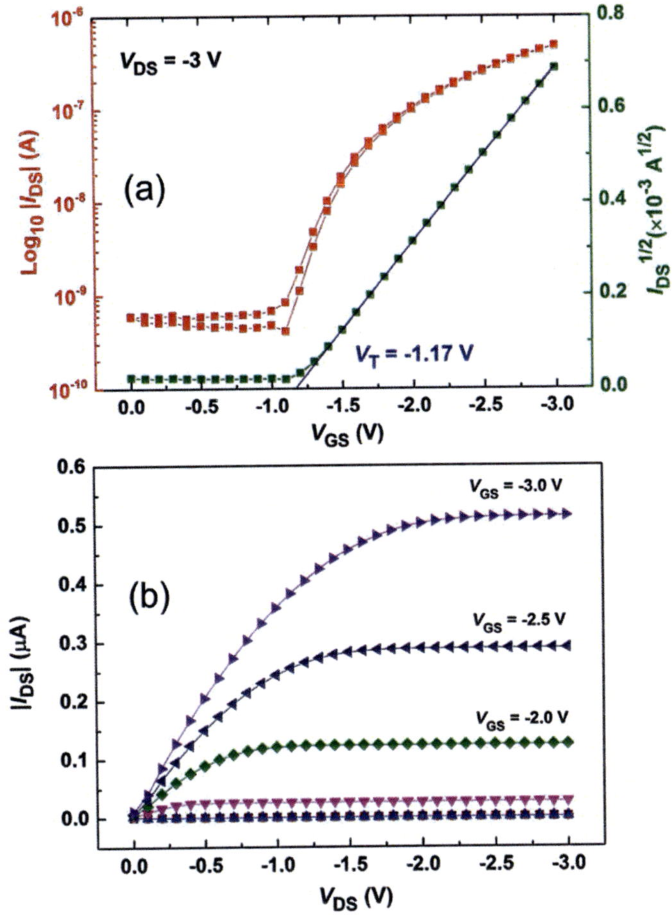

Fig. 3. (**a**) Transfer and (**b**) output curves of a typical OFET. The transfer curves were measured at a fixed source-drain voltage ($V_{DS} = -3$ V). The red curves in the transfer curve are absolute source-drain currents (I_{DS}) in the log scale (left y-axis), and the green curve is the square root of source-drain current with a linear fit (blue line) to determine threshold voltage. In the output curves, the gate-source voltage (V_{GS}) varies from 0 to -3 V with a step size of 0.5 V.

on the alcohol chain length, the devices exposed to longer chain alcohols recovered more slowly.

The OFETs also respond on exposure to the vapors of ketones such as acetone, butanone, and 2-octanone (Fig. 6a–c), and here also the percentage response increases with an increase in concentration as expected. Acetone and butanone show a rapid response and recovery to exposure, but the recovery on exposure to 2-octanone is slightly slower. Among these measured ketones, the OFETs have shown the highest sensitivity (LOD≈6 ppm) towards 2-octanone. A comparison of the sensor sensitivity (Fig. 6d) shows that the response depends on the alkyl chain length of ketones, a similar effect to that seen for the alcohol response. To further investigate the dependence of the sensitivity

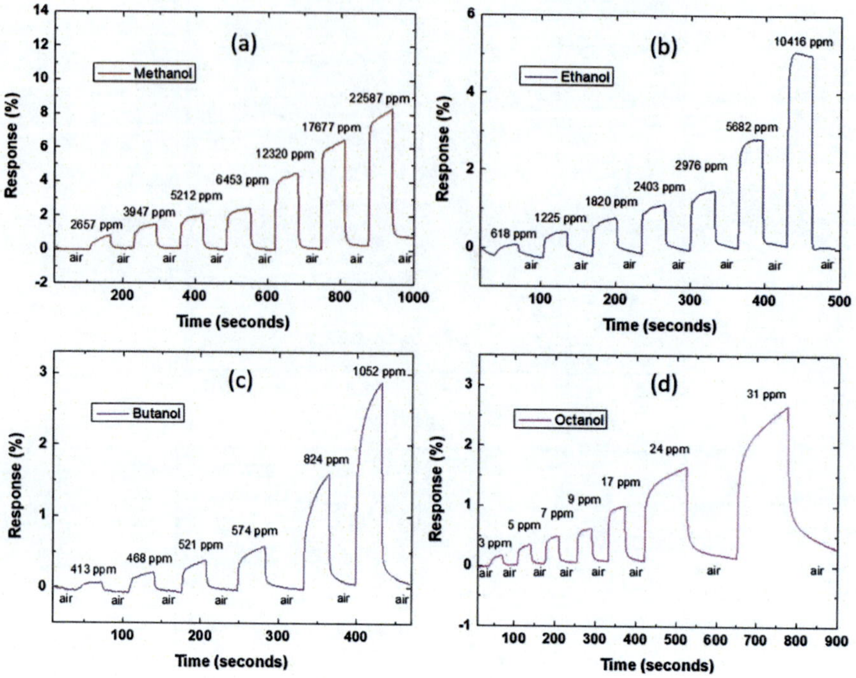

Fig. 4. Responses of OFETs to (**a**) methanol, (**b**) ethanol, (**c**) butanol and (**d**) octanol. The OFETs were exposed to alternative flows of pure air and alcohols of different concentrations as shown.

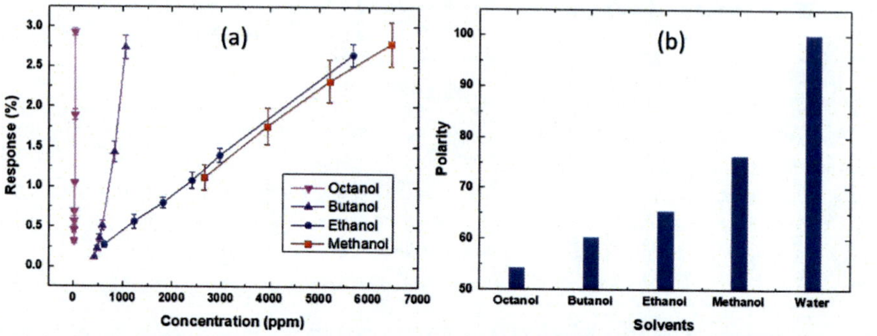

Fig. 5. (**a**) The comparison of responses of OFETs to the four alcohols and (**b**) the comparison of the relative polarities of alcohols with water set as 100. These polarity values have been taken from Smallwood [38].

on the analyte structure, the contact angle of methanol, octanol, acetone, and 2-octanone on the device surface was measured. The data collected in Table 1 indicates that the longer alkyl chain analytes have lower contact angles on this surface than shorter chain analytes. This means that the DPPT-TT surface is hydrophobic and interacts more effectively with less polar or more hydrophobic analytes. It has been reported previously that non-polar polymer semiconductors are more sensitive toward longer-alkyl chain analytes [13, 39], consistent with the observations in this paper.

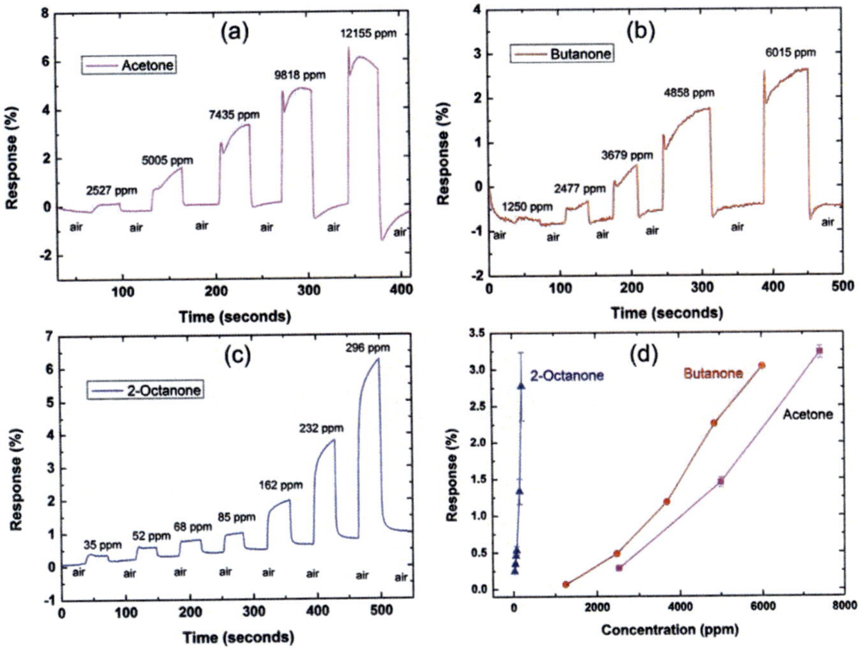

Fig. 6. Responses of OFETs to (**a**) acetone, (**b**) butanone and (**c**) 2-octanone. The OFETs have been exposed to alternative flows of pure air and ketones of different concentrations. Figure 6d shows the comparison of responses of the OFETs to the three ketones.

The OFETs show a decrease in the source-drain current on exposure to alcohols or ketones. This suggests that these devices are not charge-injection limited, and the responses are due to an interaction between the OSC layer and the analyte. These interactions are non-covalent, and can be due to hydrogen bonding, dipolar effects, and the hydrophobic effect. Absorption of polar analytes into the OSC can induce local dipoles and therefore affect the charge transport within the channel. However, it has also been shown that non-polar analytes can also induce electrical fields[40]. The dipole moments of alcohols used in this study are all the same (1.7–1.66 D) and the difference in the interaction with the OSC is presumably due to the hydrophobic nature of DPPT-TT, as shown by the contact angle measurements (Table 1) and also by the slower recovery of the responses for the longer alkyl chain analytes. It appears that the more hydrophobic and lower vapor pressure analytes partition more effectively into the OSC layer on exposure of the OFET device to the analyte vapor.

3.1 Modelling of Vapour Interactions with the Organic Semiconductor

MD simulations modeled the behavior of six VOCs (methanol, butanol, octanol, acetone, butanone, and 2-octanone) in vapour phase contact with a polymeric substrate consisting

Table 1. Contact angle measurement values of methanol, octanol, acetone, and 2-octanone on the stack of high-k (PVDF-TrFE-CFE)/low-k (PMMA)/semiconductor (DPPT-TT) films.

Analyte	Contact angle (degrees)
Methanol	23–25
Octanol	12–13
Acetone	21–23
2-Octanone	12–14

of a bottom layer of PMMA and a top layer of DPPT-TT. A typical configuration of such systems, obtained at the end of a 10 ns MD trajectory, is shown in Fig. 7.

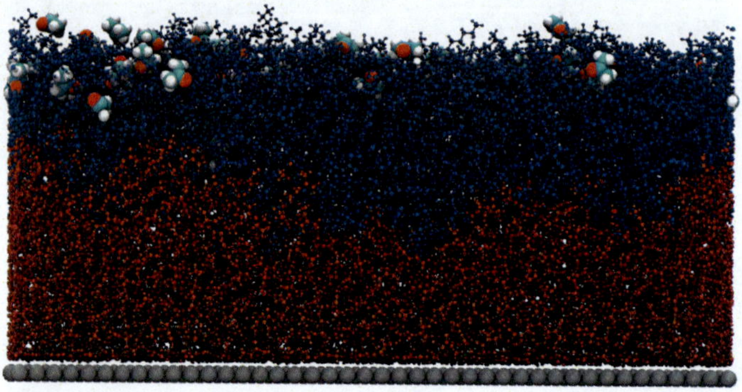

Fig. 7. Final configuration of a 10 ns MD trajectory of acetone molecules on a polymer matrix of PMMA (red layer) and DPPT-TT (blue layer). Acetone molecules have been magnified for clarity. Red, white and light blue beads in the acetone molecules represent oxygen, hydrogen and carbon atoms, respectively.

To gain an insight into the behaviour of these VOCs, their ability to diffuse into the polymer bi-layer and their two-dimensional (2D) mobility on the DPPT-TT surface were analysed. More specifically, the former was estimated by measuring the minimum distance of the analytes from the bottom of the simulation box over 10 ns, as shown in Fig. 8. The six species studied are able to diffuse into the DPPT-TT layer and can effectively reach the interface of DPPT-TT and PMMA and affect the charge transport by acting as either charge traps or interacting non-covalently at the interface between DPPT-TT and PMMA.

The analyte-polymer binding free energies were calculated by umbrella sampling (see Table 2), and these correspond to the difference between the minimum and maximum of the potential of mean force calculated at increasing distances from DPPT-TT. These results indicate that the shorter alkyl chain analyte molecules have lower binding energies. The simulation predicts that OFET sensors based on a DPPT-TT semiconductor in contact with a PMMA dielectric show a higher sensitivity towards higher alcohols

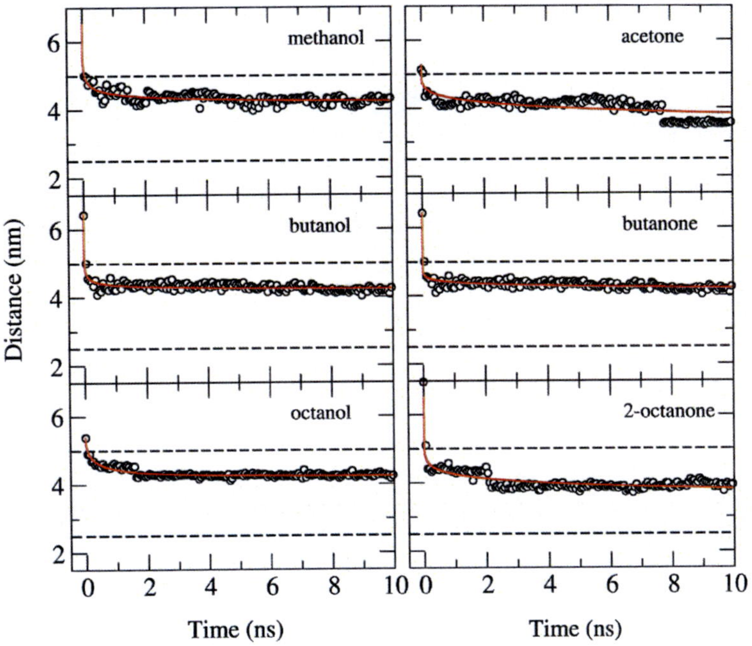

Fig. 8. Minimum distance from the bottom surface of alcohols (left column) and ketones (right column) over a 10 ns MD trajectory. The dashed lines indicate the approximate position of the PMMA/DPPT-TT and DPPT-TT/vapour interfaces at 2.5 and 5 nm, respectively. The red solid line is an exponential fit $\exp\left(-at^{b}\right)$, with a and b fitting parameters and t time.

and ketones as these show higher binding affinities to the DPPT-TT semiconductor in this system.

Table 2. Binding free energies, ΔG, between VOCs and DPPT-TT as obtained by MD simulations.

Analyte	ΔG (kcal mol^{-1})
Methanol	−0.62
Butanol	−0.57
Octanol	−0.53
Acetone	−0.80
Butanone	−0.26
2-Octanone	−0.24

4 Conclusions

OFET devices fabricated by solution processing on flexible PEN substrates with a high-k/low-k dielectric stack were capable of operation at $\leq$3 V with a high overall performance (0.1–0.2 cm^2/VS and on/off ratio of 10^3) and good ambient stability. These devices act as effective sensors for VOCs such as alcohols and ketones. The OFETS can detect ppm levels of octanol and 2-octanone and the sensitivity depends on the polarity of the analyte molecules, analytes with longer alkyl chains show higher sensitivity than shorter chain derivatives. Simulation of the analyte exposure to the organic semiconductor suggests that the sensing mechanism can be attributed to the high binding affinities of longer chain alcohols and ketones with the organic semiconductor.

References

1. Szulczyński, B., Gębicki, J.: Currently commercially available chemical sensors employed for detection of volatile organic compounds in outdoor and indoor air. Environments **4**, 21 (2017)
2. Spinelle, L., Gerboles, M., Kok, G., Persijn, S., Sauerwald, T.: Review of portable and low-cost sensors for the ambient air monitoring of benzene and other volatile organic compounds. Sensors (Basel) **17** (2017)
3. Dong, D., Zhao, C., Zheng, W., Wang, W., Zhao, X., Jiao, L.: Analyzing strawberry spoilage via its volatile compounds using longpath Fourier transform infrared spectroscopy. Sci. Rep. **3**, 2585 (2013)
4. Lonchamp, J., Barry-Ryan, C., Devereux, M.: Identification of volatile quality markers of ready-to-use lettuce and cabbage. Food Res. Int. **42**, 1077–1086 (2009)
5. Shirasu, M., Touhara, K.: The scent of disease: volatile organic compounds of the human body related to disease and disorder. J. Biochem. **150**, 257–266 (2011)
6. Fenske, J.D., Paulson, S.E.: Human breath emissions of VOCs. J. Air Waste Manag. Assoc. **49**, 594–598 (2011)
7. Parlak, O., Keene, S.T., Marais, A., Curto, V.F., Salleo, A.: Molecularly selective nanoporous membrane-based wearable organic electrochemical device for noninvasive cortisol sensing. Sci. Adv. **4**, eaar2904 (2018)
8. Bailey, A.L.P.S., Pisanelli, A.M., Persaud, K.C.: Development of conducting polymer sensor arrays for wound monitoring. Sens. Actuators, B Chem. **131**, 5–9 (2008)
9. Danesh, E., et al.: Development of a new generation of ammonia sensors on printed polymeric hotplates. Anal. Chem. **86**, 8951–8958 (2014)
10. Abraham, J.K., Philip, B., Witchurch, A., Varadan, V.K., Reddy, C.C.: A compact wireless gas sensor using a carbon nanotube/PMMA thin film chemiresistor. Smart Mater. Struct. **13**, 1045–1049 (2004)
11. Lange, U., Mirsky, V.M.: Chemiresistors based on conducting polymers: a review on measurement techniques. Anal. Chim. Act.a **687**, 105–113 (2011)
12. Pandey, S.: Highly sensitive and selective chemiresistor gas/vapor sensors based on polyaniline nanocomposite: a comprehensive review. J. Sci.: Adv. Mater. Devices **1**, 431–453 (2016)
13. Lv, A., Pan, Y., Chi, L.: Gas sensors based on polymer field-effect transistors. Sensors (Basel) **17** (2017)
14. Tate, D.J., Danesh, E., Tischler, V., Faraji, S., Majewski, L.A., Turner, M.L., Persaud, K.C.: Fully solution processed low voltage OFET platform for vapour sensing applications. In: Olfaction and electronic nose (ISOEN), 2017, ISOCS/IEEE International Symposium on, IEEE, pp. 1–3. IEEE, (Year)

15. Torsi, L., Magliulo, M., Manoli, K., Palazzo, G.: Organic field-effect transistor sensors: a tutorial review. Chem. Soc. Rev. **42**, 8612–8628 (2013)
16. Lin, P., Yan, F.: Organic thin-film transistors for chemical and biological sensing. Adv. Mater. **24**, 34–51 (2012)
17. Feng, L., et al.: Unencapsulated air-stable organic field effect transistor by all solution processes for low power vapor sensing. Sci. Rep. **6**, 20671 (2016)
18. Torsi, L., Dodabalapur, A.: Organic thin-film transistors as plastic analytical sensors. Anal. Chem. **77**, 380a–387a (2005)
19. Li, X., Jiang, Y.D., Xie, G.Z., Du, X.S., Tai, H.L., Yan, J.F., Fu, S.Q.: Ethanol gas sensors based on copper phthalocyanine thin-film transistors. Apperceiving Computing and Intelligence Analysis (ICACIA), 2010, International Conference on, IEEE, pp. 470–473. IEEE, Chengdu, China (2010)
20. Kim, Y., et al.: A composite of a graphene oxide derivative as a novel sensing layer in an organic field-effect transistor. J. Mater. Chem. C **2**, 4539–4544 (2014)
21. Sandberg, H.G.O., Bäcklund, T.G., Österbacka, R., Jussila, S., Mäkelä, T., Stubb, H.: Applications of an all-polymer solution-processed high-performance, transistor. Synth. Met. **155**, 662–665 (2005)
22. Lienerth, P., Fall, S., Lévêque, P., Soysal, U., Heiser, T.: Improving the selectivity to polar vapors of OFET-based sensors by using the transfer characteristics hysteresis response. Sens. Actuators, B Chem. **225**, 90–95 (2016)
23. Chang, J.B., et al.: Printable polythiophene gas sensor array for low-cost electronic noses. J. Appl. Phys. **100**, 014506 (2006)
24. Cheon, K.H., Cho, J., Kim, Y.H., Chung, D.S.: Thin film transistor gas sensors incorporating high-mobility diketopyrrolopyrole-based polymeric semiconductor doped with graphene oxide. ACS Appl. Mater. Interfaces **7**, 14004–14010 (2015)
25. Wedge, D.C., et al.: Real-time vapour sensing using an OFET-based electronic nose and genetic programming. Sens. Actuators, B Chem. **143**, 365–372 (2009)
26. Zang, Y., Huang, D., Di, C.A., Zhu, D.: Device engineered organic transistors for flexible sensing applications. Adv. Mater. **28**, 4549–4555 (2016)
27. Li, J., et al.: A stable solution-processed polymer semiconductor with record high-mobility for printed transistors. Sci Rep **2**, 754 (2012)
28. Oostenbrink, C., Villa, A., Mark, A.E., Van Gunsteren, W.F.: A biomolecular force field based on the free enthalpy of hydration and solvation: the GROMOS force-field parameter sets 53A5 and 53A6. J. Comput. Chem. **25**, 1656–1676 (2004)
29. Schmid, N., et al.: testing of the GROMOS force-field versions: 54A7 and 54B7 Eur. Biophys. J **40**, 843–856 (2011)
30. Martínez, L., Andrade, R., Birgin, E.G., Martínez, J.M.: PACKMOL: a package for building initial configurations for molecular dynamics simulations. J. Comput. Chem. **30**, 2157–2164 (2009)
31. Bussi, G., Donadio, D., Parrinello, M.: Canonical sampling through velocity rescaling. J. Chem. Phys. **126**, 014101 (2007)
32. Frenkel, D., Smit, B.: Understanding molecular simulation: from algorithms to applications. Elsevier (2001)
33. Chen, Y., Chen, X., Zhou, D., Shen, Q.-D., Hu, W.: Low-temperature crystallization of P(VDF-TrFE-CFE) studied by Flash DSC. Polymer **84**, 319–327 (2016)
34. Paterson, A.F., Singh, S., Fallon, K.J., Hodsden, T., Han, Y., Schroeder, B.C., Bronstein, H., Heeney, M., McCulloch, I., Anthopoulos, T.D.: Recent progress in high-mobility organic transistors: a reality check. Adv. Mater., **e1801079** (2018)
35. Sun, J., Zhang, B., Katz, H.E.: Materials for printable, transparent, and low-voltage transistors. Adv. Func. Mater. **21**, 29–45 (2011)

36. Su, Y., et al.: Low-voltage organic field-effect transistors (OFETs) with solution-processed metal-oxide as gate dielectric. ACS Appl. Mater. Interfaces **3**, 4662–4667 (2011)
37. Bharti, D., Raghuwanshi, V., Varun, I., Mahato, A.K., Tiwari, S.P.: Effect of UV irradiation on solution processed low voltage flexible organic field-effect transistors. Superlattices Microstruct. **109**, 538–544 (2017)
38. Smallwood, I.M.: Handbook of organic solvent properties. Amold, a member of the Hodder Headline Group, London, UK (1996)
39. Torsi, L., et al.: Side-chain role in chemically sensing conducting polymer field-effect transistors. J Phys. Chem. B **107**, 7589–7594 (2003)
40. Duarte, D., Dodabalapur, A.: Investigation of the physics of sensing in organic field effect transistor based sensors. J. Appl. Phys. **111**, 044509 (2012)

Thickness Dependent Chlorpyrifos Sensing Behavior of Silver Doped ZnO Nanowires

S. P. Ghosh[1,2], D. Pradhan[1], S. Pattnaik[3], and J. P. Kar[1(✉)]

[1] Department of Physics and Astronomy, National Institute of Technology, Rourkela 769008, India
karjp@nitrkl.ac.in
[2] C. V. Raman Global University, Bhubaneswar 752054, India
[3] Research Department, Manthan School, Hyderabad 502032, India

Abstract. In this research work, an attempt has been made to grow high aspect ratio zinc oxide nanowires hydrothermally by limited volume heating technique. Silver films of thicknesses ranging from 15 to 45 nm were deposited on nanowires by RF sputtering followed by thermal treatment at drive-in temperature of 475 °C in argon ambient. Morphological and structural investigation of nanowires revealed with high aspect ratio around 120 and an average length of 4.5 μm. The inter-planar lattice spacing, measured using the high-resolution transmission electron microscope is found to be around 0.26 nm, which corresponds to the wurtzite ZnO. Elemental study depicts homogeneous distribution and co-existence of Zn, O and Ag. X-ray diffraction pattern depicts the evolution of (002) orientation peak, which corresponds to the hexagonal wurtzite structure. Raman spectra displays a prominent peak at 438 cm^{-1} corresponding to the characteristic nature of wurtzite ZnO with a blue shift at higher drive-in temperature. The chlorpyrifos (CP) sensing study has shown the in/out current ratio around 68 for chlorpyrifos dosage of 3000 mg/kg at an optimized film thickness of 30 nm. The CP response and recovery time constant were found to be around 1.1 s and 1.2 s, respectively.

Keywords: ZnO · Nanowires · Silver · Chlorpyrifos · Sensor

1 Introduction

Chlorpyrifos is one of the most commonly used organophosphates (OPs), based pesticide which has a forecasted market of 576 million USD by 2025 [1, 2]. However, it is also recognized as a endocrine disruptor and may cause serious health risks even at residue level. Basically, these toxic pesticides, like chlorpyrifos (CP), can irreversibly bind to Acetylcholine esterase (AChE), a neurotransmitter and affects the nervous system for high LD_{50} [3]. The LD_{50} value for human beings for CP lies in the 2000–3000 mg/kg [4]. Farmers as well as domestic users often end up overusing toxic pesticides due to ignorance or unawareness. Thus, the reluctant intake of toxic pesticides ultimately causes severe chronic diseases and even death sometimes. Owing to its extensive use and hazardous nature, real-time detection of organophosphates like chlorpyrifos is highly

A. K. Pandey et al. (Eds.): MAMM 2022, 126, pp. 323–331, 2023.
https://doi.org/10.1007/978-3-031-20353-4_28

essential. Low dimensional semiconductors possesses very excellent material properties and therefore, currently used for chemical sensors. Among various metal-oxide based semiconductor, ZnO based one dimensional (1D) nanostructures, such as nanorods, nanowires, nanofibers, etc. have drawn significant attention in recent years towards the development of advanced micro-electromechanical (MEMS) sensors devices due to their unique charge transfer phenomena, bio-compatibility and superior electrochemical properties for the nanoscale sensing study. Moreover, metal-oxide based sensors can be fabricated using established processing techniques for semiconductors to attain miniaturized and portable devices, such as a micro-electro-mechanical systems type (MEMS-type) gas sensor. The significant advantage of ZnO over the trending carbon nanotube (CNT), is that they do not get degraded easily and are chemically and mechanically stable against oxidation and sputtering [5–7]. For the growth of ZnO nanowires three methods are widely used, such as metal organic chemical vapor deposition (MOCVD) [8], vapor liquid solid (VLS) [9] and hydrothermal methods [10]. Among these, the first two methods involve high temperature, which can only be suitable for substrates such as quartz and silicon. On the other hand, hydrothermal technique is a very cost effective approach as well as a low temperature process. In this research work a new, facile, and low cost hydrothermal approach is reported. Further, for the development of highly efficient and low cost sensors, the properties of the nanostructures need to be tailored. It has been reported that the performance of the sensors can be enhanced by UV radiation, elevated operating temperature and incorporation of metallic additives [11]. Among various metallic additives, noble metals are highly advantageous due to increase in the surface area and charge carrier concentration, which amplifies the sensor response [12]. Among various noble metallic dopants, silver (Ag) is intensively studied due to its comparatively low cost, non-toxicity, high electrical conductivity and high optical transmittance [13, 14]. It has also been reported that the physio-chemical properties of Ag incorporated ZnO varies significantly by the modulation of oxygen vacancies and crystallinity [15]. Moreover, Ag can diffuse in compound semiconductors, and thus, it can be expected that Ag atoms can residue at both the interstitial sites or the substitutional sites of the ZnO lattice [16]. Hence, the post-growth incorporation of Ag in ZnO NWs is preferred for the device fabrication. In this research, an attempt has been carried out by following a two-step post-growth process to incorporate Ag atoms in hydrothermally grown ZnO NWs by adopting pre-deposition and drive-in techniques. The pre-deposited Ag thickness dependent chlorpyrifos sensing behavior is reported in this work.

2 Experimental Work

At first, ZnO seed layer was RF sputtered on silicon [n-type, (100)] substrate. Thereafter, Ag doped ZnO NWs were processed through two steps, which include the growth of ZnO NWs by limited volume heating (LVH) method, followed by RF sputtering of the Ag film using 3 inch Ag target, (99.99%). The details of growth of ZnO NWs by LVH technique is reported in previous work [17]. The Ag film are sputtered on grow nanowires for various thickness ranging from 15 nm to 45 nm. Further, the drive-in of the sputtered atoms has been carried out at 475 °C in argon ambient. The details of the pre-deposition parameters and drive-in temperatures are mentioned in Table 1. The film thickness was measured by keeping a bare silicon substrate as reference using Stylus surface profilometer (Veeco Dektak 150). The influence of the post- heat treatment on Ag deposited ZnO NWs were investigated by field emission scanning electron microscope (FESEM:NOVA-FEI), transmission electron microscope (HRTEM, FEI TECNAI F30 G2 STWIN), and CuKα radiation ($\lambda = 1.54$ Å) of Rigaku/Ultima IV X-ray diffractometer, respectively. Besides, the elemental analysis of the samples were carried out using energy dispersive X-ray spectroscopy (EDX) technique. The microstructural properties of Ag incorporated ZnO NWs were investigated by using micro-Raman scattering system (WITecXMB3000) at an excitation wavelength of 633 nm. The composition of the chlorpyrifos was taken 5000 ppm based on the lethal dosage humans [18]. The chlorpyrifos sensing behavior of NWs were investigated by taking current-voltage (I-V) and current-time (I-t) measurement using a Keithley 6487 Picoammeter at 5V.

Table 1. Details of sputtering and thermal processing parameters for Ag incorporated ZnO NWs samples

Step—I: Sputtering (pre-deposition)		Step—2: Thermal treatment (drive-in)	
RF Power	25 W	Annealing Temperature	475 °C
Working pressure	4×10^{-3} mbar		
Substrate temperature	Room temperature		
Substrate rotation	10 rpm	Duration	1 h
Film thickness	15 nm, 30 nm, and 45 nm	Processing ambient	Argon

3 Results and Discussion

Figure 1(a–c) shows the FESEM micrographs of Ag incorporated ZnO NWs for silver film thickness ranging 15–45 nm at 475 °C. The FESEM micrographs depict s the homogeneous distribution of NWs with an average length measured to be around 4–4.5 μm. Furthermore, the morphology is also investigated from the TEM micrographs as shown in Fig. 2(a–c), which depict the with an average diameter around 35 nm and aspect ratio nearly 120. The rough surface of the NWs arises due to the larger ionic radii of Ag^{2+} (0.115 nm) in comparison to Zn^{2+} (0.074 nm). HRTEM images shown in Fig. 2(d–f) display well-resolved crystalline fringes with an interplanar spacing around 0.26 nm. It depicts the primary growth of the NWs is along the (002) plane of wurtzite ZnO. Further, the EDX analysis confirms the presence of Zn, O and Ag in the NWs with an enhancement in the Ag content at higher drive-in temperature, which is due to the availability of adequate thermal energy that facilitates the thermal diffusion of Ag atoms into the NWs lattice. Besides, the elemental mapping images shown in Fig. 3 depict the homogeneous distribution and co-existence of Zn, O and Ag, after the incorporation of Ag dopants.

Figure 4(a–c) shows the XRD pattern of ZnO NWs thermally processed at 475 °C for the Ag film thickness of 15–45 nm. The XRD pattern depicts the (002) orientated

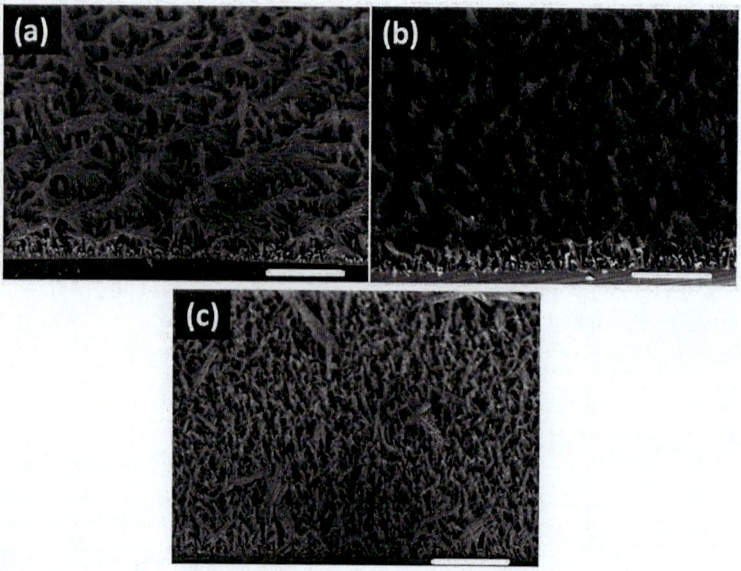

Fig. 1. FESEM micrographs of thermally (475 °C) processed ZnO NWs for Ag film thickness of (**a**) 15 nm, (**b**) 30 nm, (**c**) 45 nm. The scale bar represent 5 μm.

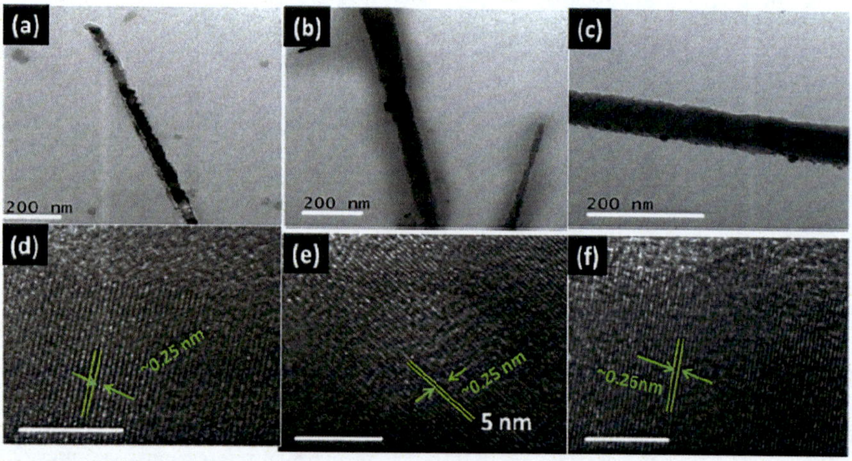

Fig. 2. Bright field TEM and HRTEM images of thermally (475 C) processed ZnO NWs for Ag film thickness of (**a**) 15 nm, (**b**) 30 nm, (**c**) 45 nm. The scale bar represent 5 nm.

crystalline peak, which corresponds to the hexagonal wurtzite structure (JCPDS file 79–0208) [19]. The variation of X-ray diffraction peak intensity is related to the combined effect of both thermal annealing and Ag incorporation. The reduction in the intensity of (002) peak intensity for higher film thickness of 45 nm, may be due to the formation of defects, which can be related to the larger ionic radius of Ag (1.26 Å) in comparison to Zn (0.74 Å) [20]. The marginal peak shift towards a lower diffraction angle, for higher

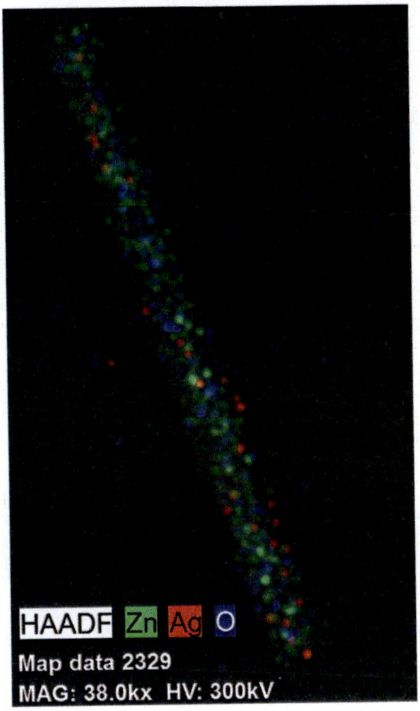

Fig. 3. Elemental mapping images of thermally (475 °C) processed ZnO NWs for Ag film thickness of 30 nm.

film thickness, depicts the substitutional incorporation of Ag [21, 22]. The shift in peak position is due to the stress arisen from the expansion of the unit cell as a result of the substitutional incorporation of larger atoms. The non-appearance of Ag related peaks in the XRD patterns depicts the crystallization of Ag atoms has not been taken place at 475 °C. Ahmed et al. have reported that the interstitial incorporation takes place for the higher concentration of Ag [23].

Figure 5 depicts the Raman spectra of ZnO NWs thermally treated at 475 °C for Ag film thickness of 15–45 nm. A prominent peak at 438 cm^{-1}, corresponding to characteristic peak of wurtzite ZnO, has been observed with a blue shift for higher film thickness. An enhancement in the intensity of E_2(H) mode along with the blue shift is may be due to the generation of compressive stress in the crystal [23]. This compressive stress may be arises due to more number of induced oxygen defects due to incorporated Ag dopants, which attributes to the E_1(LO) mode around 578 cm^{-1} [23]. The peak observed at 230 cm^{-1}, corresponds to the local vibrational modes of incorporated Ag dopants [23]. Wang et al. have reported a possible physical mechanism behind the appearance of local vibrational modes, which is the incorporation of dopant impurities into the crystal lattice [24]. The other Raman active vibrational modes are also observed at 334 cm^{-1} and 380 cm^{-1}, which corresponds to the multiphonon E_2(H)–E_2(L) and A_1(TO) modes of vibration, respectively [24].

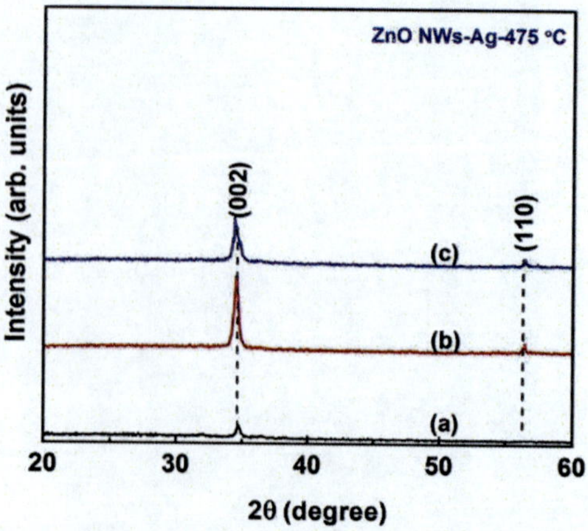

Fig. 4. XRD pattern of ZnO NWs thermally processed at (**a**) 15 nm, (**b**) 30 nm, (**c**) 45 nm thermally treated at 475 ºC.

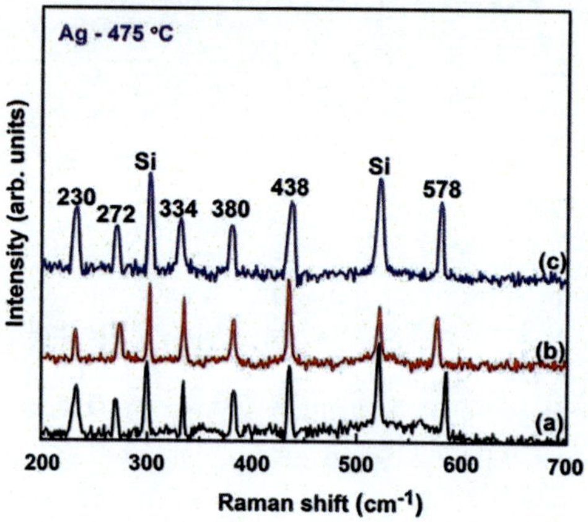

Fig. 5. Raman spectra of ZnO NWs for Ag film thickness of (**a**) 15 nm, (**b**) 30 nm, (**c**) 45 nm thermally processed at 475 °C.

Figure 6 shows the chlorpyrifos (3000 mg/kg), sensing of thermally (475 °C) response for various Ag film thickness ranging from 15 nm to 45 nm. Higher sensor response has been observed for an optimum film thickness of 30 nm with CP current in/out ratio found to be around 68. Besides, the response and recovery time of 1.1 s and 1.2 s, respectively have been obtained. The improvement in sensor response of Ag incorporated NWs is ascribed to the fact that silver acts as an acceptor in ZnO by substitution of Zn^{2+}. As a result, net donor density is decreased, which in turn facilitates the adsorption and desorption of oxygen at the NWs surface. The reason for improved response is attributed to the rise in ionic strength, which in turn boosts up the interaction at the adsorption sites. However, the reduction in CP in/out ratio for higher film thickness is might be due to the incomplete diffusion of larger size Ag dopants into the NWs lattice, thus a reduced response is observed due to the high aspect ratio and intrinsic defects. The incorporation of Ag in ZnO NWs is carried out by RF magnetron sputtering followed by thermal treatment. For higher thickness of Ag, all of the RF sputtered Ag atoms are not driven from the NWs surface into the host ZnO NWs due to larger ionic radii of Ag dopants as compared to Zn ions. As a result, the undriven amount of Ag atoms tends to agglomerate and cover the NWs surface up to some extent and act as a barrier for CP interaction.

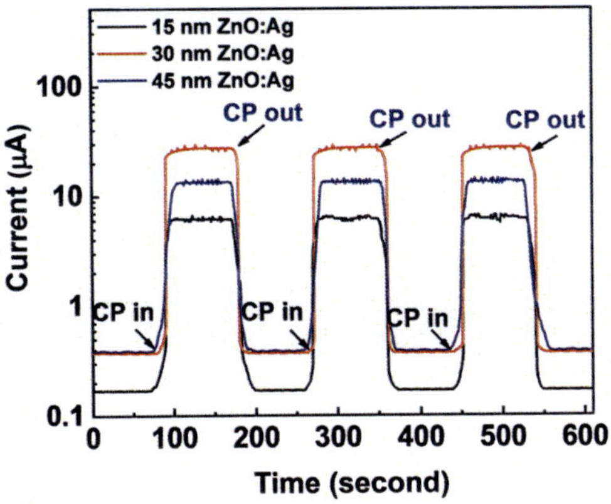

Fig. 6. Chlorpyrifos sensing response of thermally 475 °C processed ZnO NWs for Ag film thickness of (**a**) 15 nm, (**b**) 30 nm, (**c**) 45 nm for dosage of 5000 ppm.

4 Conclusion

In the this research work, the morphological and structural properties of Ag incorporated ZnO NWs were investigated for various Ag film thickness, thermally processed at a drive-in temperature of 475 °C in argon ambient. FESEM images revealed the homogeneous distribution of NWs with an average length measured to be around 4–4.5 μm. Further, the interplanar spacing estimated from the HRTEM micrographs (0.26 nm) depicts the incorporation of Ag dopants did not modulate the wurtzite structure of ZnO. Elemental mapping images further depicts the presence of Ag dopant in the host lattice. XRD analysis has revealed formation of c-axis oriented (002) peak. The Raman spectrum displays a significant peak at 438 cm^{-1}, which is associated with the $E_2(H)$ phonon mode. The sensing studies have been carried out for the NWs for CP lethal dosage of 3000 mg/kg. The NWs CP treated at 475 °C, for an intermediate Ag film thickness of 30 nm has shown better sensitivity with in/out current ratio of 68 times. Thus, the obtained result suggested that the incorporation of silver dopants into ZnO NWs can be used for the development of miniaturized, low-cost highly efficient CP sensors.

Acknowledgement. This research work was supported by the Department of Science and Technology (DST), India sponsored Indo-Korea project (INT/Korea/P-16/2013).

References

1. Cho, I.C., Kang, K.N., Yang, D.J., Yun, J.H., Park, I.K.: Localized liquid-phase synthesis of porous SnO2 nanotubes on mems platform for low-power, high performance gas sensors. ACS Appl. Mater. Interfaces **9**, 27111–27119 (2017)
2. Jagadale, S.B., Patil, V.L., Vanalakar, S.A., Patil, P.S., Deshmukh, H.P.: Preparation, characterization of 1D ZnO nanorods and their gas sensing properties. Ceram. Int. **44**, 3333–3340 (2018)
3. Wen, X., Wu, W., Ding, Y., Wang, Z.L.: Seedless synthesis of patterned ZnO nanowire arrays on metal thin films (Au, Ag, Cu, Sn) and their application for flexible electromechanical sensing. J. Mater. Chem. **22**, 9469 (2012)
4. Liu, S., Zheng, Z., Li, X.: Advances in pesticide biosensors: current status, challenges, and future perspectives. Anal. Bioanal. Chem. **405**, 63 (2013)
5. Li, Z., Wang, Y., Ni, Y., Kokot, S.: Unmodified silver nanoparticles for rapid analysis of the organophosphorus pesticide, dipterex, often found in different waters. Sensors Actuators, B Chem. **193**, 205–211 (2014)
6. He, Y., Xu, B., Li, W., Yu, H.: Silver nanoparticle-based chemiluminescent sensor array for pesticide discrimination. J. Agric. Food Chem. **63**, 2930 (2015)
7. Nanda Kumar, D., Rajeshwari, A., Alex, S.A., Sahu, M., Raichur, A.M., Chandrasekaran, N., Mukherjee, A.: Developing acetylcholinesterase-based inhibition assay by modulated synthesis of silver nanoparticles: Applications for sensing of organophosphorus pesticides. RSC Adv. **5**, 61998 (2015)
8. Lupan, O., et al.: Silver-doped zinc oxide single nanowire multifunctional nanosensor with a significant enhancement in response. Sensors Actuators, B Chem. **223**, 893 (2016)
9. Zhong, K., Mao, Y., Sun, X., Liang, C., Liu, P., Tong, Y.: Electrochemical and optical properties of ZnO nanowires modified with Ag nanoparticles by electrodeposition. J. Electrochem. Soc. **159**, 161 (2012)
10. Zhou, X., Lee, S., Xu, Z., Yoon, J.: Recent progress on the development of chemosensors for gases. Chem. Rev. **115**, 7944 (2015)
11. Wang, T., Bristowe, P.D.: Controlling Ag diffusion in ZnO by donor doping: A first principles study. Acta Mater. **137**, 115 (2017)
12. Khanna, V.K.: Nanoparticle-based sensors. Def. Sci. J. **58**, 608 (2008)
13. Bootharaju, M.S., Pradeep, T.: Understanding the degradation pathway of the pesticide, chlorpyrifos by noble metal nanoparticles. Langmuir **28**, 2671 (2012)
14. Ahmad, E., Ahmed, M.: Ahmad, The excellent photocatalytic performances of silver doped ZnO nanoparticles for hydrogen evolution. SN Appl. Sci. **1**, 327 (2019)
15. Jeong, S.H., Park, B.N., Lee, S.B., Boo, J.H.: Structural and optical properties of silver-doped zinc oxide sputtered films. Surf. Coatings Technol. **193**, 340 (2005)
16. Yildirim, Ö.A., Unalan, H.E., Durucan, C.: Highly efficient room temperature synthesis of silver-doped zinc oxide (ZnO: Ag) nanoparticles: Structural, optical, and photocatalytic properties. J. Am. Ceram. Soc. **96**, 766 (2013)
17. Ghosh, S.P., et al.: J Mater Sci: Mater Electron **26**, 7860–7866 (2015)
18. Lee, F., Lee, L.Y., Chang, Y.P.: Structural and electrical properties of silver-doped zinc oxide nanorods array. Adv. Mater. Res. **1101**, 164 (2015)
19. Poirier, L., et al.: Enzymatic degradation of organophosphorus insecticides decreases toxicity in planarians and enhances survival. Sci. Rep. **7**, 19–21 (2017)
20. Ahmad, I., et al.: Synthesis and characterization of silver doped ZnO nanoparticles for hydrogen production. J. Ovonic Res. **14**, 415 (2018)
21. Yaqoob, F., Huang, M.: Diffusion of ion implanted indium and silver in ZnO crystals. MRS Online Proc. Libr. **1394**(1), 101–107 (2011). https://doi.org/10.1557/opl.2012.696

22. Tsai, Y.T., et al.: High sensitivity of NO gas sensors based on novel Ag-doped ZnO nanoflowers enhanced with a UV light-emitting diode. ACS Omega **3**, 13798 (2018)
23. Chen, R., Zou, C., Bian, J., Sandhu, A., Gao, W.: Microstructure and optical properties of Ag-doped ZnO nanostructures prepared by a wet oxidation doping process. Nanotechnology **22**, 105706 (2011)
24. Lee, S., et al.: Solution-processed Ag-doped ZnO nanowires grown on flexible polyester for nanogenerator applications. Nanoscale **5**, 9609 (2013)
25. Muchuweni, E., Sathiaraj, T.S., Nyakotyo, H.: Effect of annealing on the microstructural, optical and electrical properties of ZnO nanowires by hydrothermal synthesis for transparent electrode fabrication. Mater. Sci. Eng. B Solid-State Mater. Adv. Technol. **227**, 68 (2018)

Investigation of $AuCl_3$ Doped MoS_2 Based IR Detector with the Variation of Annealing Temperature

Diana Pradhan[1,2], Anurag Gartia[2], Kiran K. Sahoo[2], Surya P. Ghosh[2,3], and Jyoti P. Kar[2,4,4](✉)

[1] ITER, Siksha 'O' Anusandhan Deemed to Be University, Bhubaneswar, Odisha 751030, India
[2] Department of Physics and Astronomy, National Institute of Technology Rourkela, Rourkela 769008, India
karjp@nitrkl.ac.in
[3] C V Raman Global University Bhubaneswar, Odisha 751030, India
[4] Centre for Nanomaterials, National Institute of Technology, Rourkela 769008, India

Abstract. Low dimensional transition metal dichalcogenides (TMDC) are under current investigation for the fabrication of portable and low cost IR detectors. Among the vast family of TMDC, molybdenum disulphide (MoS_2) has been considered as a potential candidate for the next-generation IR detectors due to its outstanding properties. In this research, MoS_2 thin films were grown by rapid thermal process at an elevated temperature for a shorter duration with the flow of inert (argon) and reducing (hydrogen) gas. In order to modulate the post-growth characteristics of MoS_2 films, spin coated $AuCl_3$ layer was used as a source of p-type dopant. Thereafter, rapid thermal annealing of the AuCl3 treated MoS2 films was carried out at various temperatures ranging from 200 °C to 500 °C. FESEM images have revealed the modulation in the morphology beyond 300 °C. XRD studies depicted the appearance of (002) characteristic peak, whose intensity was found to be increased with annealing temperature up to 300 °C. Raman spectra have shown the evolution of $E^1{}_{2g}$ and A_{1g} active mode peaks. The Raman peak position was found to be shifted with the variation in post-treatment annealing temperature. Using Mott-Schottky analysis, the dependence of carrier concentration of MoS_2 films and built-in potential across MoS_2/Si heterojunction on the annealing temperature were estimated. The ideality factor of MoS_2/Si heterojunction was investigated from current-voltage characteristics. The IR detection behaviour of MoS_2/Si heterojunction was studied.

Keywords: Molybdenum disulfide · Thin films · Heterojunction · Gold chloride solution · Annealing temperature · IR detection

1 Introduction

Near infrared (NIR) light detectors are widely used in CCTV, virtual reality headsets, 3D printing, optical interconnects, biometrics for iris recognition and finger-vein reader,

A. K. Pandey et al. (Eds.): MAMM 2022, 126, pp. 332–341, 2023.
https://doi.org/10.1007/978-3-031-20353-4_29

defence and in medical diagnosis and treatment [1–6]. Hence, there is a need to detect NIR light. Recently, microelectronic devices are realized using TMDC due to their reduced dimensions, improved functionalities as compared to conventional semiconductors [7]. Among various types of TMDC, molybdenum disulphide (MoS_2) has drawn tremendous attention for applications in MEMS to be used as photodetector, photovoltaic, sensor, transistor and memristor due to tunable bandgap, effective electron mobility varying from 1 to 480 $cm^2V^{-1}s^{-1}$, interplanar and interlayer resistivity is found to be 10 and 2000 Ω.cm, respectively [8–12]. Moreover, MoS_2 has better ability to absorb light from a wide range of spectrum (visible to IR) in comparison to Si and GaAs [13, 14]. Although an extensive study has been carried on MoS_2 based photodetectors, there exists a limitation in the performance of these devices due to poor quality p-n heterojunction. Hence, the MoS_2 based detectors have been considered as the potential candidate for numerous applications, which demands for monolithic integration with silicon for achieving low cost, high performance miniaturized device. Researchers have shown higher photoresponse for MoS_2 heterojunction based detectors [15–18]. However, those suffer from complexity and reproducibility in the fabrication steps. Hence, an attempt has been made to fabricate MoS_2/Si heterojunction. In order to fabricate a p-n heterojunction, the prerequisite is the choice of an appropriate p-type dopant and doping methods, which will pave way for p-type MoS_2 to be used in Si compatible heterojunctions.

Chemical doping is being highly adopted for modifying the electronic properties of MoS_2 because of its numerous advantages like, the structure of host material remains unaltered, the electron acceptor dopants create states near the valence band and reduce the contact resistance at electrode/MoS_2 interface. In comparison to other chemical doping techniques, spin coating is easier, simpler, cheaper and reliable chemical deposition process. It allows preparation of molecular level homogeneous film and easy formation of complex hybrid composites. Several precursors such as, gold (III) chloride ($AuCl_3$) [19], gold nanoparticles (Au NPs) [20], Tris(4-bromophenyl)ammoniumyl hexachloroantimonate [21], 7,7,8,8-tetracyanoquinodimethane (TCNQ) [22], polyethylene imine (PEI) [23], tetracyanoethylene (TCNE) [24] have been adopted to synthesize p-type materials. Among these, $AuCl_3$ possesses large reduction potential (~1.4 V) and has strong charge transfer (electron extraction) efficiency in low dimensional materials [25]. In order to explore the $AuCl_3$ doping in TMDC, several research groups have considered mechanically exfoliated MoS_2 film as the host material [26, 27], which is not suitable for IC processing. Therefore, MoS_2 thin films grown by rapid thermal technique in this research work, which has low thermal budget. Thereafter, an attempt has been made to study the effect of post-treatment annealing on morphological and microstructural properties of $AuCl_3$ treated MoS_2 films. In addition, IR detection study has been carried out by taking current-time measurement of MoS_2/Si heterojunction.

2 Experimental Work

Molybdenum (Mo) thin films and sulfur (S) films were deposited on Si by RF sputtering and thermal evaporation techniques, respectively. S/Mo/Si substrates were placed in such a way that both the S films on each substrate touch each other. The details of MoS_2 thin film growth, using rapid thermal process (RTP), is discussed elsewhere [28].

Thereafter, post-deposition treatment of $AuCl_3$ solution on RTP processed MoS_2 films were carried out with 6 number of coats using spin coating technique. Thereafter, rapid thermal annealing of 15 mM $AuCl_3$ solution treated MoS_2 films were carried out at different temperatures (200–500 °C). The annealing was done in vacuum ambient for 10 min at a ramp rate of 20 °C/s. The methodology adopted in this study is shown schematically in Fig. 1.

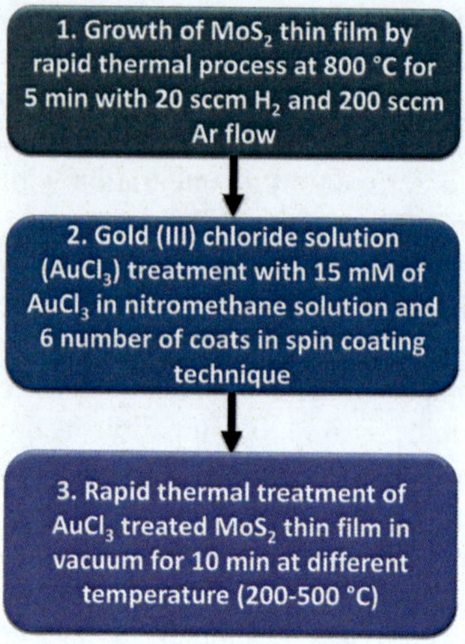

Fig. 1. Process flow for AuCl3 solution treatment of MoS_2 thin films

Morphological studies of MoS2 thin films were carried out using Field Emission Scanning Electron Microscope (FESEM, Nova NanoSEM 450). Atomic Force Microscopy (AFM, Park Systems) was used to study the RMS roughness of MoS_2 thin films. The microstructural characterizations were performed using X-ray diffraction (XRD, Rigaku Ultima IV) system, whereas Raman spectra were obtained from the Witec Alpha 300 Raman Spectrometer at excitation wavelength of 532 nm. Prior to the electronic measurements, aluminum (Al) film was thermally evaporated to fabricate electrodes across the MoS_2/Si heterojunction. Capacitance-voltage (C-V) characteristics of the heterojunction was investigated by Agilent E4980A LCR meter, Keithley 6487 Picoammeter/Voltage source was used to study the current-voltage (I-V) and current-time (I-t) measurements of MoS2/Si heterojunction.

3 Results and Discussion

The morphological characteristics of MoS_2 films were investigated from FESEM images as shown in Fig. 2. As observed from SEM images, the surface morphology is not significantly varied upto 200 °C. This suggests that the thermal energy, provided for a short period, is not sufficient to modulate the surface morphology such as grain growth. Beyond 300 °C, the modulation in $AuCl_3$ treated MoS_2 film morphology can be attributed to the enhancement in the incorporation of Au particles into the MoS_2 films through the grain boundaries. At higher annealing temperature the thermal energy is sufficient for the grain growth, as a result the morphology changes from smooth to coarse. The calculated RMS roughness of the $AuCl_3$ treated MoS_2 thin films at various annealing temperatures are shown in Fig. 3. The decrease in the roughness with the increase in temperature upto 300 °C can be attributed to the incorporation of Au into MoS_2 films and lateral spreading of Au particles with increased density. The further rise in annealing temperature does not significantly modulate the surface roughness.

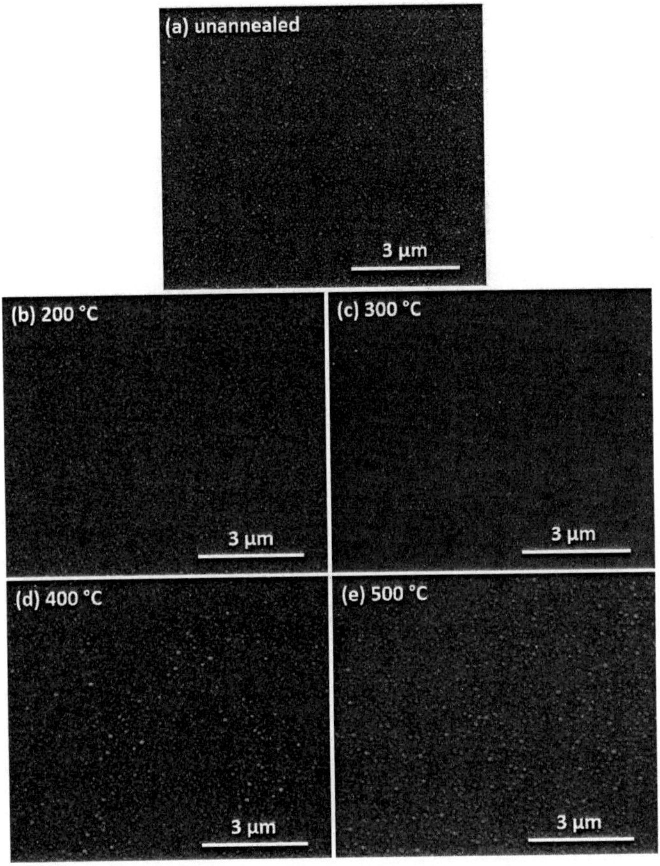

Fig. 2. FESEM images of $AuCl_3$ treated MoS_2 thin films annealed at different temperatures

The XRD patterns, shown in Fig. 4, reveal the prominent peak around $2\theta = 14.1°$, which corresponds to the hexagonal MoS_2 structure. The decrease in the intensity of cubic Au peak at $2\theta = 38.1°$ for higher annealing temperature is observed. This can be ascribed the presence of Au aggregates till 300 °C and with the further rise in temperature these particles get incorporated into the MoS_2 through grain boundaries and/or desorption from MoS_2 surface, thereby the intensity of Au peak reduces. The crystallite size is calculated from Scherer analysis as 6.4 nm, 6.3 nm, 6.0 nm, 5.8 nm and 5.77 nm for unannealed, annealed at 200 °C, 300 °C, 400 °C and 500 °C, respectively. These values are in well agreement with the reported values [29].

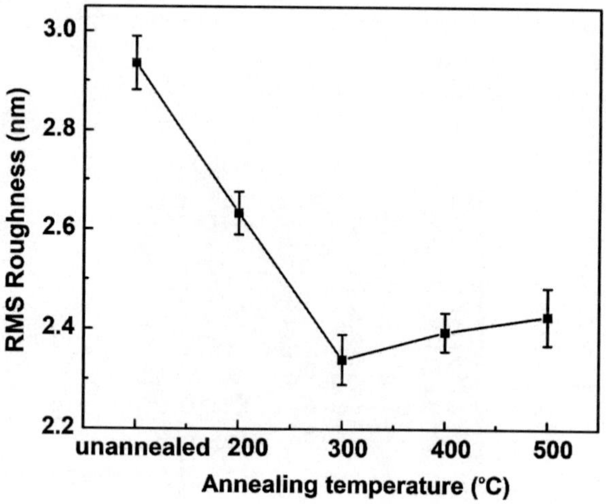

Fig. 3. RMS roughness plot of $AuCl_3$ treated MoS_2 thin films at different annealing temperatures

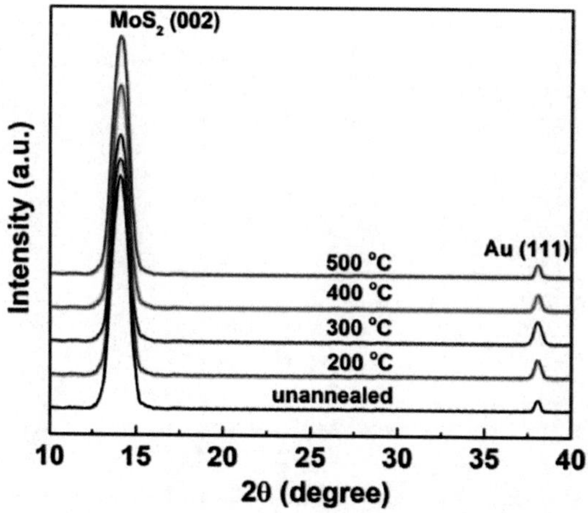

Fig. 4. XRD pattern of $AuCl_3$ treated MoS_2 thin films at various annealing temperatures

In Fig. 5, a slight variation in the Raman peak position has been observed. The shifting of A_{1g} peak towards higher wavenumber till 300 °C is ascribed to the p-type doping and reverse shift beyond 300 °C is ascribed to de-doping process. Yoon et al. reported the formation Au^0, $AuCl^-$ and Cl_2 by thermal decomposition of $AuCl_3$ at 200 °C [30]. Beyond 200 °C, the reduction of Au^{3+} is assisted by thermal energy. So, the transfer of electrons from MoS_2 thin films is restricted at elevated annealing temperatures. This decrease in the charge transfer is termed as de-doping process. In addition, the presence of electronegative species ($AuCl_2^-$ and $AuCl_4^-$) transfers electron to MoS_2 and deteriorating the p-type characteristics of MoS_2 thin films. But, the transfer of electrons between $AuCl^-$ and MoS_2 and evaporation of Cl_2 gas can be facilitated at temperature beyond 300 °C [31]. This describes the p-type doping process persists till the annealing temperature of 200 °C by the reduction of Au^{3+} with the transfer of electron from MoS_2. At temperatures beyond 300 °C, the de-doping nature initiates in MoS_2 thin films.

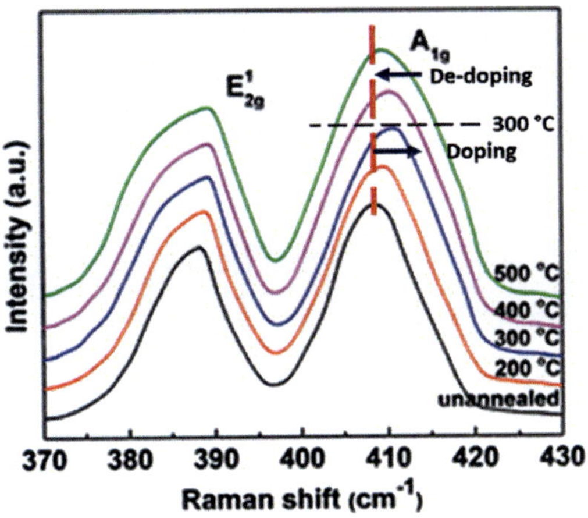

Fig. 5. Raman spectra of $AuCl_3$ treated MoS_2 thin films at various annealing temperatures

The I-V characteristics across $AuCl_3$ treated MoS_2/n-Si heterojunction shown in Fig. 6 are found to be rectifying in nature. The ideality factor of p-MoS_2/n-Si heterojunction is found to be 1.36, 1.42, 1.35, 1.28 and 1.23 for unannealed, annealed at 200 °C, 300 °C, 400 °C and 500 °C, respectively. Figure 7 shows the decrease in the carrier concentration (p-type) beyond 200 °C, which is ascribed to the de-doping process at higher post-treatment annealing temperature. The built-in potential across p-MoS_2/n-Si heterojunction is estimated as 0.48 V, 0.53 V, 0.43 V, 0.38 V and 0.35 V for unannealed, annealed at 200 °C, 300 °C, 400 °C and 500 °C, respectively.

The IR detection study of MoS_2 thin films, treated with 15 mM, 6 coats of $AuCl_3$ solution and annealed at 200 °C, was carried out for 850 nm exposure. A pronounced current On/Off ratio is depicted from the current-time measurements with IR exposure. Upon IR illumination, the charge separation takes place at the heterojunction [32]. The diffusion of carriers across the p-n heterojunction results in the formation of uncompensated acceptors (by diffusion of holes) and donors (by diffusion of electrons) near the interface [33]. The presence of positive space charge and negative space charge near n-type semiconductor and p-type semiconductor, respectively, induces an E-field. This E-field facilitates the separation of the photoexcited charges on exposure to illumination, thereby allowing the movement of electrons to the conduction band of n-type semiconductor and vice-versa for holes, thereby leading to the flow of photocurrent across the heterojunction. In addition to the majority carriers, minority carriers also diffuse to the

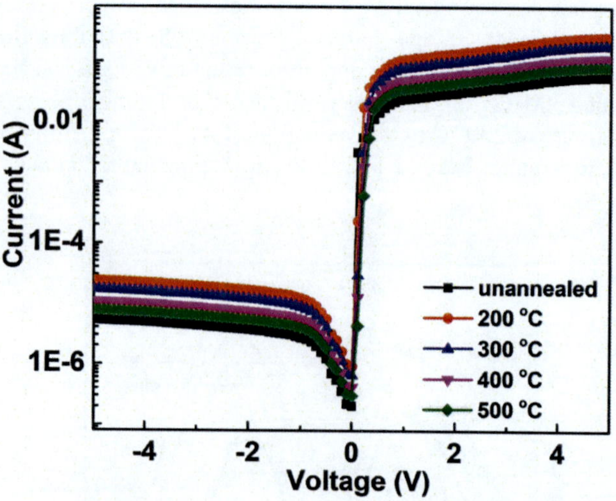

Fig. 6. I-V curves of $AuCl_3$ treated MoS_2 thin films at various annealing temperatures

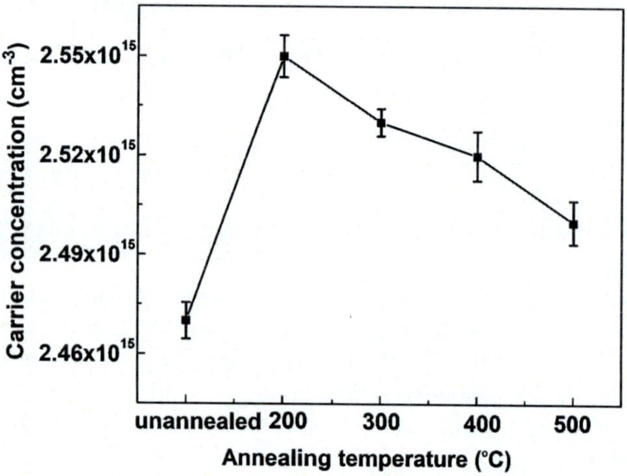

Fig. 7. Variation in carrier concentration with annealing temperature of $AuCl_3$ treated MoS_2 thin films

space region and are separated by the built-in voltage in heterojunction. Figure 8 shows the current-time curves illustrating the fitting of rise and fall time of p-MoS_2/n-Si heterojunction with 850 nm IR illumination. The current On/Off ratio, rise time and fall time are found to be 350, 1.2 s and 1.4 s, respectively.

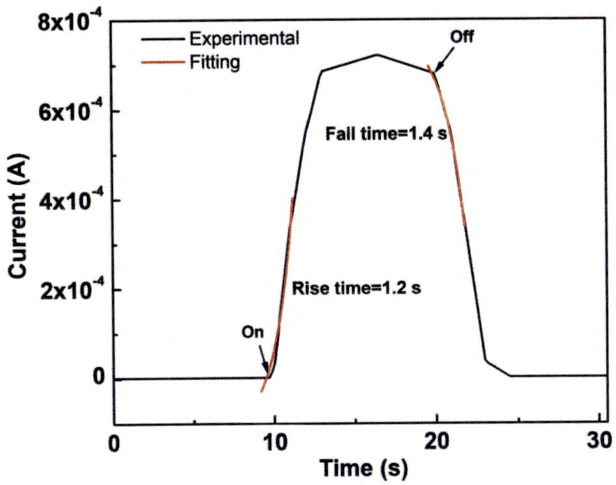

Fig. 8. Current-time characteristics across p-MoS_2/n-Si heterojunction with 850 nm IR illumination.

4 Conclusion

In this work, the effect of post-treatment annealing temperature on the morphological, microstructural and electronic properties of $AuCl_3$ treated MoS_2 films have been discussed elaborately. FESEM images revealed the modulation in the morphology beyond 300 °C, which can be attributed to the enhancement in incorporation of Au particles into the MoS_2 films through the grain boundaries. Moreover, the RMS roughness is found to be reduced with the increase in temperature upto 300 °C, which can be attributed to the lateral spreading of Au particles with increased density. Along with MoS_2 XRD characteristic peak ($2\theta = 14.1°$), an appearance of additional peak ($2\theta = 38.1°$) corresponds to (111) plane of cubic Au. The increase in the intensity of cubic Au peak can be ascribed the presence of Au aggregates till 300 °C. The XRD results are corelated with morphological results. The shifting of A_{1g} peak beyond 300 °C was illustrated using de-doping process. The dependence of carrier concentration of MoS_2 films and built-in potential across MoS_2/Si heterojunction on the annealing temperature were evaluated from C-V measurements. The ideality factor of MoS_2/Si heterojunction was investigated from I-V characteristics. The response time across p-MoS_2/n-Si heterojunction upon illumination of 850 nm IR source were found to be 1.2 s (rise time) and 1.4 s (fall time). The study of variation in the properties of MoS_2 films with the post-treatment annealing temperature paves way towards the integration of MoS_2 based devices in silicon technology.

Acknowledgement. This work was supported by Defence Research and Development Organisation (DRDO), India sponsored Extramural Research and Intellectual Property Rights (ERIP) project (ERIP/ERJ201701014/M/01/1748).

References

1. Cooperstein, I., Indukuri, S.R.K.C., Bouketov, A., Levy, U., Magdassi, S.: 3D printing of micrometer-sized transparent ceramics with on-demand optical-gain properties. Adv. Mater. **32**, 1–8 (2020). https://doi.org/10.1002/adma.202001675
2. Lu, S., Li, R., Jiao, J., Kang, J., Zhao, N., Li, M.: An eye gaze tracking method of virtual reality headset using a single camera and multi-light source. J. Phys. Conf. Ser. **1518**, 1–8 (2020). https://doi.org/10.1088/1742-6596/1518/1/012020
3. Heerlein, J., Behringer, M., Jäger, C.: Near-infrared power LED for emerging security and defence applications, Electro-Optical Remote Sensing, Photonic Technol. Appl. V. **8186**, 81860O1–81860O9 (2011). https://doi.org/10.1117/12.897992
4. Fard, M.M.P., Williams, C., Cowan, G., Liboiron-Ladouceur, O.: A 35 Gb/s silicon photodetector for 850 nm wavelength applications. In 2016 IEEE Photonics Conference IPC 2016, pp. 53–54, IEEE (2017). https://doi.org/10.1109/IPCon.2016.7830950
5. Reddy, P.V., Kumar, A., Rahman, S.M.K., Mundra, T.S.: A new antispoofing approach for biometric devices. IEEE Trans. Biomed. Circuits Syst. **2**, 328–337 (2008). https://doi.org/10.1109/TBCAS.2008.2003432
6. Stevenson, D.J., Lake, T.K., Agate, B., Gárcés-Chávez, V., Dholakia, K., Gunn-Moore, F.: Optically guided neuronal growth at near infrared wavelengths. Opt. Express. **14**, 9786–9793 (2006). https://doi.org/10.1364/oe.14.009786
7. Lemme, M.C., Wagner, S., Lee, K., Fan, X., Verbiest, G.J., Wittmann, S., Lukas, S., Dolleman, R.J., Niklaus, F., van der Zant, H.S.J., Duesberg, G.S., Steeneken, P.G.: Nanoelectromechanical sensors based on suspended 2D materials. Research. **2020**, 1–25 (2020). https://doi.org/10.34133/2020/8748602
8. Cui, F., et al.: Tellurium-assisted epitaxial growth of large-area, highly crystalline ReS_2 atomic layers on mica substrate. Adv. Mater. **28**, 5019–5024 (2016). https://doi.org/10.1002/adma.201600722
9. Pezeshki, A., Shokouh, S.H.H., Nazari, T., Oh, K., Im, S.: Electric and photovoltaic behavior of a few-layer α-$MoTe_2$/MoS_2 dichalcogenide heterojunction. Adv. Mater. **28**, 3216–3222 (2016). https://doi.org/10.1002/adma.201504090
10. Guo, Q., et al.: Black phosphorus mid-infrared photodetectors with high gain. Nano Lett. **16**, 4648–4655 (2016). https://doi.org/10.1021/acs.nanolett.6b01977
11. Wang, W., et al.: MoS_2 memristor with photoresistive switching. Sci. Rep. **6**, 1–11 (2016). https://doi.org/10.1038/srep31224
12. Kalantar-Zadeh, K., Ou, J.Z.: Biosensors based on two-dimensional MoS_2. ACS Sensors. **1**, 5–16 (2016). https://doi.org/10.1021/acssensors.5b00142
13. Wang, L., et al.: MoS2/Si heterojunction with vertically standing layered structure for ultrafast, high-detectivity, self-driven visible-near infrared photodetectors. Adv. Funct. Mater. **25**, 2910–2919 (2015). https://doi.org/10.1002/adfm.201500216
14. Bernardi, M., Palummo, M., Grossman, J.C.: Extraordinary sunlight absorption and one nanometer thick photovoltaics using two-dimensional monolayer materials. Nano Lett. **13**, 3664–3670 (2013). https://doi.org/10.1021/nl401544y
15. Lee, Y.H., et al.: Synthesis of large-area MoS_2 atomic layers with chemical vapor deposition. Adv. Mater. **24**, 2320–2325 (2012). https://doi.org/10.1002/adma.201104798
16. Kwak, J.Y., Hwang, J., Calderon, B., Alsalman, H., Spencer, M.G.: Long wavelength optical response of graphene-MoS_2 heterojunction. Appl. Phys. Lett. **108**, 0911081–0911085 (2016). https://doi.org/10.1063/1.4943169
17. Xue, F., et al.: P-type MoS_2 and n-type ZnO diode and its performance enhancement by the piezophototronic effect. Adv. Mater. **28**, 3391–3398 (2016). https://doi.org/10.1002/adma.201506472

18. Li, Y., Xu, C.Y., Wang, J.Y., Zhen, L.: Photodiode-like behavior and excellent photoresponse of vertical Si/monolayer MoS_2 heterostructures. Sci. Rep. **4**, 1–8 (2014). https://doi.org/10.1038/srep07186
19. Li, H.M., et al.: Ultimate thin vertical p-n junction composed of two-dimensional layered molybdenum disulfide. Nat. Commun. **6**, 1–9 (2015). https://doi.org/10.1038/ncomms7564
20. Remškar, M., Škraba, Z., Stadelmann, P., Lévy, F.: Structural stabilization of new compounds: MoS2 and WS2 micro- and nanotubes alloyed with gold and silver. Adv. Mater. **12**, 814–818 (2000). https://doi.org/10.1002/(SICI)1521-4095(200006)12:11%3c814::AID-ADMA814%3e3.0.CO;2-0
21. Tarasov, A., et al.: Controlled doping of large-area trilayer MoS_2 with molecular reductants and oxidants. Adv. Mater. **27**, 1175–1181 (2015). https://doi.org/10.1002/adma.201404578
22. Mouri, S., Miyauchi, Y., Matsuda, K.: Tunable photoluminescence of monolayer MoS_2 via chemical doping. Nano Lett. **13**, 5944–5948 (2013). https://doi.org/10.1021/nl403036h
23. Du, Y., Liu, H., Neal, A.T., Si, M., Ye, P.D.: Molecular doping of multilayer mos2 field-effect transistors: reduction in sheet and contact resistances. IEEE Electron Device Lett. **34**, 1328–1330 (2013). https://doi.org/10.1109/LED.2013.2277311
24. Dey, S., Matte, H.S.S.R., Shirodkar, S.N., Waghmare, U.V., Rao, C.N.R.: Charge-transfer interaction between few-layer MoS_2 and tetrathiafulvalene. Chem. An Asian J. **8**, 1780–1784 (2013). https://doi.org/10.1002/asia.201300174
25. Li, X., Zhu, M., Dai, B.: $AuCl_3$ on polypyrrole-modified carbon nanotubes as acetylene hydrochlorination catalysts. Appl. Catal. B Environ. **142–143**, 234–240 (2013). https://doi.org/10.1016/j.apcatb.2013.05.031
26. Ren, J., Teng, C., Cai, Z., Pan, H., Liu, J., Zhao, Y., Liu, B.: Controlled one step thinning and doping of two-dimensional transition metal dichalcogenides. Sci. China Mater. **62**, 1837–1845 (2019). https://doi.org/10.1007/s40843-019-9461-8
27. Bellani, S., Najafi, L., Capasso, A., Del Rio Castillo, A.E., Antognazza, M.R., Bonaccorso, F.: Few-layer MoS_2 flakes as a hole-selective layer for solution-processed hybrid organic hydrogen-evolving photocathodes. J. Mater. Chem. A. **5**, 4384–4396 (2017). https://doi.org/10.1039/c6ta10572f
28. Pradhan, D., Ghosh, S.P., Gartia, A., Sahoo, K.K., Bose, G., Kar, J.P.: Modulation of microstructural and electrical properties of rapid thermally synthesized MoS2 thin films b the flow of H_2 gas. Superlattices Microstruct. **145**, 1–10 (2020). https://doi.org/10.1016/j.spmi.2020.106598
29. Nandhagopal, P., Mohan, D.B.: Optimisation study on few layer formations of MoS_2 thin films by a novel sulfurization method. AIP Conf. Proc. **2162**, 020135 (2019). https://doi.org/10.1063/1.5130345
30. Yoon, S.M., et al.: Thermal conversion of electronic and electrical properties of AuCl3-doped single-walled carbon nanotubes. ACS Nano **5**, 1353–1359 (2011). https://doi.org/10.1021/nn103055u
31. Kim, S.M., et al.: Role of anions in the AuCl3-doping of carbon nanotubes. ACS Nano **5**, 1236–1242 (2011). https://doi.org/10.1021/nn1028532
32. Dhyani, V., Das, S.: High-speed scalable silicon-MoS_2 P-N heterojunction photodetectors. Sci. Rep. **7**, 2 (2017). https://doi.org/10.1038/srep44243
33. Hao, L., Liu, Y., Gao, W., Han, Z., Xue, Q., Zeng, H., Wu, Z., Zhu, J., Zhang, W.: Electrical and photovoltaic characteristics of MoS2/Si p-n junctions. J. Appl. Phys., **117** (2015). https://doi.org/10.1063/1.4915951

Investigation of Structural and Electrical Properties of Ta_2O_5 Thin Films with Sputtering Parameters for Microelectronic Applications

Kiran K. Sahoo[1], D. Pradhan[1,2], S. P. Ghosh[1,3], A. Gartia[1], and J. P. Kar[1,4](✉)

[1] Department of Physics and Astronomy, National Institute of Technology Rourkela, Rourkela 769008, India
karjp@nitrkl.ac.in
[2] Department of Electronics and Communication Engineering, ITER, Siksha 'O' Anusandhan Deemed to Be University, Bhubaneswar, Odisha 751030, India
[3] C. V. Raman Global University Bhubaneswar, Odisha 751030, India
[4] Centre for Nanomaterials, National Institute of Technology, Rourkela 769008, India

Abstract. Tantalum oxide (Ta_2O_5) thin film is considered as an alternative dielectric layer in both microelectronics and MEMS devices due to its high dielectric constant, high breakdown field and low leakage current density. In this research, radio-frequency magnetron sputtering was used to deposit Ta_2O_5 thin films on p-type Si (100) substrates. During the film deposition, the RF power and Ar/O_2 gas flow ratio were kept constant while the sputtering pressure and substrate temperature were varied. The films were annealed in the air for an hour at 900 °C after the deposition. The structural, morphological, and electrical properties of the films were studied with various sputtering parameters. Orthorhombic β—phase structure of Ta_2O_5 films is observed from XRD investigation. The crystallinity of the films was found to be improved with the increase in the sputtering pressure and substrate temperature. The films, deposited at higher working pressure, became rough, whereas the films deposited at higher temperature became smooth. The Capacitance-voltage and current-voltage techniques were used to study the electrical properties of the thin films. Low oxide charge density of 6.5×10^{11} cm^{-2} and 3.1×10^{12} cm^{-2} observed at sputtering pressure of 8.0×10^{-3} mbar and substrate temperature of 300 °C, respectively.

Keywords: High-k dielectric · Ta_2O_5 · Thin film · Sputtering · C-V · I-V

1 Introduction

In the last few decades, microelectronics has become the most essential driving factor behind all technological advancements [1, 2]. In microelectronic systems such as the central processing unit (CPU), metal oxide semiconductor field effect transistor (MOSFET), dynamic random-access memory (DRAM) and flash memory, the dielectric layer is a key component [3, 4]. Although the dielectric material's primary role is to improve capacitive coupling between nearby metals and semiconductors, it also inhibits leakage

A. K. Pandey et al. (Eds.): MAMM 2022, 126, pp. 342–350, 2023.
https://doi.org/10.1007/978-3-031-20353-4_30

current between electrodes, which reduces energy consumption (in CPUs and DRAMs) and long-term dependability (in flash memory) [5]. Other than microelectronic systems, dielectric material is also used in Micro-electromechanical Systems (MEMS) specifically in RF MEMS switches for satellite reconfigurable antennas to achieve lower insertion loss and lower power consumption than semiconductor switches [6–8].

In microelectronics, silicon dioxide (SiO_2) has been utilised as an ideal dielectric material for decades because of its ease of growth, low-charge density and better interface with silicon. During downscaling of SiO_2 layer, the leakage current caused by the quantum tunnelling phenomenon become prominent [9], causing major difficulties with power consumption and device performance. To overcome the scaling restriction of microelectronic device, conventional dielectric layers are replaced with advanced high permittivity (high-k) materials [10–12]. Various high k materials, such as barium strontium titanate (BST, $BaSrTiO_3$), tantalum oxide (Ta_2O_5), titanium oxide (TiO_2), hafnium oxide (HfO_2), zirconium oxide (ZrO_2), silicon nitride (Si_3N_4), and aluminium oxide (Al_2O_3), aluminium nitride (AlN) are being investigated to meet the need. Among various metal oxide, Ta_2O_5 is the most promising gate insulator and also appropriate candidate for use of memory dielectric in storage capacitors in place of SiO_2, because of its high dielectric constant, high breakdown field, high amorphous-to-crystalline transition temperature and better chemical stability [11–15]. In addition, Ta_2O_5 has high bandgap (4.5–5 eV) to achieve a sufficient barrier height to reduce the leakage current. Because of the above properties, Ta_2O_5 thin film has drawn tremendous attention for high-density dynamic random-access memory (DRAMs) and metal–oxide–semiconductor field-effect transistors (MOSFETs). On the other hand, the charge density of the dielectric layer needs to be systematically investigated since the constraints imposed by the charging mechanisms in the dielectric material will impact the collapse of bridges and cantilevers of RF MEMS [16]. Charges are injected and subsequently trapped in the dielectric layer in electrostatically actuated MEMS switches, when the bridge comes into contact with the dielectric under the pull-in voltage. This charge injection creates a change in the actuation voltage, which eventually leads to a stiction effect, which severely limits the device performance [17]. The improvement of the fabrication process as well as the exploration of alternate materials for dielectric layers is necessary to overcome the limits of RF MEMS reliability. Ta_2O_5 is a promising option for the use of dielectric layer in RF MEMS switches because of its excellent material properties [18, 19]. The dielectric constant of Ta_2O_5 is significantly greater than that of the conventional dielectric material utilised in RF MEMS switches. This feature has the critical benefit of the enhancement of capacitance ratio in RF MEMS based capacitive switches using Ta_2O_5 as the dielectric layer by improvement in the switching performance [20]. Hence, the systematic study of the electrical behavior of Ta_2O_5 films is required prior to its microelectronic applications.

Several physical and chemical deposition strategies for fabrication of Ta_2O_5 thin films have been carried out, whereas sputtering is found to be a low-temperature, non-toxic, CMOS compatible technology that produces homogeneous smooth thin films of better quality [21, 22]. Regardless of the Ta_2O_5 synthesis process, the deposition parameters have a significant impact on the modulation of morphological and electrical properties of the sputtered films. Hence, the effect of sputtering pressure and substrate temperature on the structural and electrical properties of Ta_2O_5 thin films is investigated in this research work.

2 Experimental Details

Tantalum oxide thin dielectric films were deposited on p-type silicon [(100), 1–10 Ω cm] wafers using an RF magnetron sputtering technique using a pure tantalum target (99.95%) in the presence of high-quality oxygen and argon gas. During the sputtering process, the sputtering power and Ar/O_2 gas flow ratio were fixed as 300 W and 3:2, respectively. The sputtering pressure was varied from 4×10^{-3} mbar to 1×10^{-2} mbar and the substrate temperature was varied from room temperature to 300 °C. During the variation of the sputtering pressure, the substrates were kept at room temperature and while varying the substrate temperature the sputtering pressure was fixed at 6.0×10^{-3} mbar. After the deposition, all the films were undergone post-deposition annealing treatment. The annealing was carried out at 900 °C for 1 h in air ambient. The thickness of the films was kept around 50 nm. X-ray diffraction (XRD) (Rigaku ultima IV), Atomic force microscope (AFM) (Park System XE7), were used to investigate the structural, morphological properties of the thin film. To study the electrical properties of Ta_2O_5 films, metal oxide semiconductor (MOS) structure was fabricated using thermal evaporation of aluminum. The capacitance-voltage (C–V) measurement technique was used to calculate the interface charge density, oxide charge density, and dielectric constant, while the current–voltage (I–V) studies were used to estimate the leakage current of the films. The C-V and I–V behaviour were investigated using an Agilent E4980 precision LCR meter.

3 Results and Discussion

Figure 1 shows the XRD patterns of the Ta_2O_5 thin film, deposited at various sputtering pressure and substrate temperatures, on Si substrate. There was not evolution of crystalline peaks for as-deposited Ta_2O_5 thin films. When the films are annealed at 900 °C, β—phase of orthorhombic structures is obtained. The position of the diffraction peaks is in agreement with the JCPDS data (Card No. 25–0922) [23]. The crystallinity of the film increases with increase in the sputtering pressure as shown in Fig. 1a. The increase in sputtering pressure produces sufficient sputtering species (Ar) and reacting species (O_2). Hence, the film of better stoichiometry is produced during sputtering. Figure 1b indicates the crystallinity of Ta_2O_5 at various substrate temperatures. As the annealed Ta_2O_5 film is polycrystalline in nature with orthorhombic structure, with increase in substrate temperature the crystallinity of the film improved slightly. The increase in substrate temperature has provided sufficient energy to the adatoms to migrate to the favorable sites to preliminarily produce dense and homogeneous structure. However, the temperature and duration of deposition is not enough for the orientation of ad-atoms. Therefore, post-deposition annealing is required for structural improvement.

Figures 2 and 3 depicts the AFM micrographs of Ta_2O_5 thin films for various sputtering pressure and substrate temperature respectively. Surface roughness, estimated from AFM images, is found to be increased from 4 nm to 4.8 nm with increased sputtering

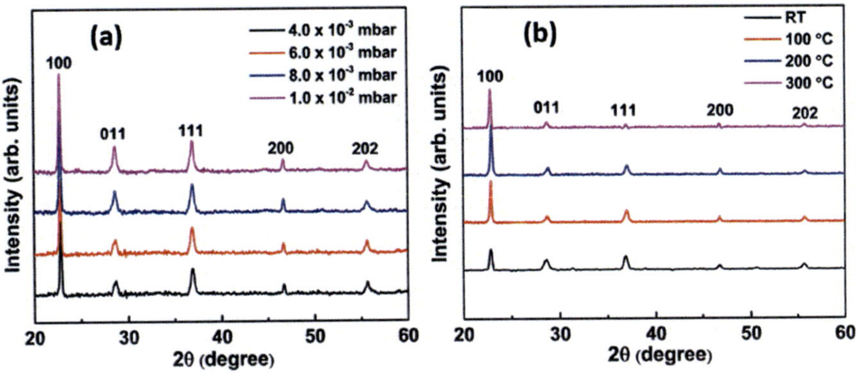

Fig. 1. XRD pattern of Ta_2O_5 thin films deposited at different (**a**) sputtering pressure and (**b**) substrate temperature followed by 900 °C annealing.

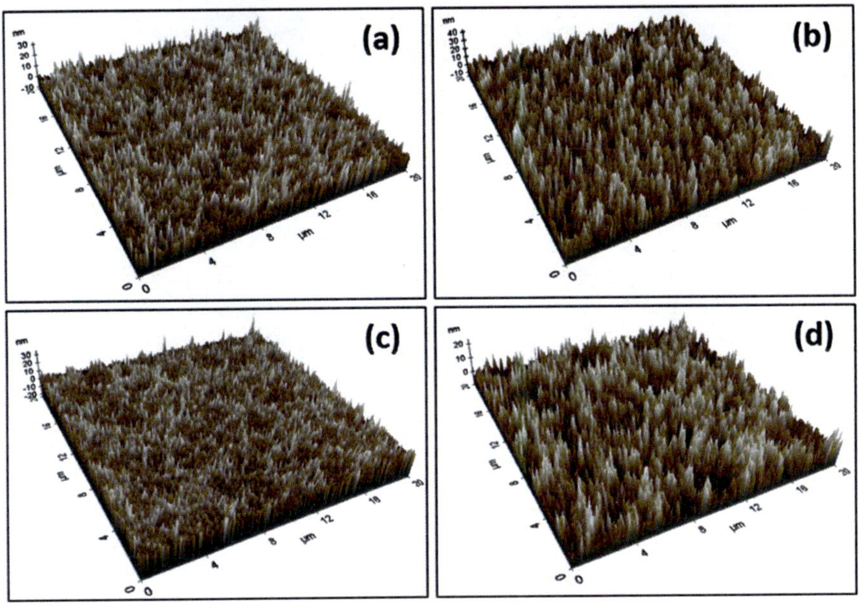

Fig. 2. AFM micrographs of Ta_2O_5 thin films for different sputtering pressure (**a**) 4.0×10^{-3} mbar (**b**) 6.0×10^{-3} mbar (**c**) 8.0×10^{-3} mbar (**d**) 1.0×10^{-2} mbar

pressure from 4.0×10^{-3} mbar to 1.0×10^{-2} mbar, as shown in Fig. 4a. The increase in roughness of the sputtered film with pressure is due to the grain growth as a result of rise in crystallinity as seen from XRD plots. In addition, the average distance between molecule collisions rises because there are fewer gas molecules in the chamber at lower sputtering pressure and the chance of particle collision decreases and sputtered particles have enough energy to homogeneously organise themselves on the substrate. As the sputtering pressure rises, the bombardment of the target atoms by the sputtering Ar^+ increases, and as a result of the huge number of target atoms ejected, the scattering of the sputtered species improves, lowering the kinetic energy of the impacting ions. As a result, at greater pressures, adatom mobility on the substrate surface is reduced and the roughness increases. On the other hand, the surface roughness decreases, when the deposition temperature rises [Fig. 4b]. At higher substrate temperature, sputtered particles have increased mobility, which gives rise to homogeneous nucleation and grain development. As a result, the atoms uniformly distributed to produce smoother surface.

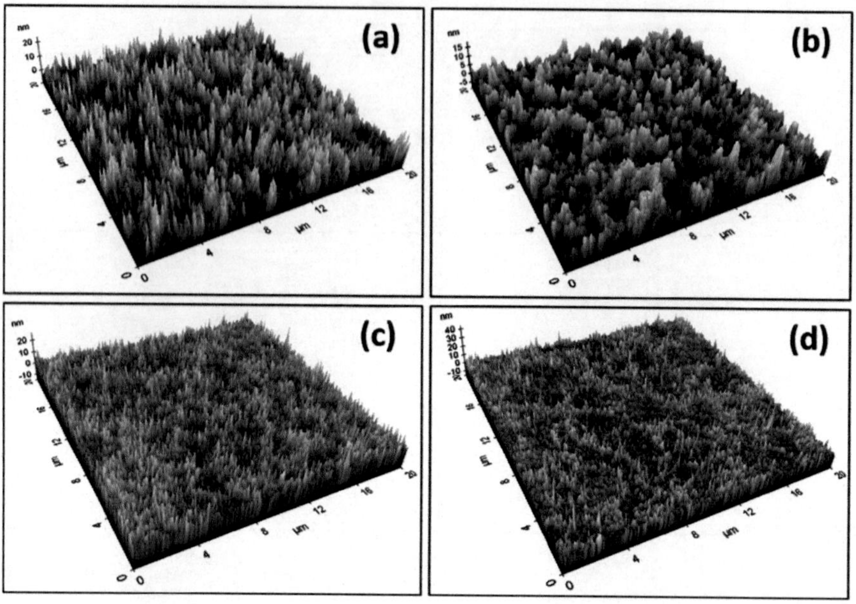

Fig. 3. AFM micrographs of Ta_2O_5 thin films for different substrate temperature (**a**) room temperature (**b**) 100 °C (**c**) 200 °C (**d**) 300 °C

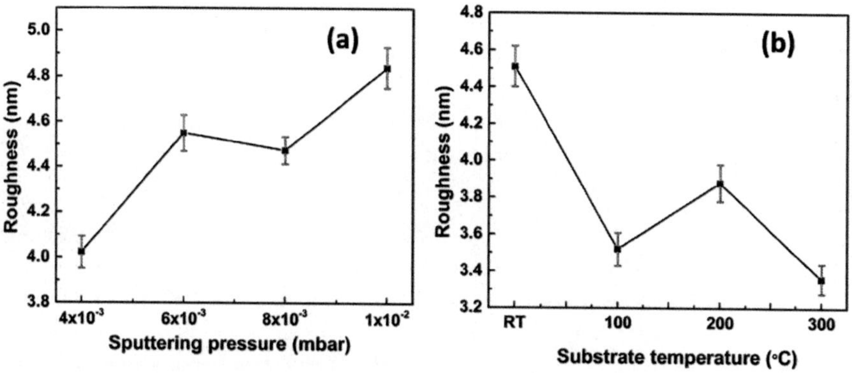

Fig. 4. AFM surface roughness of Ta_2O_5 thin films for different (**a**) sputtering pressure and (**b**) substrate temperature

Capacitance–voltage measurements of Ta_2O_5 based MOS structures are carried out, with a sweep in bias voltage from −5 V to 5 V, to investigate the electrical characteristics. Figure 5 depicts the normalized high frequency (1 MHz) C–V curve of the Al/Ta_2O_5/Si MOS structure for various sputtering pressures and substrate temperatures. The oxide charge density (Q_{ox}), interface charge density (D_{it}), and dielectric constant of the Ta_2O_5 thin film were determined from the C–V curve. With a bias voltage sweep, the accumulation, depletion, and inversion regions are clearly observed. Here, Cox represent the oxide capacitance per unit area.

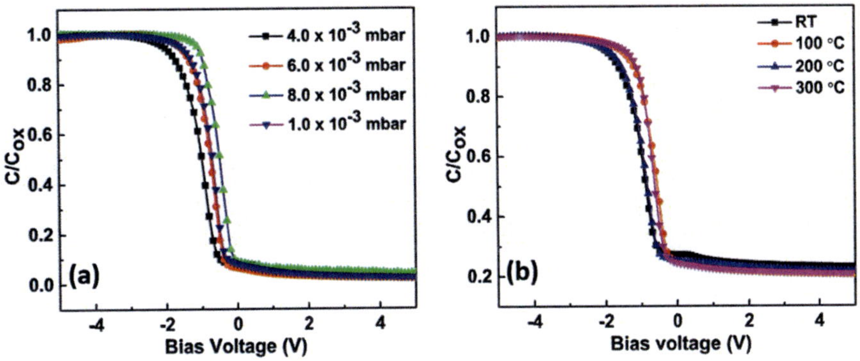

Fig. 5. Capacitance–voltage characteristics of the Ta_2O_5 thin film deposited at different (**a**) sputtering pressure and (**b**) substrate temperature

The accumulation capacitance is used to compute the dielectric constant, which is found to be increased from 14.8 to 18.8 with the increase in sputtering pressure from 4.0×10^{-3} mbar to 8.0×10^{-3} mbar and from 17.2 to 19.1 with the rise in substrate temperature from room temperature to 300 °C, respectively. The dielectric constant increases with the increase in the sputtering pressure because the crystallinity of the film improved with increase sputtering pressure and hence, at the accumulation region the capacitance of the film increases [24]. The shift in C–V curves, indicates a reduction in oxide charge density as the sputtering pressure rises. Figure 6 shows the Q_{ox} and D_{it} values for various sputtering pressures and substrate temperatures. When the sputtering pressure is increased from 4.0×10^{-3} mbar to 8.0×10^{-3} mbar, the interface trap density found to be decreased. Similarly, the oxide charge density decreases with increase in substrate temperature. This may be due to the modulation is grain size and their boundaries. Defects, voids, unsaturated bonds, and impurities are commonly found along grain boundaries, which act as trap centres. Films of smaller grains have larger grain boundary area, which means there are more trapping centres accessible for collecting free carriers. With the rise in granular dimensions, the surface area of the grain boundaries is comparatively annihilated and therefore, the defect density has decreased. The interface charge density (D_{it}) is found to be maximum at lower sputtering pressure, which may be due to the high-energetic atom bombardments as the mean free path of the atomic species are comparatively larger during sputtering. The films grown at moderate substrate temperature yield lower D_{it}.

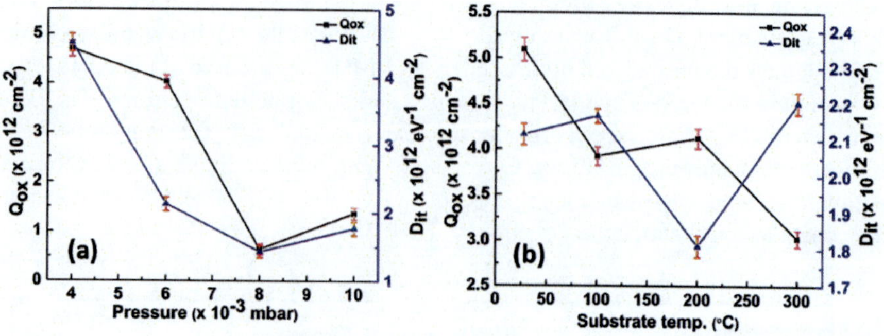

Fig. 6. Oxide (Q_{ox}) and interface (D_{it}) charge densities of Ta_2O_5 thin films deposited at various (**a**) sputtering pressure and (**b**) substrate temperature

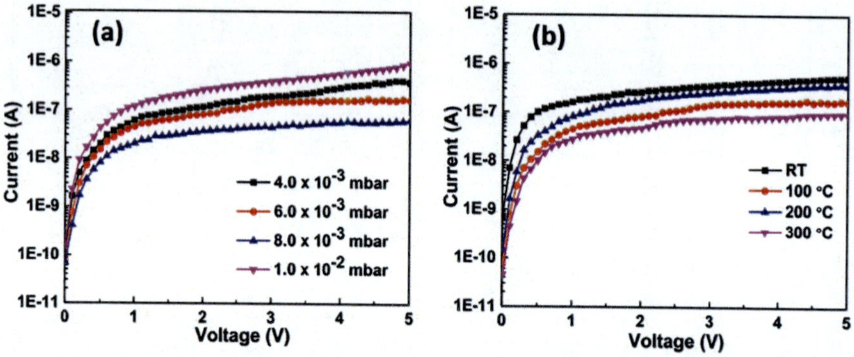

Fig. 7. Current–voltage characteristics of the Ta_2O_5 thin films deposited at various (**a**) sputtering pressure and (**b**) substrate temperature

Current–voltage (I–V) curves for Al/Ta_2O_5/Si structures for various sputtering pressures and substrate temperatures are shown in Fig. 7. A rapid rise in current occurs at lower bias voltages, and then the current becomes quasi-saturated. The lower leakage current is achieved for a sputtering pressure of 8.0×10^{-3} mbar. The decrease in leakage current with increasing sputtering pressure is related to rise in grain size, with fewer grain boundaries accessible to offer a current conducting channel [25]. Similarly with increase in substrate temperature the leakage current decreases and found to be minimal for sample prepared at 300 °C. This may be due to the improvement of film quality by the reduction of voids, unsaturated bonds with the enhancement of crystallinity. Hence, the Ta_2O_5 films grown at 8.0×10^{-3} mbar and substrate temperature of 300 °C found suitable for microelectronic applications.

4 Conclusion

Ta_2O_5 thin films were deposited on p-type silicon wafer using RF sputtering technique to study its structural and electrical properties. The sputtering pressure and substrate temperature have been varied to optimize the growth parameters of Ta_2O_5 thin films. The sputtering pressure was varied from 4.0×10^{-3} mbar to 1.0×10^{-2} mbar and the substrate temperature was varied from room temperature to 300 °C. From XRD study it was found that the Ta_2O_5 film is polycrystalline in nature and possesses β—phase of orthorhombic structure. With increase in sputtering pressure and substrate temperature the crystallinity of the Ta_2O_5 thin film is found to be increases. The roughness of the film is found to be increased with increase in sputtering pressure and decreased with increase in substrate temperature. From the C-V and I-V measurements, the dielectric constant, oxide charge density, interface charge density and leakage current are estimated. The oxide charge density and leakage current for film prepared at 8.0×10^{-3} mbar sputtering pressure is found to be minimum with the enhancement of dielectric constant. The Ta_2O_5 film, deposited at 8.0×10^{-3} mbar sputtering pressure and 300 °C of substrate temperature, has possessed high dielectric constant, lower leakage current and low oxide charge. Hence, this film can be used as an alternative dielectric layer in microelectronics and MEMS devices.

References

1. Iwai, H.: Future semiconductor manufacturing: challenges and opportunities. In IEDM Technical Digest. IEEE International Electron Devices Meeting 2004, pp. 11–16. IEEE (2004)
2. Iwai, H.: CMOS downsizing toward sub-10 nm. Solid-State Electron. **48**(4), 497–503 (2004)
3. Cheng, C.H., Chin, A.: Low-leakage-current DRAM-like memory using a one-transistor ferroelectric MOSFET with a Hf-based gate dielectric. IEEE electron device letters 35(1), 138–140 (2013)
4. Atanassova, E., Paskaleva, A.: Challenges of Ta_2O_5 as high-k dielectric for nanoscale DRAMs. Microelectron. Reliab. **47**(6), 913–923 (2007)
5. Xu, W., et al.: Anomalous bias-stress-induced unstable phenomena of InZnO thin-film transistors using Ta_2O_5 gate dielectric. J. Phys. D Appl. Phys. **45**(20), 205103 (2012)
6. Vinoy, K.J., Vardan, V.K.: Design of reconfigurable fractal antennas and RF-MEMS for space-based systems. IOP Smart Mater Struct **10**(6), 1211–1223 (2001)
7. Yao, Z.J., Chen, S., Eshelman, S., Denniston, D., Goldsmith, C.: Micromachined low-loss microwave switches. IEEE J Microelectromech. Syst. **8**, 129–134 (1999)
8. Muldavin, J.B., Rebeiz, G.M.: High-Isolation CPW MEMS switches: part 1: modeling. IEEE Trans Micro Tech **48**(6), 1045–1052 (2000)
9. Kingon, A.I., Maria, J.P., Streiffer, S.K.: Alternative dielectrics to silicon dioxide for memory and logic devices. Nature **406**, 1032–1038 (2000)
10. Sayan, S., Garfunkel, E., Suzer, S.: Soft x-ray photoemission studies of the $HfO_2/SiO_2/Si$ system. Appl. Phys. Lett. **80**(12), 2135–2137 (2002)
11. Intern. Techn. Roadmap for Semiconductor (ITRS) 2004 edition, http://public.itrs.net
12. Manchanda, L., et al.: Multi-component high-K gate dielectrics for the silicon industry. Microelectron. Eng. **59**(1–4), 351–359 (2001)
13. Das, K.C., et al.: Evolution of microstructural and electrical properties of sputtered HfO2 ceramic thin films with RF power and substrate temperature. Ceram. Int. **42**(1), 38–145 (2016)
14. Kar, J.P., Bose, G., Tuli, S.: Influence of rapid thermal annealing on morphological and electrical properties of RF sputtered AlN films. Mater. Sci. Semicond. Process. **8**(6), 646–651 (2005)

15. Ezhilvalavan, S., Tseng, T.Y.: Preparation and properties of tantalum pentoxide (Ta2O5) thin films for ultra-large-scale integrated circuits (ULSIs) application–a review. J. Mater. Sci.: Mater. Electron. **10**(1), 9–31 (1999)
16. Wibbeler, J., Pfeifer, G., Hietschold, M.: Parasitic charging of dielectric surfaces in capacitive microelectromechanical systems (MEMS). Sens. Actuators, A **71**(1–2), 74–80 (1998)
17. Melle, S., et al.: Reliability modeling of capacitive RF MEMS. IEEE Trans. Microw. Theory Tech. **53**(11), 3482–3488 (2005)
18. Ezhilvalavan, S., Tseng, T.Y.: Conduction mechanisms in amorphous and crystalline Ta_2O_5 thin films. J. Appl. Phys. **83**(9), 4797–4801 (1998)
19. Tsai, K.C., Wu, W.F., Chao, C.G., Wu, C.C.: Improving Electrical Characteristics of Ta/Ta_2O_5/Ta Capacitors Using Low-Temperature Inductively Coupled N_2O Plasma Annealing. J. Electrochem. Soc. **154**(6), 512 (2007)
20. Lisec, T., Huth, C., Wagner, B.: Dielectric material impact on capacitive RF MEMS reliability. In 34th European Microwave Conference, 2004. Vol. 1, pp. 73–76. IEEE (2004)
21. Sahoo, A.K., Wu, G.M., Liu, C.Y., Lu, H.C.: Improved amorphous indium gallium zinc oxide thin film transistors by low power RF-sputtering deposition using Ta_2O_5 dielectric. Nanosci. Nanotechnol. Lett. **7**(3), 193–199 (2015)
22. Sertel, T., Sonmez, N.A., Cetin, S.S., Ozcelik, S.: Influences of annealing temperature on anti-reflective performance of amorphous Ta_2O_5 thin films. Ceram. Int. **45**(1), 11–18 (2019)
23. Perez, I., et al.: Evidence for structural transition in crystalline tantalum pentoxide films grown by RF magnetron sputtering. J. Alloy. Compd. **712**, 303–310 (2017)
24. Kar, J.P., Bose, G., Tuli, S.: Correlation of electrical and morphological properties of sputtered aluminum nitride films with deposition temperature. Curr. Appl. Phys. **6**(5), 873–876 (2006)
25. Das, K.C., et al.: Surface and interface studies of RF sputtered HfO_2 thin films with working pressure and gas flow ratio. J. Mater. Sci.: Mater. Electron. **26**(8), 6025–6031 (2015)

Optimization of Controllable Pulsed LASER Deposition Parameters for the Fabrication of Lead Free $Ba(Zr_{0.15}Ti_{0.85})O_3$ Thin Films

M. L. V. Mahesh[1,2](✉), Prem Pal[2], V. V. Bhanuprasad[1], and A. R. James[1]

[1] Ceramics and Composites Group, Defence Metallurgical Research Laboratory, Kanchanbagh, Hyderabad 500058, India
mahesh.dmrl@gov.in

[2] Department of Physics, Indian Institute of Technology, Sangareddy, Kandi, Telangana 502285, India

Abstract. Owing to the superior energy and power storage properties, the Zr doped $BaTiO_3$ (BZT) material system is reportedly used in numerous applications. In this work, the optimal process parameters viz., substrate temperature, oxygen gas pressure, laser fluence and target-substrate distance for deposition of lead-free thin films using pulsed laser deposition (PLD) system were evaluated. The deposition of BZT thin films were carried out based on design of experiments by adopting the Taguchi method. The thickness and dielectric properties of deposited thin films were analyzed to assess the influence of each parameter on the growth of thin films using analysis of variance (ANOVA) method. Optimal process conditions based on grey relational analysis were evaluated. The phase and morphology analysis on BZT films indicate possibility of domain engineering by choosing appropriate process variables. BZT thin films deposited at optimal process conditions resulted in superior properties comparable to properties of single crystal of the same material system.

Keywords: Ferroelectrics · Dielectric · Optimization · Taguchi method

1 Introduction

1.1 Taguchi Method

$Ba(Zr_{0.15}Ti_{0.85})O_3$ (BZT) material system shows pinched phase transition and thus exhibits enhanced dielectric and ferroelectric properties [1]. The enhanced energy and power storage properties of these materials assist in realizing numerous applications in electronics and electrical industries [2]. The influence of process parameters on the properties of lead-free BZT thin films deposited using pulsed LASER Deposition (PLD) system was evaluated using the Taguchi method, which is extensively useful in designing and improving process conditions [3]. The improvement in process conditions lead to improved properties. The classical method of parameter optimization is complex and

A. K. Pandey et al. (Eds.): MAMM 2022, 126, pp. 351–356, 2023.
https://doi.org/10.1007/978-3-031-20353-4_31

needs lot of experimentation. Taguchi method helps in systematic evaluation of optimum process conditions with fewer experiments when compared to the classic method.

In the present work, four crucial and impactful factors viz., substrate temperature, oxygen partial pressure, LASER fluence, duration of deposition were chosen for deposition of thin films. For each factor, three different levels are considered. The details of each variable and levels are tabulated in below.

Process variable	Level 1	Level 2	Level 3
Substrate temperature (^{o}C)	600	650	700
Oxygen partial pressure (mTorr)	100	200	300
LASER Fluence (J/cm^2)	1.0	1.5	2.0
Target-substrate distance (cm)	05	06	07

1.2 Analysis of Variance (ANOVA)

The analysis of variance (ANOVA) is used to obtain quantitative information about the percentage influence of each factor on the deposition characteristics of thin films. The ANOVA model determines the dependence of process parameters on the growth and performance of thin films.

2 Results and Discussion

2.1 Signal to Noise Ratio

The optimized conditions were assessed using signal-to-noise (S/N) ratio using Eq. 1, usually classified into three categories i.e., the lower-the-better, the higher-the-better, and the nominal-the-better. The optimal process parameters were chosen based on highest S/N ratio corresponding to thickness data in this study.

$$\eta = -10 \log \frac{1}{n} \sum_{i=1}^{n} \frac{1}{y_n^2} \tag{1}$$

where η denotes the calculated value of S/N ratio in decibels (dB), y_n represents the obtained experimental values viz., thickness and dielectric constant and n stands for the number of trials carried out.

The S/N response corresponding to different input parameters for growth conditions of BZT thin films is shown in Fig. 1. The optimal process variables are chosen based on ANOVA results.

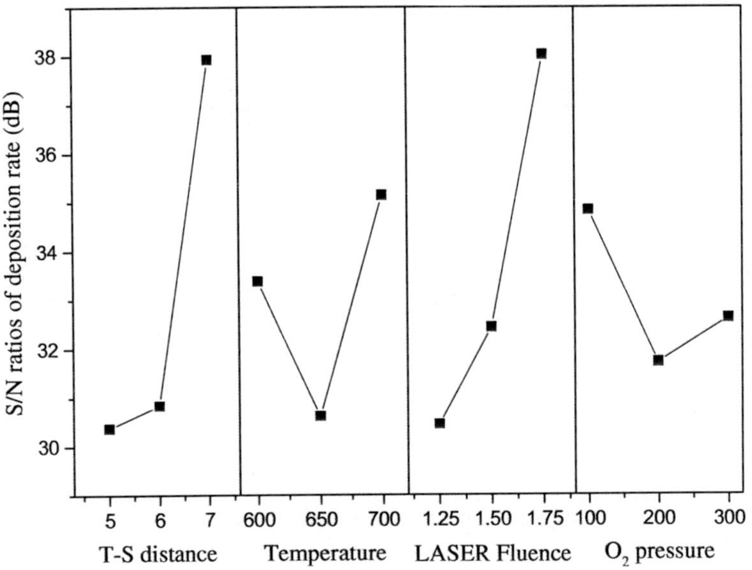

Fig. 1: S/N response graph corresponding to different input parameters for growth conditions of $Ba(Zr_{0.15}Ti_{0.85})O_3$ thin films.

2.2 Phase Analysis

The results of X-ray diffraction (XRD) studies carried out for phase analysis confirmed the presence of single-phase perovskite structure of BZT thin films (refer Fig. 2). The phase analysis reveals the possibility of formation of perovskite structure with preferred (110) orientation growth of BZT thin films.

2.3 Microstructural Studies

The scanning electron microscope (SEM) analysis of BZT thin films shown in Fig. 3 reveals different surface morphologies viz., formation of clusters of target species, pointed flakes, granular morphology and cracked surfaces depending on the process parameters applied for growth conditions of thin films. The films grown at optimal process variables exhibited maximum dielectric constant of ~380 at zero electric field. The values obtained are higher than the BZT films of same composition deposited by different methods [4] and (100) oriented BZT thin films of different composition as reported in the literature [5].

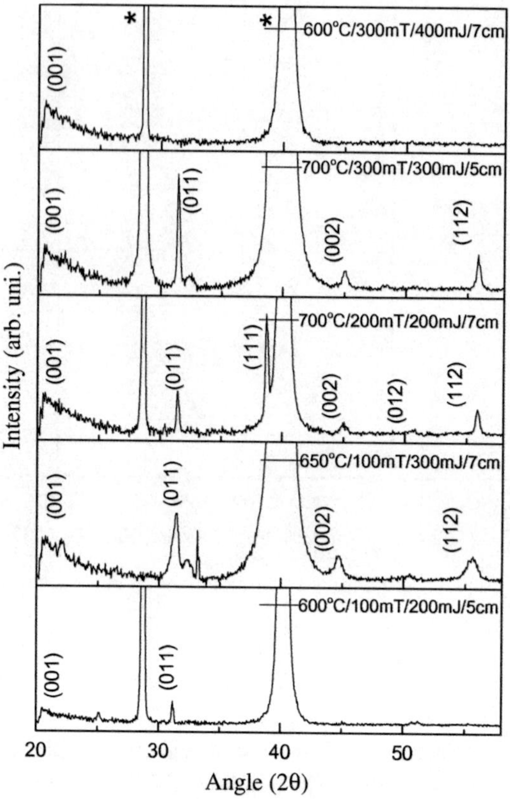

Fig.2. XRD spectra of the $Ba(Zr_{0.15}Ti_{0.85})O_3$ thin films deposited at various process parameters.

2.4 Evaluation of Contributing Factors Using ANOVA

The dominant and significant process parameters affecting the growth of thin films were evaluated in this study.

Process variable	Degrees of freedom	Sum of squares	Variance	Contribution (%)
Substrate temperature (°C)	2	31.18	15.6	12.48
Oxygen gas pressure (mTorr)	2	15.03	7.51	6.01
LASER Fluence (J/cm^2)	2	95.7	47.88	38.3
Target-substrate distance (cm)	2	107.9	53.98	43.2

The results indicate that the target-substrate distance and LASER fluence play a 'dominant' role which are primarily responsible and are the contributing factors to the thickness. The other two parameters viz., substrate temperature and oxygen partial pressure play a 'significant' role on the thickness.

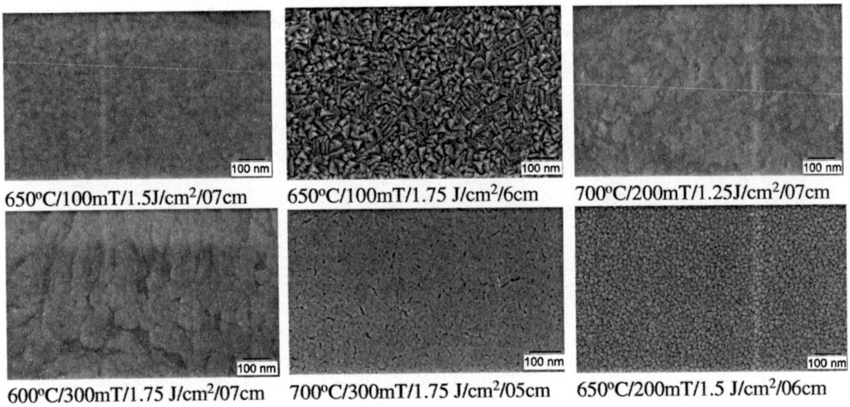

Fig. 3. SEM analysis of BZT films at different growth conditions.

3 Conclusions

This work highlights the importance of optimization of process parameters for growth of thin films. The methodology for obtaining oriented films was discussed. The possibility of domain engineering of surface morphology of BZT thin films was discussed. The factors responsible for growth of ferroelectric thin films were evaluated using ANOVA results. The films grown at optimal process variables i.e., at a substrate temperature of 650 °C, oxygen gas pressure of 200mT, energy fluence of 1.5 J/cm^2 and target-substrate distance of 06cm exhibited electrical properties comparable to oriented single crystals of same material.

Acknowledgements. The authors acknowledge the financial support from the Defence Research and Development Organization, Ministry of Defence, New Delhi for carrying out the present work. Also, the authors would like to thank the Director of Defence Metallurgical Research laboratory (DMRL) for permitting publication of this work.

References

1. Mahesh, M.L.V., Bhanu Prasad, V.V., James, A.R.: Enhanced dielectric and ferroelectric properties of lead-free $Ba(Zr_{0.15}Ti_{0.85})O_3$ ceramics compacted by cold isostatic pressing. J. Alloys Compounds **611**, 43–49 (2014)
2. Veerapandiyan, V., Benes, F., Gindel, T., Deluca, M.: Strategies to improve the energy storage properties of perovskite lead-free relaxor ferroelectrics: a review. Materials **13**(24), 5742–5789 (2020)
3. Peace, G.S.: Taguchi Methods: A hands-on approach. Addison-Wesley, Boston (1993)

4. Gao, C., Zhai, J., Yao, X.: Preparation and dielectric properties of $Ba(Zr_xTi_{1-x})O_3$ thin films grown by a sol-gel process. Integr. Ferroelectr. **74**(1), 147–153 (2005)
5. Tang, X.G., Wang, X.X., Wong, K.H., Chan, H.L.W.: Effects of texture on the dielectric properties of $Ba(Zr_{0.2}Ti_{0.8})O_3$ thin films prepared by pulsed laser deposition. Appl. Phys. A **81**(6), 1253–1256 (2005)

Resistive Switching Behavior of TiO_2 Nanostructures Grown by Dip Coating Process

S. Roy[1,2], P. K. Sahu[1], D. Pradhan[3,4], N. Tripathy[5], and J. P. Kar[3](✉)

[1] Department of Electrical Engineering, National Institute of Technology, Rourkela 769008, India
[2] Department of Electronics and Communication, Madanapalle Institute of Technology and Science, Madanapalle 517325, India
[3] Department of Physics and Astronomy, National Institute of Technology, Rourkela 769008, India
karjp@nitrkl.ac.in
[4] ITER, Siksha O Anusandhan Deemed to be University, Bhubaneswar, Odisha 751030, India
[5] Department of Humanities and Sciences, Vardhaman College of Engineering, Shamshabad, Hyderabad 501218, India

Abstract. In recent years, titanium dioxide (TiO_2) has drawn tremendous attentions due to its excellent electronic properties, nontoxic nature, high chemical stability, low cost, and huge availability in nature. TiO_2 nanostructures were grown on boron doped, (100) oriented, 1–10 Ω-cm silicon substrates by using wet chemical method, where the concentration of titanium n-butoxide was varied from 0.24 ml to 0.60 ml. The chemical process was conducted at 140°C for 2 h using an autoclave. The morphological and structural properties of TiO_2 nanostructures were investigated by scanning electron microscope (SEM) and X-ray Diffraction (XRD) techniques. The nano flowers like structures were observed by the SEM study and rutile phase of TiO_2 was depicted by the XRD analysis. The nanostructures were increasing with the titanium butoxide concentration that was confirmed by SEM study. In precursor concentration of 0.60 ml, the nanostructures of width 256 nm and length 1.54 μm were obtained. Resistive switching study was showing the memory capacity of the nanorod structures.

Keywords: Nano flowers · Hydrothermal process · SEM · XRD · Resistive switching

A. K. Pandey et al. (Eds.): MAMM 2022, 126, pp. 357–364, 2023.
https://doi.org/10.1007/978-3-031-20353-4_32

1 Introduction

Dimensionality has a huge contribution in the performance of the nanostructure based microelectronic devices [5,19]. Introduction of 1-D nanomaterials such as nanorods or nanowires have become an attractive approach both in technical as well as scientific research areas. 1-D nanomaterials can offer significant larger surface area to volume ratio than thin films [2,12]. Therefore, 1-D nanomaterials have excellent features such as it has unique physical and chemical properties, and other applications properties can be achieved from their mechanical to electrical domain [11,16].

There are various fabrication processes that have been established for the synthesis of 1-D nanomaterials such as template assisted, vapor-liquid-solid (VLS), chemical vapor deposition (CVD), anodic oxidation method, solvothermal and hydrothermal methods etc. In this research, hydrothermal process is used to fabricate the TiO_2 nanorods due to its simplicity, low cost, chemical stability and solubility, higher reaction velocity, better electron conducting properties such as diffusion, sizes and phase conversions at low temperature, high yield, higher nucleation as well as control of surface morphologies [8,18]. Hydrothermal method is quite well established method because of its versatility and the process can be performed easily on different types of substrate like silicon wafer, FTO, ITO etc. In addition, hydrothermal process can easily suppress the huge crystallographic defects even at a lower temperature with controlling particle dimension and their spatial location of particles [17].

The formation of TiO_2 nanorods can be expressed in step wise process. It starts from nucleation, attachment to the oriental surface, and nanorod growth, which takes place in the acidic medium. Dissolution and recrystallization of the nanoparticle affect the nucleation process strongly in high acidic medium of the hydrothermal process [4]. The crystallite size largely depends on the reaction time and the growth direction of the nanorods is depended on the nucleation sites [13]. Surface energy and thermodynamic stability have huge effect on the formation of the nanowires and hydrothermally grown TiO_2 nanorod arrays is a an cost effective process for the application of resistive switching [6]. In the present work, the fabrication and characterization of TiO_2 nanorod arrays with various aspect ratio by hydrothermal process has been reported. Finally, the resistive switching behavior of the TiO_2 nanorods has been studied.

2 Experimental Procedure

TiO_2 nanorod arrays were grown on the p-type (100) Silicon substrate (1–10 Ω.cm) oriented using hydrothermal process. In the typical experiment, 25 ml of deionized water was mixed with 25 ml hydrochloric acid followed by stirring at the ambient temperature for 5 min. Afterwards, various concentration of titanium butoxide such as 0.24–0.60 ml was added with the mixture in order to prepare different precursor. After stirring another 15 min, the mixture was transferred to a Teflon lined stainless steel autoclave (100 ml volume). The hydrothermal synthesis was conducted at 140 °C for 2 h in regular laboratory air oven.

After synthesis, the autoclave was cooled down to room temperature. The silicon substrate was taken out from the autoclave and rinse extensively with the deionized water and allowed to dry in the ambient air.

The surface and structural morphologies of the samples were characterized by the scanning electron microscopy (JEOL JSM- 6480LV Scanning Electron Microscope) and X-ray diffractometer (Rigaku Ultima IV). For fabricating Al/TiO_2/ p^{++} Si MIM structures, Aluminum electrodes were deposited on dip coated TiO_2 film by thermal evaporation method. Top electrodes were deposited by using a shadow mask, whereas globally deposited Al film were used as bottom electrode. Resistive switching behavior of TiO_2 nanorod were studied by using Keithley 2410 source meter.

3 Results and Discussion

Figure 1 depicts the SEM micrographs, where the TiO_2 nanorods are synthesized at 140 °C for 2 hrs with various precursor concentrations such as (a) 0.24 ml (b) 0.36 ml (c) 0.48 ml (d) 0.60 ml. The morphologies of the nanorods have shown randomly aligned structure with a length of 400–1.54 m and width varies approximately 210–280 nm. The SEM images describe that the nanorods are becoming wider with the increased concentration of titanium precursor and at 0.60 ml concentration, the average length of the nanorod becomes 1.5 m, which shows that precursor concentration has a significant effect on the width and length of the

Fig. 1. SEM images of the TiO_2 nanorods grown at 140 °C for 2 hrs with various precursor concentration of Titanium butoxide (**a**) 0.24 ml (**b**) 0.36 ml (**c**) 0.48 ml (**d**) 0.60 ml.

nanorods. During growth of TiO_2 nanorods, the molecule of the precursor reacts with the aqueous solution during the hydrolysis process and the vapor is used to form during reaction in the autoclave which creates a certain amount of pressure [3].

The overall morphology such as width and length of nanorod are determined by the nucleation and growth mechanism. During the chemical reaction, the atomic species are attached to the nucleation site for a thermodynamical stable orientation, which depends on the parameters such as growth temperature, precursor concentration, substrates, growth time. The controlled supply of precursor atoms for subsequent growth of nanorods at an optimized condition is highly important for achieving the reproducible nanostructures. The nucleation and agglomeration of TiO_2 nanoparticles is related with the hydrolysis process and as a result, higher density nanorods are obtained at higher precursor concentration of 0.60 ml where flower like morphology is observed. The orientation of the nanorods growth is governed by surface energy. The growth mechanism of nanorods can be explained in a two-step process. First, due to hydrothermal method, TiO_2 nanoparticles are formed and later anisotropic agglomeration of nanoparticle leads to formation of nanorods reducing the surface energy of nucleation sites. It is shown that initial reaction conditions have significant influence on the nanorod length as well as width. So, the hydrothermal process is highly dependent on the amount of titanium source and are very suitable for industrial applications.

Table 1. Average TiO_2 nanorod dimensions with the variation of precursor concentration

Conditions of hydrothermal process		Average nanorod dimensions	
Temperature (°C)	Precursor concentration (ml)	Length (μm)	Width (nm)
140	0.24	0.4	210
	0.36	0.8	251
	0.48	1.6	284
	0.60	1.5	256

Figure 2 depicts the XRD plot of the TiO_2 nanorods grown on the p-type silicon substrate. The XRD peak intensity is found to be increased with the precursor concentration which can be attributed to the enhancement in the density and dimension of the nanorods. Diffraction peak obtained at the 27.47° indicates the rutile (100) phase of TiO_2 [9]. The nucleation process of the nanoparticles is largely dependent on the interfacial energy. Rutile phase is more stable as compared to the anatase phase of TiO_2 at any calcination temperature and pressure [1]. The reason behind the stability is based on the lower free energy of rutile phase [10,14] and anatase phase is more stable for extremely small crystallites due to high surface area and anatase phase containing lower surface energy [15].

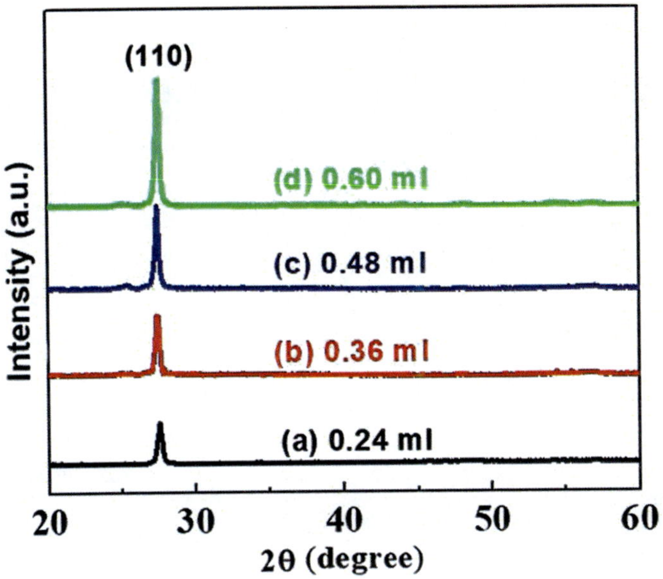

Fig. 2. XRD patterns of the TiO_2 nanorod with 140 °C for 2 hrs with various precursor concentrations of nanorod growth.

It is shown that multidimensional TiO_2 nanorods are formed dandelions structure, which is observed in the random orientation of the nanorod growth and rutile phase is observed.

Resistive switching phenomenon of TiO_2 nanorods are shown in Fig. 3. Nanorods are excited by oxygen molecule and make an oxygen deficiency center easily because TiO_2 itself is an oxygen deficient material. At slow hydrolysis, the morphology has not changed but Ti^{3+} species and oxygen vacancies are formed at the shell. The geometrical structure of the nanorods have large surface area so it can absorb O_2 molecule on the defective area of nanorods as trapping the charge carriers and accelerate the oxidation process. Thereafter, the resistance of TiO_2 nanorods are increased in the form of leading conductivity of TiO_2 nanorods surface [5,7]. In the SET process, the oxygen vacancies are diffused from the anode, where voltage is applied to the cathode, which is the ground electrode and conducting filaments are formed after connecting the top and bottom electrode and switching is occurred between high resistance state (HRS) to low resistance state (LRS). A high and low resistance states are switched back and forth based the set voltage and reset voltage. In this research compliance current is fixed at 0.1 mA, which is used to prevent permanent breakdown of the device. The forming voltage is applied at 3 V to initiate the electron conductance process. For the SET process of the on/off ratio become 10 for 0.36 ml precursor concentration and at 0.48 the on/off ratio was around 18. The resistive switching behavior is bipolar in nature. Due to the multifunctional behavior of TiO_2 nanorods such

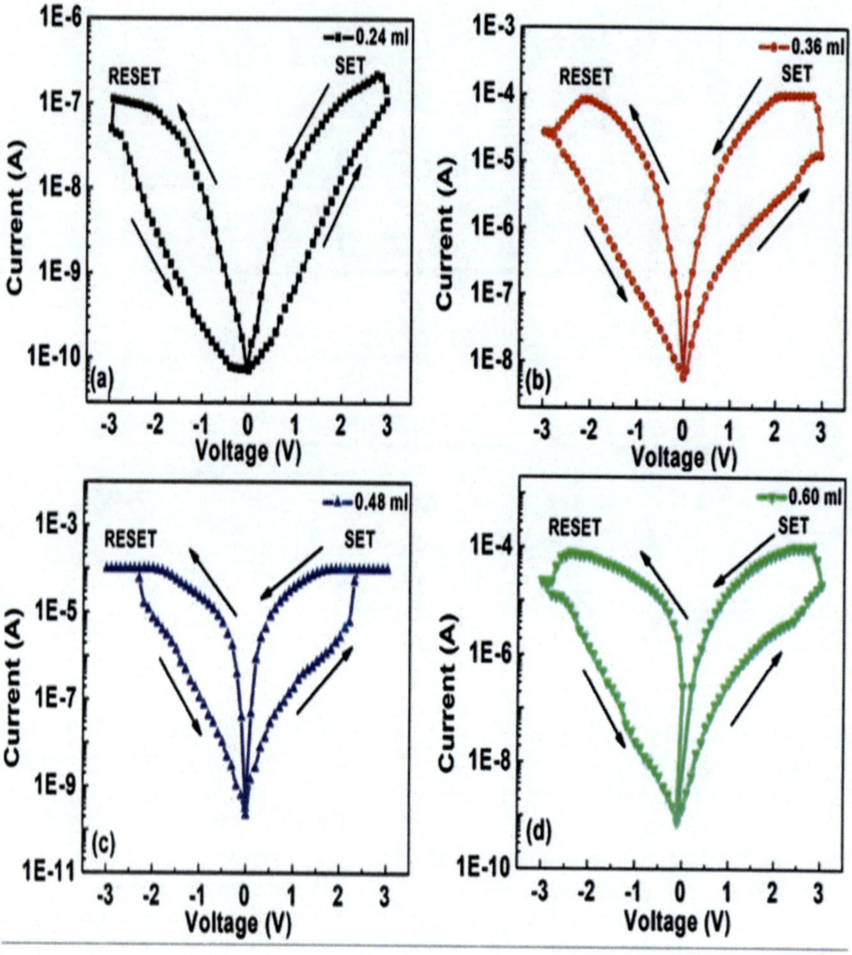

Fig. 3. Resistive switching behaviors of the TiO_2 nanorod with 140 C for 2 hrs with various precursor concentrations of nanorod growth

as high surface to volume ratio as well as resistive switching, these devices can be used in MEMS industries as functional materials and sensors.

4 Conclusion

TiO_2 nanorods were grown on the p-type silicon substrate using hydrothermal process. The SEM micrograph has shown the formation of scattered nanorod structures. The 0.60 ml precursor concentration has resulted flower shaped TiO_2 nanorods. The XRD investigation has shown the rutile phases of the TiO_2. Resistive switching behaviour was also studied to use in the memory as well as switching devices and process condition at 140°C, 2 hrs with 0.48 ml precursor concentration is showing better memory window with On/Off current ratio of 18 for nanorods.

Acknowledgment. Authors are thankful to Prof. Pitamber Mahanandia, NIT Rourkela for the resistive switching measurements.

References

1. Attar, A.S., Ghamsari, M.S., Hajiesmaeilbaigi, F., Mirdamadi, S., Katagiri, K., Koumoto, K.: Sol-gel template synthesis and characterization of aligned anatase-tio_2 nanorod arrays with different diameter. Mater. Chem. Phys. **113**(2–3), 856–860 (2009)
2. Cao, G.: Growth of oxide nanorod arrays through sol electrophoretic deposition (2004)
3. Dey, S., Roy, S.C.: Designing tio_2 nanostructures through hydrothermal growth: influence of process parameters and substrate position. Nano Express **2**(1), 010028 (2021)
4. Grillo, F., La Zara, D., Mulder, P., Kreutzer, M.T., Ruud van Ommen, J.: Oriented attachment and nanorod formation in atomic layer deposition of tio_2 on graphene nanoplatelets. J. Phys. Chem. C **122**(34), 19981–19991 (2018)
5. Hu, J., Odom, T.W., Lieber, C.M.: Chemistry and physics in one dimension: synthesis and properties of nanowires and nanotubes. Accounts Chem. Res. **32**(5), 435–445 (1999)
6. Huang, C.H., Chou, T.S., Huang, J.S., Lin, S.M., Chueh, Y.L.: Self-selecting resistive switching scheme using tio_2 nanorod arrays. Sci. Rep. **7**(1), 1–9 (2017)
7. Huang, T.H., Yang, P.K., Lien, D.H., Kang, C.F., Tsai, M.L., Chueh, Y.L., He, J.H.: Resistive memory for harsh electronics: immunity to surface effect and high corrosion resistance via surface modification. Sci. Rep. **4**(1), 1–5 (2014)
8. Iraj, M., Nayeri, F.D., Asl-Soleimani, E., Narimani, K.: Controlled growth of vertically aligned tio_2 nanorod arrays using the improved hydrothermal method and their application to dye-sensitized solar cells. J. Alloys Compounds **659**, 44–50 (2016)
9. Khalid, N.S., Fazli, F.I.M., Hamed, N.K.A., Napi, M.L.M., Fhong, S.C., Ahmad, M.K.: Biocompatibility of tio_2 nanorods and nanoparticles on hela cells. Sains Malaysiana **45**(11), 1675–1678 (2016)
10. Kim, K.J., Benkstein, K.D., van de Lagemaat, J., Frank, A.J.: Characteristics of low-temperature annealed tio_2 films deposited by precipitation from hydrolyzed ticl4 solutions. Chem. Mater. **14**(3), 1042–1047 (2002)
11. Lee, S.M., Cho, S.N., Cheon, J.: Anisotropic shape control of colloidal inorganic nanocrystals. Adv. Mater. **15**(5), 441–444 (2003)
12. Limmer, S., Chou, T., Cao, G.: A study on the growth of tio_2 nanorods using sol electrophoresis. J. Mater. Sci. **39**(3), 895–901 (2004)
13. Meng, X., Shin, D.W., Yu, S.M., Park, M.H., Yang, C., Lee, J.H., Yoo, J.B.: Formation mechanism of rutile tio_2 rods on fluorine doped tin oxide glass. J. Nanosci. Nanotechnol. **14**(11), 8839–8844 (2014)
14. Park, N.G., Van de Lagemaat, J., Frank, A.J.: Comparison of dye-sensitized rutile- and anatase-based tio_2 solar cells. J. Phys. Chem. B **104**(38), 8989–8994 (2000)

15. Patra, S., Davoisne, C., Bouyanfif, H., Foix, D., Sauvage, F.: Phase stability frustration on ultra-nanosized anatase tio$_2$. Sci. Rep. **5**(1), 1–10 (2015)
16. Peng, X., Manna, L., Yang, W., Wickham, J., Scher, E., Kadavanich, A., Alivisatos, A.P.: Shape control of cdse nanocrystals. Nature **404**(6773), 59–61 (2000)
17. Scheepens, A., van de Waarenburg, M., van den Hove, D., Blanco, C.E.: A single course of prenatal betamethasone in the rat alters postnatal brain cell proliferation but not apoptosis. J. Physiol. **552**(1), 163–175 (2003)
18. Tan, C., Zhang, H.: Wet-chemical synthesis and applications of non-layer structured two-dimensional nanomaterials. Nat. Commun. **6**(1), 1–13 (2015)
19. Yuan, Z.Y., Colomer, J.F., Su, B.L.: Titanium oxide nanoribbons. Chem. Phys. Lett. **363**(3–4), 362–366 (2002)

Correction to: Detection of Volatile Organic Compounds Using Solution Processed Organic Field-Effect Transistors

Suresh Kumar Garlapati[1,2,3](✉), Sheida Faraji[1,2], Daniel Tate[1], Aiman Rahmanudin[1], Palaniappan Valliappan[2], Alessandro Patti[2], Krishna Persaud[2], and Michael Turner[1](✉)

[1] Department of Chemistry, OMIC, University of Manchester, Manchester M13 9PL, UK
gsuresh@msme.iith.ac.in, michael.turner@manchester.ac.uk
[2] Department of Chemical Engineering, University of Manchester, Manchester M13 9PL, UK
[3] Department of Materials Science and Metallurgical Engineering, Indian Institute of Technology Hyderabad, Hyderabad 502285, India

Correction to: Chapter "Detection of Volatile Organic Compounds Using Solution Processed Organic Field-Effect Transistors" in: A. K. Pandey et al. (Eds.): MAMM 2022, 126, pp. 310–322, 2023. https://doi.org/10.1007/978-3-031-20353-4_27

In the original version of the book, the author's "Alessandro Patti" first name was inadvertently published with a typo. The name has now been corrected from "Alessanndro Patti" to "Alessandro Patti" in the chapter "Detection of Volatile Organic Compounds Using SolutionProcessed Organic Field-Effect Transistors" updated version.

The chapter and book have been updated with the changes.

The updated original version of this chapter can be found at https://doi.org/10.1007/978-3-031-20353-4_27

A. K. Pandey et al. (Eds.): MAMM 2022, 126, p. C1, 2023.
https://doi.org/10.1007/978-3-031-20353-4_33